UFOs and Water

UFOs and Water

Physical Effects of UFOs on Water Through Accounts by Eyewitnesses

Carl W. Feindt

To order additional copies of this book, contact:
Xlibris
1-888-795-4274
www.Xlibris.com
Orders@Xlibris.com
580738

In Memory of Richard H. Hall

While still working on this project, I was saddened to learn, along with many in the UFO field, that Richard passed away on July 17, 2009, after a long battle with cancer.

Richard was a man of utmost dignity and distinction in the field of ufology, and his thoughts have been expressed on many UFO e-mail lists. He was a man whom I greatly respected as a serious UFO researcher. That he would accept to help me with my book during his illness speaks volumes to his character and dedication to ufology. I hope that my book will add to the knowledge contained in his two-volume work entitled *The UFO Evidence*. I hold him in the highest esteem, and his life's contribution to the study of UFOs serves as a fitting legacy to us all. He was a pioneer in ufology and he will be missed.

Rest in peace and with our admiration, Mr. Hall.
December 25, 1930 - July 17, 2009

Preface to the Second Edition

As those of you who read the First Edition of my book may know, it has been several years since it was published in August 2010. My main goal in writing the Second Edition of *UFOs and Water* is to add some newfound information in order to provide more proof of the UFO's surrounding field and its rotation and solve some other mysteries that were previously unexplainable. Accompanying this are extra eyewitness accounts and a few supplementary sketches. Additionally, every chapter in this new publication has been thoroughly revised and updated.

The following is a list of **new** topics in this edition:

Introduction now includes an excerpt from Robert J. Low's "trick" memorandum concerning the Condon committee's study of UFOs (pp. xxviii-xxix).

Chapter 2 now contains a section on ionization in general and as it relates to UFOs (pp. 39-41). The idea of an on-off switch controlling the UFO's field is discussed (pp. 41-43). A new case dated 10-15-1979 has been added with different terminology for the bulge seen in the water (p. 49). I also talk about our technology versus alien technology (pp. 77-78).

Chapter 3 has been expanded to provide more proof of the rotating field in reverse with sections on pyramid- or inverted-cone-shaped waves (pp. 87-94) as well as the Roswell gouge versus underwater gouges (pp. 94-97). In addition, I offer some thoughts on the electrical energy of UFOs and the use of a black hole to accumulate energy (pp.

109-112). Then I revisit Roswell to see what might have caused the Roswell crash (p. 112).

Chapter 4 examines the phenomenon of engraved ice circles (pp. 152-158). I also introduce some eminent physicists, including a Nobel Prize winner, who discuss the reality of UFOs (pp. 164-173).

Chapter 5 includes an additional extensive abductee account (pp. 195-199).

Chapter 11 delves into missing water where mysterious disappearances of large quantities of water have occurred (pp. 362-374). In conjunction with this, a discussion of camouflage as another possible purpose for the water that UFOs take on is presented (pp. 374-378).

Chapter 19 proposes an explanation for how the UFO's field can have the effect of flattening surface waves, giving them the appearance of a sheet of glass (pp. 448-451).

Chapter 20 was previously Chapter 19 (pp. 452-472).

Finally, an index is now included, in chronological order, of all cases presented in this Second Edition.

Contents

Dedication v

Preface to the Second Edition.............................. vii

Acknowledgments... xv

Foreword.. xix

Explanation of Case Dates xxiii

Introduction.. xxv

Chapter 1: Water-UFO History 1
1.1   In the Beginning...of Ufology (June 24, 1947) 1
1.2   Some Acronyms That Describe Water-UFOs—
       And the Futility of Some Acronyms........................... 6
1.3   A Question of History—How Long Water-UFOs
       Have Been Seen and by Whom 7
1.4   Verbal Testimony as Anecdotal Tales.........................12
1.5   Case Examples of Out of, Into, Hovering Over,
       and Floating on the Water 12
       1.5 a.   Coming Out of Water 12
       1.5 b.   Entering Water....................................16
       1.5 c.   Hovering over Water 20
       1.5 d.   Floating on Water 20
       1.5 e.   Out Of and Into or Into and Out Of Water 24
1.6   Summary 27

Chapter 2: Physical Influences of a UFO on Water......... 29
2.1   Physical Effects on Water—Tying It Together...............33
2.2   Ionization in General and As It Relates to UFOs...........39

2.3   The On-Off Switch...41

2.4   UFO Coming Up—Water Bulges 46

2.5   UFO Just above the Surface—Water Depression 53

2.6   Above the Surface but Close—Water Mounding 56

2.7   Above the Surface—a Waterspout 58

2.8   What's It All About?...66

2.9   A Function of Height..67

2.10   Water Nests (Crop Circles) 71

2.11   From Water to Land 72

2.12   How Tough Is This Field 75

           2.12 a.   Guess What, Dr. Haines,

                      I've Found a Bigger Gun 75

           2.12 b.   Our Technology versus Their Technology 77

2.13   Tilting UFO—Loss of the Vacuum Cleaner—

           Vector Analysis..79

Chapter 3: The Field in Reverse 82

3.1   Operation of the Field in Reverse...............................82

3.2   Now the Important "One"...85

3.3   Yet More Proof of the Rotating Field in Reverse—

           Tiny Pyramid- or Inverted-Cone-Shaped Waves...........87

3.4   The Roswell Gouge versus Underwater Gouges—

           More Evidence of the Field in Reverse 94

3.5   Energy Transfer between UFOs Using

           the Reversed Field 97

3.6   Similar Exchange As Seen by an Abductee

           in a Non-Water Landed Craft 102

3.7   Fuel 109

3.8   Let's Play Roswell Again 112

Chapter 4: Heat and Water-Related UFOs......................113

4.1   How Hot Is UFO Radiation?.....................................114

4.2   Explanation of Terms...116

4.3   On or Close to the Water's Surface 119

4.4   Bubbles and Boiling—Separate Causes? 121

4.5   Close Encounters above the Surface 127

4.6   In the Rain...130

4.7   Through the Clouds.................................134

4.8   In the Snow 140

4.9   Through the Ice 145

4.10   Engraved Ice Circles................................152

4.11   Flipping the Field Switch to Off—UFO Makes Ice 158

4.12   Eminent Physicists Discuss the Reality of UFOs 164

Chapter 5: Abductees—Water and Bases 175

5.1   What We Can See through the Eyes of Others............175

5.2   Abductees and Water 175

5.3   "Beam Me Up, Scotty" 179

5.4   Another Case History................................184

5.5   Abductees That Know They Are Underwater..............186

5.6   Abductees and Underwater Bases...........................199

5.7   Amazing the Information We Have at Hand.
If Only We Would Listen................................210

Chapter 6: Radar and Sonar 212

6.1   Show Us Your Instruments' Documentation 212

6.2   Radar...213

        6.2 a.   *Radar Only (operating and detecting)* 213

        6.2 b.   *Radar Only (operating and NOT detecting)* ...217

        6.2 c.   *Radar Only (rendered inoperable)* 219

        6.2 d.   *Radar Can Illuminate Plankton*................222

6.3   Sonar...224

        6.3 a.   *Sonar Only (operating and detecting)* 225

        6.3 b.   *Sonar Only (operating and NOT detecting)* ...228

6.4   Radar and Sonar.......................................229

6.5   Now You See Them (on Radar or Sonar),
Now You Don't..231

Chapter 7: Aircraft Carriers and Other Large Ships.......234

7.1   Fear in Wartime of an Unknown Enemy 235

7.2   Closer to the End of the War................................237

7.3   Radar Only (operating and detecting) 238

7.4   Radar Only (operating and NOT detecting) 252
7.5   Electromagnetic (EM) Effects 255
7.6   Can They Travel Fast Underwater? 260

Chapter 8: Secrecy about UFOs 263
8.1   "UFOs Do Not Constitute a Threat to National Security." ... 263
8.2   Early On—Before UFO Secrecy 265
8.3   But Then... 266
8.4   The JANAP-146 Documents 267
8.5   The MERINT Document 272
8.6   The Veil of Secrecy Tightens 274
8.7   The U.S. Navy and Flying Submarines 283

Chapter 9: Fish and Animal Reactions to UFOs 292
9.1   Reactions of Fish 292
9.2   More than Fish 296
        9.2 a.   *Birds in Water Cases* 297
        9.2 b.   *Dogs in Water Cases* 299
        9.2 c.   *A Cat in a Water Case* 303
        9.2 d.   *Animals in General in Water Cases* 303
9.3   Lightning Strikes 305
9.4   Dead in the Water 306
        9.4 a.   *Dead Fish* 308
        9.4 b.   *Dead Ducks and Seagulls* 312
        9.4 c.   *Dead Professional Divers* 312

Chapter 10: Negative Aspects of Water Ufology 314
10.1   General Introduction 315
10.2   Misrepresentation (*Eltanin*—"Thing" on the
      Ocean Floor) 315
        10.2 a. *This Is the Start of the Whole Thing* 315
        10.2 b. *As Promised: The Truth of the Matter*
                 after Years of Speculation 321
10.3   Anomaly Attachment (Water Wheels) 324
10.4   Gross Supposition (Missing Ships) 332
10.5   Honest Mistakes (Misidentifications) 335
10.6   Malicious Intent (Hoaxes) 346

Chapter 11: UFOs—Taking on Water by Conventional Means 352

11.1 For What Purpose? 352

11.2 Encounters with Aliens Fetching Water 354

11.3 Missing Water 362

11.4 Taking on Water for Camouflage? 374

Chapter 12: Beam of Light...Into a Body of Water 379

Chapter 13: Burning Objects or Water-UFOs? 387

13.1 Very Simple Reports 388

13.2 The Similarity between a Fiery, Glowing UFO and a Burning Aircraft 389

13.3 The Similarity between a Fiery, Glowing UFO and a Burning Boat 392

13.4 A Meteor? 395

13.5 Descriptions as Resembling "Fire" 395

Chapter 14: Sea Serpents or Water-UFOs? 398

Chapter 15: Physical Influences of a UFO on a Boat's Motion 409

Chapter 16: Totally Submerged 413

16.1 Hooked onto Something 413

16.2 Submarines? 416

16.3 Hit by Unseen Metal Craft (UFO or Submarine?) 419

16.4 The Reverse—Ship Hits Something Unknown 421

16.5 Submarines That Could Not Have Been Subs 423

16.6 Actually Seen Underwater 424

Chapter 17: Underwater "Solid" Lights 429

17.1 Glowing into Water and Continuing to Glow Underwater 429

17.2 Glowing Underwater and Continuing to Glow out of the Water 431

17.3   In and Out or Out and In 432
17.4   Totally Submerged 434

Chapter 18: Shag Harbour: The "Water Roswell" 440
18.1   An Early Account of the Sighting 440
18.2   How Are They Going to Fix It Underwater? 444

Chapter 19: The Sea Was Like a Sheet of Glass 448

Chapter 20: Up Close and Personal............................... 452
20.1   The Fear Factor 453
20.2   From Our Government Which Does
       Not Believe UFOs Exist 456
       20.2 a. From the CIA 456
       20.2 b. From Naval Intelligence 460
20.3   Two UFO Stories from before the UFO Era 462
20.4   Electromagnetic (EM) Effects after 1947.................... 466

Conclusion... 473

References 477

Index of Cases in Chronological Order......................... 525

About the Author.. 539

Acknowledgments

Writing a book is a considerable endeavor, especially for someone like me whose writing experience is quite limited. However, I was determined to make known my ideas. In undertaking this book, it has been not only unavoidable but entirely appropriate that I draw upon the work of so many others, to whom I am greatly indebted.

I would like to start by thanking all my mentors in ufology, past and present, who inspired me and whose written texts from 1960 onward have contributed to my knowledge of UFOs and have made this book possible.

Thanks to Jan Aldrich who asked for help in researching newspapers of the fifty states of the United States in the early 1990s. It was the right time and the right place as I had just retired from my occupation in aviation. I volunteered to search for UFO articles about my state, Delaware, which started my active participation in ufology.

Ted Phillips's work in trace cases inspired me to pursue the idea of water as a means of discovering a UFO's attributes.

My friend Tony Rullán's enthusiasm pushed me to get the ball rolling in 1998 while at a MUFON Symposium in Washington, D.C. (Thanks to Robert and Susan Swiatek of the Fund for UFO Research for hosting it.) Although he was busy with both his own research and a demanding work schedule, he encouraged me in what I was trying to accomplish and furnished me with photocopies of relevant texts he was able to acquire.

I would like to take this opportunity to express my thanks to those who helped in so many ways in compiling this book. Special appreciation is due those who read various chapters of it in draft form. Their expertise, corrections of errors, and constructive comments improved this volume immeasurably. They generously volunteered their valuable time, and it has been a privilege to learn from them all.

First and foremost, I wish to thank Richard Hall who passed away on July 17, 2009. In spite of his health problems, he very graciously accepted my e-mail request to edit the introduction and first chapter of this book, and I feel privileged to have been able to work with him. To those of us who study UFOs, his work will endure and continue to inspire us.

Bruce Maccabee is of long standing in the UFO field and understood what I, a non-scientist, was thinking in Chapters Two, Three, and Four. His scholarship and excellent advice contributed substantially to this book's accuracy and content. I appreciate all his valuable feedback.

Joan Woodward, with her experience in the field of animal reactions, helped me to add fish and ducks and other animal reaction cases that had a water connection to Chapter Nine.

Don Ledger and Chris Styles of Nova Scotia, Shag Harbour case researchers, helped me with Chapter Eighteen, and I welcomed their expertise with the facts of the case.

I offer thanks to my son Jeffrey who, before this book began, took my garbled copy of "Physical Influences of a UFO on Water" and worked it into a paper that won me second prize at the 2006 MUFON Symposium in the contest called "Best Physical Evidence." That text has served as the cornerstone of this book.

I am immensely grateful to my son Andrew who visited me frequently on weekends to work with me chapter by chapter on phraseology, helping me over many hurdles, including my "New Yorkese" articulation.

My sister Helen and my brother-in-law Bill Jatzen read the entire first draft and offered constructive criticism, support, and encouragement. They have also helped me over the years in correcting my transcriptions of taped interviews and case investigation reports.

Many other people helped me bring this book to fruition. And while I fear that it is impossible to mention everyone who has helped by sending me cases, this book would not exist without them. A hearty thank-you to Marco Bianchini in Italy for more than 280 water cases, Jean-Luc Rivera in France, Peter Hassall in New Zealand, Bill Chalker in Australia, Anders Liljegren in Sweden, Edoardo Russo in Italy, Brian Vike in Canada, Albert Rosales, Chris Aubeck, Peter Davenport, George Filer, Larry Hatch, Donald Johnson, Keith Chester, Barry Greenwood, Mark Rodeghier and Frank Reid at CUFOS, Richard Haines, and Timothy Good. I consider myself very fortunate to have so many friends in the field of ufology who work with me in a true sense of camaraderie. And, of course, I am beyond grateful to all the eyewitnesses who reported their UFO sightings. Without these cases, my work would not have been possible.

And finally, I need to mention Helene Jouan, who painstakingly read, corrected, and improved my writing. Her thoughtful suggestions, meticulous and conscientious work, unfailing interest, enthusiasm, support, and dedication to this project were unparalleled. I am grateful to her for her time and for giving me the benefit of her experience in the areas of language and grammar. She remains an indispensable colleague whom I have nicknamed the

Feindt-N-Stein Monster (meant solely as a pun) because I think I am making a ufologist out of her. I hope she forgives me for that!

But with that said, any and all remaining errors and omissions are, of course, my own.

Thanks to you all.
Carl—The **Aquarian**, <u>February 6</u>, 1938

Foreword

The Earth is covered by water to the extent of some three-fourths of its surface, making this water planet unique in our solar system. It has often been said that many mysteries still lie undiscovered beneath its vast surface, and it is still true today that we know more about the surface of Mars than we do about our own oceans. There have been reports of sea monsters and strange goings-on in and on the world's oceans for hundreds of years, in fact as long as man has sailed its surface.

One aspect of these high strangeness reports comes in the form of what appear to be aerial objects, often referred to as unidentified flying objects (UFOs), entering, being recorded submerged, and leaving the ocean depths. This is an extension of the UFO phenomenon and what Carl Feindt refers to as water-UFOs, rejecting the term USO (underwater submerged object) since he feels this should apply only to a totally submerged object, not one that goes into or comes out of the water. This text, however, is not just another UFO book, as the author has assembled some of the most compelling water-related UFO cases collected to date. Carl is a meticulous recorder and investigator of the water-UFO phenomenon, and this volume is the culmination of many years of research by him, with the number of reports gleaned from his own files approaching 1,200 cases [1,700 cases in 2014]. While physical traces left by UFOs on land have been frequently studied and documented, for the most part the UFO-water connection has been overlooked by ufologists. *UFOs and Water* is unique in that it deals specifically with how UFOs affect the water with which they have contact. From the wealth of eyewitness testimony that he has collected, the author has been able to analyze the data and has found a commonality among what has been observed by people around the world. Now, thanks to him, we have a valuable new resource with hundreds of eyewitness reports, insightful comments, plus sketches he has drawn which clearly illustrate his theories on the water-UFO connection.

Though we had frequented Errol Bruce-Knapp's internet list, "UFO Updates," for a couple of years, it was quite by accident that Carl and I met at a MUFON Conference in Rochester, New York, in July 2002. We became fast friends. One of the reasons why I am writing this foreword is because he considers the Shag Harbour incident in Nova Scotia to be one of the best recorded water-UFO events in the UFO phenomenon. First investigated by Chris Styles and then later by me, Chris and I co-authored a book entitled *Dark Object* about the Shag Harbour UFO incident which helped to familiarize interested readers around the world with this particular occurrence.

In the opening chapter of *UFOs and Water*, the author briefly touches on the Kenneth Arnold sighting in June 1947, the breakout case by a pilot that began the modern investigation into the UFO phenomenon which still has people struggling to this day in an attempt to explain these things. Carl slides back almost a thousand years and reveals that 1947 was not the first time that people took notice of these occurrences but simply the year that the mass media became interested. It is well known among researchers of the air- and water-UFO phenomena that witnesses tend to describe their sightings in the vernacular of the time, bound by its technology, an example being that in the fifteenth and sixteenth centuries, unidentified objects were often referred to as ships or galleons. Sometimes, however, witnesses stick to the facts when they describe what they see as was the case in 1067 over the North Sea off northeastern England: *"In this year people saw a fire that flamed and burned fiercely in the sky. It came near the earth and for a little time brilliantly lit it up. Afterwards, it revolved, ascended on high, and then descended into the sea."* This case—which incidentally was probably not the first but rather the one that actually made it into ancient records—sets the tone for what is to come.

Carl is one of only a few researchers who has actually tackled the subject as to how UFOs move through air and water and how they influence these mediums. With his special interest in water-UFOs, he has not remained static in his cataloging of waterborne anomalies. In Chapter 2, titled "Physical Influences of a UFO on Water," and continued in Chapters 3 and 4, he proposes a theory about how water-UFOs interact with water or influence it by their presence, providing numerous accounts and drawings to support his ideas. He draws on papers by Dr. Richard Haines (retired NASA research scientist and co-founder of NARCAP) and Larry Lemke (aerospace engineer with NASA) to support his hypothesis. Water-UFOs have been seen diving into, rising out of, hovering over, and traveling just under the surface of oceans and lakes, creating a swell-like effect on the surface. Indeed, a schoolteacher once approached me and Paul Kimball, documentary and TV director, while we were discussing UFOs in a small restaurant. He related an event which happened to him when he was seven years old and living right on the coast in Lockport, Nova Scotia. He was alone and became frightened when at about three o'clock in the afternoon, he witnessed some object just under the surface which created a large hump in the water and which traveled toward him causing him to run for the safety of his home. Both Paul and I were not inclined to think—given the description of this event—that this was some marine animal, such as a whale.

In succeeding chapters, a variety of subjects are discussed. Carl talks about the abductee phenomenon where it relates to water-UFOs. Since ships have reported these objects by the hundreds through the years, he shares some of these accounts as well as numerous cases involving sonar and radar. He also examines what people called "sea monster" sightings, some dating back hundreds of years, as possible misidentifications of water-UFOs, comparing various features of the "monsters" to what we know about UFO craft today and showing how what was mistaken as a biological monster could have been a metallic hard-bodied monster, a UFO, in its own right. However, he does not restrict his book to physical objects and goes on to describe UFOs taking on water (even getting permission from witnesses to do so in some cases), luminescence, water wheels, underwater solid lights, government secrecy, and other fascinating topics.

I must encourage the reader to move on into the book that follows. Some of the cases herein will make your hair stand on end. *UFOs and Water* by Carl Feindt is an honest and detailed investigation into an area of the UFO phenomenon largely ignored in the field of UFO research. The more than one thousand cases he has studied and the hundreds he has included here make this an especially interesting and readable book. I heartily recommend it to anyone interested in this phenomenon and to researchers of the same. It has been worth the wait, Carl.

Don Ledger
Author
Molega Lake, Nova Scotia,
Canada
January 2010

Explanation of Case Dates

All of the cases in this book can be found in their entirety on my website (http://www.waterufo.net), provided my website survives my eventual demise.

The dates shown in the cases are formatted as Month-Day-Year (MM-DD-YEAR) and have letters substituted where the date was not remembered. The following abbreviations should help make the connection easier. Note that although a single "E" could suffice for "Early," in order to keep a uniform format, I used two E's. Therefore, all dates should generally follow a XX-XX-XXXX format.

EE	Early (in the month or year)
LL	Late (in the month or year)
MM	Mid (month or year)
??	Unknown period (month, day, or both)
04?	Unsure of day or month or year. "It was sometime around the 4th" or "in April."
06~08	Between the 6th and 8th, day or month or year.
1968/9	Either 1968 or 1969
196.	The period signifies '60s. "It happened sometime in the '60s."
UNDATED-#	No listed date of the sighting in the publication. The date of the sighting would have to be prior to the publication date of the source in which it appeared.

In place of a month, sometimes the only thing that can be remembered is the season:

SP	Spring
SM	Summer
FF	Fall (Autumn)
WW	Winter

This would be a good page to dog-ear for further reference.

Introduction

When you consider the fact that a large part of our planet is covered with water, it is not surprising that reports of UFO sightings are not just confined to our skies. Sightings of these objects coming up from or going down into our oceans, rivers, streams, lakes, and ponds have been described by witnesses for quite some time. When a frightened witness reports something that is so far in advance of our technology that it reeks of magic, charlatanism, liquor, or ignorance, we ufologists, conscious of our own reputation, don the cloak of skepticism and review the case to try to understand the facts behind the witness's account. As more and more eyewitness accounts come in, we become aware that the housewife, the sailor, the janitor, and the company executive have all observed some physical characteristics common within their stories. As time goes by, these characteristics grow not only in number but in types as well. What was unbelievable in an individual account now becomes plausible given the similarities among collective accounts.

To this end, we owe a deep debt of gratitude to those who came forward in the past with their stories and continue to come forward, often enduring the ridicule, harsh skepticism, and disregard of others. We also thank the investigators who picked up this challenge without remuneration or recognition, for the most part, and keep on doing so, defining the parameters of this mystery that is still in the process of being explored.

Our First Ufologist (Well...Anomalist)

For our purposes, the history of UFOs goes back to an 18-year-old who, in 1892, left home and spent a year in the newspaper industry. Years later, after receiving a small inheritance, he savored the option of doing what he enjoyed, which was writing about mysteries he had discovered for which science had no respect. He had no love of "orthodox" scientists and their proclivity to rapidly dismiss the strange and unusual as nonexistent trash. The gentleman's name was Charles Fort. He might be considered the first real ufologist, although the term was unknown back then, and his interests were in many varieties of the unknown. His first book had the religious-sounding title of *The Book of the Damned*, [001] but it was in fact a book about what had been observed by local witnesses. Their accounts had appeared in newspaper reports as well as in respected journals as unexplained mysteries which were damned to obscurity by scientists of the period.

Mr. Fort had a great curiosity about the above mysteries that were reported and were dismissed with ridicule. He, as a researcher, I would imagine, had stumbled on a particular oddity on a few occasions that stimulated his drive to find other peculiarities of like nature. He fell into reading other newspapers and scientific journals to satisfy that curiosity, and it led to many cigar boxes (his pigeon-hole boxes) of various strange occurrences. These included, in part, fish falling from the sky, spontaneous human combustion, ghosts, and strange lights in the sky that could not be accounted for as comets or meteors. Keeping in mind that he lived from August 6, 1874, to May 3, 1932, the chances of these newspaper accounts being airplanes flying at night, prior to or even shortly after the 1903 flight of the Wright brothers, were slim to nonexistent.

A few of his "cigar boxes" are specifically of interest to us in this book: light wheels rotating on or just under the water, strange objects that floated but were not the whales or fish with which we are familiar, things that submerged, and above all, a few objects that appeared to come out of the water and fly away. I have checked all of the water-related cases in his books and most are Water Wheels, which will be discussed in Chapter 10 that deals with "Negative Aspects of Water Ufology," ufology being the study of UFOs. Following are a couple of examples from Charles Fort's books which illustrate areas of interest.

THROUGHOUT THIS BOOK, ONLY SPECIFIC CASES OR EXTRACTS ARE PRESENTED, WITH TEXT PUT IN BOLDFACE TYPE BY ME FOR EMPHASIS.

10-28-1902   *Ship* – Fort Salisbury, *Gulf of Guinea, Africa*

Zoologist, 4-7-38—that, according to the log of the steamship *Fort Salisbury*, the second officer, Mr. A. H. Raymer, had, Oct. 28, 1902, in Lat. 5° 31′ S., and Long. 4° 42′ W., been called, at 3:05 a.m., by the lookout, who reported that there was a huge, dark object bearing lights in the sea ahead. Two lights were seen. **The steamship passed a slowly sinking bulk of an estimated length of five or six hundred feet**. Mechanism of some kind—fins, the observers thought—was making a commotion in the water. "A scaled back" was slowly submerging.

One thinks that seeing for such details as "a scaled back" could not have been very good at three o'clock in the morning. So doubly damned is this datum that the attempt to explain it was in terms of the accursed Sea Serpent.

Phosphorescence of the water is mentioned several times, but that seems to have nothing to do with two definite lights, like those of a vessel. The captain of the *Fort Salisbury* was interviewed. "I can only say that he (Mr. Raymer) is very earnest on the subject, and has, together with the lookout and helmsman, seen something in the water of a huge nature, as specified." [002]

There was a series of occurrences in the summer of 1910. Early in July, the crew of the French fishing smack *Jeune Frédéric* reported having seen, in the sky off the coast of Normandy, a large, black, bird-like object. **Suddenly it fell into the sea, bounded back, fell again, and disappeared, leaving no findable traces**. Nothing was known of the flight of any terrestrial aircraft by which to explain. [003]

Mr. Fort was not in awe of scientists, so, while speaking of scientists...

Rejection of Reality

It is a bit painful when our respected gods of science turn their noses up at us as if we were lepers. At the end of 1968, a book sponsored by the U.S. Air Force was published called the *Final Report of the Scientific Study of Unidentified Flying Objects,* [004] which was conducted under the direction of physicist Edward Condon and based on case investigations performed by scientists at the University of Colorado. Ufologists, of course, had their fingers crossed for good luck, but even before the book was published, the cat was out of the bag, so to speak.

A scientist disputed the study's results, saying the assessment was a sham and coauthored the book *UFOs? Yes! Where the Condon committee went wrong.* [005] Robert J. Low, project coordinator of the University of Colorado study, wrote a memo with some damning evidence which was included in this book. The memo showed just how low the project administrators had sunk in their attempt to make people believe that the Condon committee was conducting an objective investigation:

The trick would be, I think, to describe the project so that, to the public, it would appear a totally objective study but, to the scientific community, would present the image of a group of nonbelievers trying their best to be objective but having an almost zero expectation of finding a saucer. [005]

Nevertheless, the book (*Scientific Study...*) was published with the conclusion by the director, Dr. Condon, at the very front of the book instead of at the end, relating that further investigation into UFOs would be unproductive and "unlikely to yield major scientific discoveries." Well, I read the report, and roughly one-third of the cases' conclusions read "unexplainable." Obviously Dr. Condon didn't listen to his own group of scientists. So much for sour grapes. Science has a bad reputation that goes back much further.

In the 17th century, there were visual sightings of rocks falling from the skies. However, at that time many scientists could not conceive of such an idea, but we are all familiar with meteors and meteorites today.

In my research of UFOs going into the water in the 17th and 18th centuries (yes, some could have been those same meteors), another strange event reared its ugly head. It was the occurrence of giant waves (now called rogue waves) that smashed those wooden sailing ships down to Davy Jones' locker. But if a ship survived such a wave, far be it that the captain or crew would report it, for everyone knew that such a thing did not exist. They were only reported by drunken sailors or captains looking for excuses for lost cargo or damage to the ship. Sounds very much like seeing a UFO these days. But again, science came to the rescue in the 1990s in the form of two German satellites which found that rogue waves do exist. With the help of modern technology, scientists continue to study ways to predict their occurrence and warn ships of the danger.

Given the two previous examples, heaven forbid that anyone would take any anecdotal accounts seriously (I am being facetious here), especially when the tales are about unidentified flying objects and corroborate each other, as they do in ufology. As an enlightened scientist, trying to educate his colleagues, so aptly titled his book, *UFOs? YES!*

Water, water, every where,
And all the boards did shrink;
Water, water, every where,
Nor any drop to drink.

—From *"The Rime of the Ancient Mariner"*
by Samuel Taylor Coleridge (1772-1834)

In our newly acquired, scientific paradigm of the solar system, if not the universe as well, there is, indeed, "water, water, every where," but only one humble little planet in this solar system is known to accommodate it in liquid form on its surface. *Why...that's us.*

Why Study Water?

We study water because it restricts our technology. Water is denser than the atmosphere in which we fly and slows the forward motion of ships for the same amount of horsepower in respect to aircraft. Given its higher density, it is also harder to penetrate than air. A good example of this is a dive from a ten-foot high diving board and landing flat on your stomach (known as red belly in my day). Another example would be how a wide-bladed knife cuts water like butter when the edge "cuts" water. However, rotate the knife 90 degrees so that the flat surface is moving forward, and more effort is needed to push the knife.

Propulsion in water is another issue. For a surface ship, this is a very small problem as the engines are above the surface and have a constant supply of air which is necessary for ignition to occur. Likewise, a submarine carries its own oxygen and fuel, while a nuclear sub manufactures oxygen by division through electrolysis of its abundant water supply. Of course, as efficient as a nuclear sub is, it is also very, very heavy due to the shielding required to protect the crew from radiation. In either case, neither type of ship flies.

This brings us to aircraft and water. We know that they can fly over water and have done so for quite some time now. Certain aircraft can also land on water, given that the water is free from floating debris, because the forward speed hitting this debris could cause damage to the aircraft's skin and structural components. Two well-known examples of these aircraft are the Consolidated PBY Catalina and the Martin M-130 Clipper, both of WW2 vintage. There have been several of these "flying boats" in use in both the military and civilian areas. But can they submerge into the water? No, they cannot. First, there is no engine that can perform in both mediums (air and water) without being totally contained, and second, the aircraft structure would have to be greatly strengthened in order to "crash" through the water's surface. This would add a great deal of weight to the aircraft and restrict its airborne ability. "You can't have one without the other" (from the song "Love and Marriage"). During the 1960s, the U.S. Navy studied the feasibility of a combination seaplane-submarine. Details of this are discussed in Chapter 8, "Secrecy." It goes without saying that the Foreign Technology Division (FTD) of the U.S. Air Force at Wright-Patterson Air Force Base in Ohio would also be interested in a craft that could function in both the air and the water.

So given all these restrictions to *our* technology, why should we study *UFO* technology? The answer is obvious: The aliens and their craft have overcome these limitations and they can move rapidly between both mediums. But why should a ufologist study UFOs *and* water? Well, by examining what a UFO does physically *to* that medium, which is not as evident in the atmosphere, we can pick up clues as to its operation. While the UFO-water connection has been largely ignored by other ufologists, I will explore this avenue more closely in the coming chapters.

Therefore, this book, based on the investigations of thousands of eyewitness accounts that ufologists have gathered through the years, is my small contribution to the further study of UFOs. I do this, not for fame or fortune, but in hopes that this text will endure and encourage others who share my interest in the UFO-water connection or their own specific interests in ufology.

Chapter 1

Water-UFO History

1.1   In the Beginning...of Ufology (June 24, 1947)

Many books start the UFO era with the famous Kenneth Arnold sighting, which sparked what became the modern UFO phenomenon. Much has been written about this sighting, so rather than reviewing an evergreen, I would like to show how UFOs were regarded as time progressed and what it meant eventually to their relationship with water.

Since UFOs were recognized by relatively few but forward-looking mentalities, even these people were restricted in their conception of what craft made by some superior intelligence could do. The conception of a machine that could outperform our highest scientific technology was already an embarrassing event, not only to us, but to the scientific community which could not conceive, even today, of another intelligence that could be superior to ours. In short, if we could not conceive of it, how could anyone in the evidently dim-witted universe build one? Well, they did. Therefore, when these things were bouncing around our skies back then, it was one thing to say they were machines created by a superior intelligence of unknown origin, but it was quite another to say that the source was extraterrestrial (ET). So the early ufologists, already considered nut cases by those who were not acquainted with the tales of witnesses like we are today, were denigrated for their ideas. Like a Monday-morning quarterback reviewing what happened, I can only humbly appreciate what those who preceded us went through.

Resistance rose against the concept that such a craft could really exist. We talk about holding concepts at arm's length to show that we are not ready to grasp strange ideas, and in doing so keep the illusion of being open-minded.

However, even ufologists can be arm's-length intellectuals. So where is the craft that goes into water? There were a few of them; however, despite minimal study of such things, some thought it possible that a mechanical craft without a crew—an Unmanned Aerial Vehicle (UAV)—*might* accomplish this. Although we could fathom the possibility of a mechanical ability that could do strange things, we were still at arm's length to accept pilots and crew in these craft. After all, a machine of superior technology is one thing, but to have occupants in these craft seemed impossible because what these craft did in their travels would make jelly of a human being on any wall within a UFO (according to our laws of physics). But, more and more cases came in with them being seen, so eventually we had to admit craft members to the fold. The proverbial arm shrank in one instance but not in others. Then came the abductees, who were considered bizarre, until reports came in from all corners of the globe and shared a commonality with each other. The hand came yet closer to the body (acceptance).

As time advanced, more and more cases were put together in various publications: general-interest magazines such as *Life* and *Look*, ufological journals like *Flying Saucer Review*, *A.P.R.O. Bulletin*, and NICAP's *UFO Investigator*, as well as newsletters of various other organizations. Acceptance? Yes, but in a very limited way. Of what importance is water to a UFO? Probably none, and who cares anyway? It is what is above our heads that is important, right?

So arm's length works yet again, but not as strongly as the other controversial aspects of ufology. Then, in 1968, a book was published with the title *UFOs over the Americas* [006] by Jim and Coral Lorenzen, with one chapter of ten pages on UFOs that went into and came out of water. This was a start in the recognition of water-UFOs, but it was not enough to satisfy me. My frustration grew. After that, in 1970, Ivan T. Sanderson, a biologist and spy who had worked for the British government, wrote a book called *Invisible Residents*.[007] At first I was delighted, but my joy changed to disappointment. The book contains all water-related UFO cases with the premise that an alien civilization coexists with us on this planet. It never addresses *how* they are able to enter or exit water, which is something about which we inquisitive individuals cannot seem to find a solution. Back to the drawing board.

Nevertheless, that stayed on my shelf until I was able to retire. At that point, some thirty years later, I saw that no one else had tackled the water-UFO question. Having gained some friends in the UFO community, I was able to begin my quest for a broader understanding of the water-UFO connection. Sanderson had done a good job for the period up to 1970; there were 262 references to water-related UFOs. I currently have almost 1700 cases (dated from 06-09-0597 through 02-05-2013), and I have made a connection pertaining to how the UFO *may* affect water, which will be elaborated upon in the coming chapters.

When one thinks of UFOs, the image is of a flying object. This is already a misconception of what these craft actually do. In all their shapes and sizes, none have an airfoil shape necessary for lift in our atmosphere and, therefore, do not fly like birds, from which we derived our concept of lift for our own aircraft. Indeed, why should they? A true spaceship should be able to travel in any environment it encounters and not rely on conditions at the time of arrival. Would we expect a glider to function on the airless moon? Of course not.

In grammar school, I remember seeing a drawing depicting a scene from over 1,000 years ago of a well-dressed Asian gentleman sitting in a chair that had rockets attached to the back of it, all fused together for the obvious purpose of transportation. Missing were the means of flight control, so that if this experiment had gone forward, the results could have been disastrous. Yet where are we today? Three chairs on one very large rocket (three stages) with directional control. Progress? Of sorts.

So why should we get excited about some water-related reports of our ever-popular UFO cases? Because in these cases we can see what we must accomplish in order to actually travel to the stars: a new propulsion system in addition to an independently operating field that surrounds the craft to protect it from micrometeoroid hits, acidic climates, hostile gunfire, or whatever else the craft may encounter.

The word "field" might be unfamiliar to most except in the context of a sports field or a plowed field as in farming. However, its use in the succeeding chapters would be better served by the dictionary definition: "*Physics* **a** A portion of space at every point of which force is exerted. **b** The force exerted therein: the magnetic field, a field of force." [008]

In ufology, the term electromagnetic is frequently used in relation to a field surrounding a UFO. Unknown, however, is how the field used by a UFO operates and what it is that comprises said force. Since we are dealing with an unknown, I would like to view this field as a clear beach ball with the UFO centered within it. Although this does not satisfy the scientific concept of a "force field," or "field of force," we have seen this idea used in motion pictures such as *Star Wars* (1977) and *Independence Day* (1996) to show a protective wall surrounding a UFO. This science-fictional force field may be defined as "an invisible barrier of exerted strength or impetus." [009] So I beg the forgiveness of the scientific community because, due to the nature of the term "force field," and not necessarily the scientific principles behind it, I have no recourse but to rely upon the implied meaning as opposed to the specific meaning of this term until a better word or phrase can be instituted or our knowledge of what it is exactly that surrounds the UFO is identified.

In the upcoming chapters, we will see that the craft has in essence a three-way toggle switch of sorts: 1) On, in "Normal" mode, 2) Off, and 3) Reverse. At night the on or off state is rather obvious as the craft appears as either a glowing ball (on) or it is visible or barely visible by itself (off). During the daytime, the sun's rays obscure the colored field, so the UFO is seen as itself regardless of whether its field is on or off.

In some eyewitness accounts, there are large discrepancies in the amount of observed facts about a craft.

As an example, people at a beach with their back to the water suddenly hear a loud splash and turn around to see what has happened. In this time frame, the splash has indicated that something has already entered the water (therefore nothing is seen), or they see the blur of a craft as it disappears after having come out of the water. Distance becomes yet another factor here, as the witness is on the beach, and the UFO could be at a great distance from him or her. The horizon also comes into play because if the object or meteor goes beyond it, the UFO would appear to be going into the sea, whereas it could, in actuality, be traveling in a straight line back into space on the other side of the planet.

In one case, fishermen had their boat rocked as a UFO started to ascend to the surface. The rocking motion set them looking for the cause. Distance in this case was quite close. This then led to an observation of the water conditions as the UFO came even closer to the surface and caused more effects on the surface of the water. The men who looked at the place where the UFO was to come out were now visually geared to observe all of its aspects. The object not only exited the water, but also hovered briefly, allowing further observation of the craft. All this would make the tabulation of characteristics of water cases almost meaningless unless cases of similar natures are considered and grouped.

In the following chapters I hope to show some of the physical effects that have been seen and recorded by people around the world: men, women, and children of all ages, nationalities, and religious beliefs. Witnessed effects are virtually all we have in the way of physical proof, as these craft are too advanced to lose their proverbial tailpipe here. Physical traces left on land have been pursued by UFO researcher Ted Phillips. They include broken tree limbs caused by UFOs descending, calcified soil in a ring, and burnt grass roots. Mr. Phillips has an advantage because in the cases that he investigates, samples of the affected items can be taken to a laboratory and studied. In water-related cases, would-be evidence drifts away or is diluted. However, many of the water characteristics can be seen from the wealth of witness testimony. Heat and pressure of the field surrounding the UFO act on both environments equally.

This is why my interest is in water-related UFOs. Through its changes in form, water is the standard we can use to see the effects of these craft. A rock placed on the water's surface will rapidly descend to the bottom while a block of wood will float. Observing what selected materials do to water gives us a clue as to their physical nature, just as the influence of a craft on water will reveal information about a craft.

1.2 Some Acronyms That Describe Water-UFOs— And the Futility of Some Acronyms

USO	Unidentified Submerged Object #1 in use
AFO	Amphibious Flying Object
UAO	Unidentified Aquatic Object
UNO	Unidentified Nautical Object
UUO	Unidentified Underwater Object
UWO	Unidentified Water Object

"Etcetera, etcetera, etcetera"

—*Yul Brynner in the movie* The King and I *(1956)*

Now consider describing the sighting of an unidentified aircraft (A) and its various aspects as we do for a UFO. We could have:

ULA	Unidentified Landed Aircraft
UAA	Unidentified Ascending Aircraft
UDA	Unidentified Descending Aircraft
UCA	Unidentified Crashed Aircraft
UTA	Unidentified Turning Aircraft

In all the above situations, we know that the object is an aircraft. Therefore, to fasten a group of letters to describe what it is or is doing is pointless. In the case of a UFO, we are aware that the object, which has *not* been identified after a period of research and exhibits extraordinary flight characteristics, is an Unconventional Flying Object, as explained in Paul Hill's wonderful book by the same name. The acronym still applies. Therefore, I feel it is unnecessary to alter the phraseology when we talk about the craft's nautical characteristics, inasmuch as both air and water are governed by the one principle in physics known as "fluid dynamics."

1.3   A Question of History—How Long Water-UFOs Have Been Seen and by Whom

In our haste to get the latest information on current UFO cases, we ufologists are asked by well-meaning people why there were none of these reports prior to 1947, the supposed beginning of the UFO era. We scratch our collective heads and agree that there must be some record of this phenomenon somewhere if it is a continuous phenomenon. Some researchers, having broad interests, bring their knowledge of these cases to the forefront, while others engaged in historical research, with only a passing interest in UFOs, come up with other cases. The database begins to expand.

The following may not be viewed as credible UFO sightings, but in light of today's knowledge of the subject, the nagging question remains: Could they have been UFOs?

One of the earliest cases I have goes back to the year 1067:

??-??-1067   *North Sea off northeastern England, UK*

The year is 1067 A.D. We learn in Geoffroy Gaimar's *L'Estoire des Engleis*: "In this year people saw **a fire that flamed and burned fiercely in the sky. It came near the earth and for a little time brilliantly lit it up. Afterwards, it revolved, ascended on high, and then**

descended into the sea. In several places it burned woods and plains, and in the County of Northumberland, this fire showed itself in two seasons of the year." [010]

And some other early cases:

??-??-1361   *Sea of Japan off western Japan*

In 1361, **a flying object** described as being "shaped like a drum about twenty feet in diameter" **emerged from the inland sea off western Japan**. [011]

Twenty feet is the size of the front of my house, and since the dimension is given as a diameter, a circle is implied. A submarine in that time period? A circular submarine at *any* time? I don't think so.

In the following, although the case involving Christopher Columbus cannot easily be identified with a water-UFO, his ship would not be in a position to see land, or anything else, over the horizon until the next morning. Therefore, the source of light which he sees is unexplained. Enjoy a piece of history:

10-11-1492   *Atlantic Ocean off San Salvador*

Thursday, 11 October. Steered west-southwest and encountered a heavier sea than they had met with before in the whole voyage. Saw pardelas and a green rush near the vessel. The crew of the *Pinta* saw a cane and a log; they also picked up a stick which appeared to have been carved with an iron tool, a piece of cane, a plant which grows on land, and a board. The crew of the *Nina* saw other signs of land and a stalk loaded with rose berries. These signs encouraged them, and they all grew cheerful. Sailed this day till sunset, twenty-seven leagues.

After sunset steered their original course west and sailed twelve miles an hour till two hours after midnight, going ninety miles, which are twenty-two leagues and a half; and as the *Pinta* was the swiftest sailer, and kept ahead of the Admiral, she discovered land and made the signals which had been ordered. The land was first seen by a sailor called Rodrigo de Triana, although the Admiral at ten o'clock that evening standing on the quarter-deck **saw a light, but so small a body that he could not affirm it to be land;** calling to Pero Gutierrez, groom of the King's wardrobe, he told him he saw a light, and bid him look that way, which he did and saw it; he did the same to Rodrigo Sanchez of Segovia, whom the King and Queen had sent with the squadron as comptroller, but he was unable to see it from his situation. The Admiral again perceived it once or twice, **appearing like the light of a wax candle moving up and down,** which some thought an indication of land. But the Admiral held it for certain that land was near; for which reason, after they had said the Salve which the seamen are accustomed to repeat and chant after their fashion, the Admiral directed them to keep a strict watch upon the forecastle and look out diligently for land, and to him who should first discover it, he promised a silken jacket, besides the reward which the King and Queen had offered, which was an annuity of ten thousand maravedis. At two o'clock in the morning the land was discovered, at two leagues' distance; they took in sail and remained under the square-sail lying to till day, which was Friday, when they found themselves near a small island, one of the Lucayos, called in the Indian language Guanahani. [012]

The next case came to light because of a reason I had mentioned before, a non-believer who was a historian. However, he was evidently intrigued enough to write the following (name intentionally left out to give him some privacy):

"As a doctoral student in the humanities, I've been spending much time reading 18th-century British periodicals. In one of them, I have found a report of an unexplained phenomenon. I leave it to your judgment and experience to decide its merits as a UFO sighting. Extract of a letter from Edinburgh, Sept. 8, 1767, follows:" [013]

09-08-1767   *River Isla near Coupar Angus, Scotland, UK*

"We hear from Perthsire [sic-Perthshire] that **an uncommon phaenomenon was observed on the water** of Isla, near Cupor [sic] Angus, preceded by a thick, dark smoke, which soon dispelled, and discovered a large, luminous body, like a house on fire, but presently after took a form something pyramidal, and rolled forwards with impetuosity till it came to the water of Erick, up which river it took its direction, with great rapidity, and disappeared a little above Blairgowrie. The effects were as extraordinary as the appearance.

"In its passage, it carried a large cart many yards over a field of grass; a man riding along the high road was carried from his horse, and so stunned with the fall as to remain senseless a considerable time. It destroyed one half of a house and left the other behind, undermined, and destroyed an arch of the new bridge building at Blairgowrie, immediately after which it disappeared." [014]

It must be noted here and now, before we attribute the physical destruction to a meteor, that in order for this "meteor" to go from "the water of Isla, near Cupor Angus to the water of Erick, up which river it took its direction," a course change had to be made, something not associated with meteors. We are further informed that "it carried a large cart many yards over a field of grass." Not very exciting (except to those of that period), as we have cases in ufology where cars have been lifted and, in a particular instance, transported over national borders. (The occupants had passport problems because they had not been prepared to travel outside their country.)

06-11-1881   *Between Sydney & Melbourne, Australia*

King George V of England

In the *Cruise of the Bacchante*, a work compiled from the journals of the late King George V of England (then Duke of York) and his brother Prince Albert Victor, is reported the sighting of a phantom ship. The brothers served as midshipmen on H.M.S. *Bacchante*'s round the world voyage between 1879 and 1882.

At 4 a.m. on July 11, 1881 [should be June, not July], while the vessel was sailing to Sydney from Melbourne, Australia, the late King's diary says **an eerie red light was noticed**. His account follows:

In the midst of the red light, the masts, spars and sails of a brig two hundred yards distant stood out in strong relief as she came up on the port bow. The lookout in the forecastle reported her as close to the bow, while also the officer of the watch from the bridge clearly saw her. So did the quarterdeck midshipman, who was sent forward at once to the forecastle, but on arriving, there was no vestige or sign of any material ship. The night was clear and the sea calm.

Thirteen persons altogether saw her. Two other ships of the squadron, the *Tourmaline* and the *Cleopatra*, who were sailing off our starboard bow, asked whether we had seen the strange red light. [015]

??-??-1947   *Pacific, Humboldt Current off South America*

Thor Heyerdahl (*Kon-Tiki*)

"We saw the shine of phosphorescent eyes drifting on the surface on dark nights, and on one single occasion, **we saw the sea boil and bubble while something like a big wheel came up and rotated in the air,** while some of our dolphins tried to escape by hurling themselves desperately through space." [016]

So here a discoverer of a new continent, a researcher, and a king of a powerful state all observed unexplained phenomena. U.S. presidents Jimmy Carter and Ronald Reagan also observed UFOs. Even though all these were important, influential people, they saw the same things that a host of people of various statuses in society have also seen throughout history, though some closer to them and in more detail.

1.4   Verbal Testimony as Anecdotal Tales

Stories are given, in many cases, with the assurance that the witness will testify in a court of law to the actuality of what has been observed. Skeptics and debunkers, however, would raise the specter of misidentification in all cases, yet they have no way to renounce the witness other than by suggestion of error via the implication of substance abuse, delirium, or misconception. What we fail to realize is that the testimony given by sane, sober, and reputable people who witness a crime may condemn a person to death, yet in regard to UFO phenomena, these same witnesses are called into question, and their testimony is treated as a joke by both the news media, in many instances, and by skeptics and debunkers of questionable repute.

1.5   Case Examples of Out of, Into, Hovering Over, and Floating on the Water

Many people do not know about the reality of UFOs, much less about them going into, coming out of, or hovering above water. Therefore, I will introduce you to these types of cases through eyewitness accounts.

1.5 a.   Coming Out of Water

This is possibly the most terrifying apparition to someone who is on a body of water. It is so totally unexpected, and its arrival, causing a great disruption of the water, also makes the witness fear for his life even before the craft is in view. The boiling, bubbling mass of water could be signaling the exit of a large whale or even the eruption of an underwater volcano. Nevertheless, these cases provide the best evidence of the craft because the only other "things" that exit water are flying fish and the Polaris missile, which had its first launch on July 20, 1960, and, therefore, cannot account for those cases seen prior to that date. Even those that came after that date would be hard-pressed to explain the majority of the crafts' actions that cannot be applied to the normal flight of a missile.

06-EE-1950   *Atlantic Ocean off several Argentinean cities*

Suddenly and breaking the nocturnal silence, he heard **something similar to a loud noise of water violently disturbed.** Several sheep, who were sleeping in a field, jumped up alarmed and hurriedly ran away. There was no wind, storm, or thunderstorm that might have explained the event. Immediately **a luminous object of oval form appeared emerging from the sea about 500 meters from the shore.** It rose up vertically to a certain altitude and then made a turn of ninety degrees and disappeared towards the northeast in the direction of Argentine territory. [017]

07-LL-1955   *Lake Ontario near Niagara Falls, NY, USA*

Witnesses were seven in total, two families. The families were on the beach on an outing when **an object suddenly rose out of the water and hovered about 12-14 feet in the air**. The families moved toward their cars when the object rose out of the water, and it then moved toward them, hovering right overhead when they were by their vehicles. The object was 100 feet across, disk-shaped with a dome, and a dull metallic surface. They tried to start their cars but the motors wouldn't work at all. [018]

11-13-1965   *Off Rugged Islands, New Zealand*

"We were about half a mile off Rugged Islands at the northwest of Stewart Island about 11:30 on Saturday morning when **we saw this thing come out of the water**," Mr. Hanning said.

"It rose out about 15 ft., a tapered object perhaps 12 ft. long at the waterline and 5 ft. at the top. Then about 30 ft. away from it was a box-shaped object, about 10 ft. long and 5 ft. high.

"There was no sign of any periscope or railings—nothing but the tower and box were visible."

The water was smooth and the object was in clear view only about 300 yards away, Mr. Hanning said.

"We had it in sight for 10 to 11 seconds, then I would say we were either seen or heard.

"**There was a great surging of water, like a tide boil,** then the objects disappeared." [019]

SM-??-1972   *Lake Erie close to Buffalo, NY, USA*

One afternoon in the summer of 1972, this man (who has requested that his name not be used) decided to do some fishing. He took his boat out into the eastern end of Lake Erie, close to Buffalo, and started to fish. This day, however, the fish were not cooperating, and the only thing disturbing the otherwise calm surface was the plop of his lure. But the lake did not remain placid for long. Without warning, **what appeared to be a large, silver, disk-like object burst from under the water and shot straight up into the summer sky with such a force that the wave it created almost swamped the fisherman's boat!** [020]

03-05-1979   *Atlantic Ocean near the Canary Islands*

The photographer took five shots of the splendid spectacle in the sky, then took some pictures of the fishing boats that were sailing before the coast. He was about to dismount his camera when **he suddenly saw something emerging**

from the ocean. "The light was not at all white, more like old ivory, and very shiny. **It came out of the water, absolutely. I can swear to that. It rose into the sky at a tremendous speed**. The big ball of light looked like suddenly released energy. On the top the UFO was shaped like a pyramid." [021]

02-08-1981   *Isla Cristina in the Gulf of Cadiz, Spain*

Two boys studying in BUP [BUP stands for Bachillerato Unificado y Polivalente = Unified and Multipurpose Degree] observed an object that emerged from the water on the night of Sunday, February 8. According to their statements, **they suddenly saw how a light was coming up from the depths of the sea about 300 or 400 meters (1000-1300 feet) from the beach** (in the area of Critina [sic—Cristina] Island, province of Huelva). "This light was getting brighter and brighter, which made us think of a submarine, but **suddenly it came out of the water and stayed still at a height of about 500 meters** (1600 feet), disappearing without a trace soon after." [022]

09-03-1989   *Caspian Sea off Sari, Iran*

The eyewitness (an employee of our correspondent) was walking with some friends on the beach of the Caspian, near Sari, at 6:30 p.m., when suddenly this **brilliant orange-coloured object**, about 8-12 m in diameter, **rose up out of the Caspian Sea at a distance of some 250 m from them**. There were, of course, also other people on the shore at the time, and the eyewitness says the UFO generated such a general panic that, in a matter of just a few seconds, the entire beach was empty! [023]

07-25-2006   *Waterhen Lake, Saskatchewan, Canada*

He then quickly added, "What was that?" I asked him what he had seen and he described it as a yellow, round object consisting of two disks with an orange-colored center. **He said that it had come out of the water** and flown rather erratically in a vertical and sometimes horizontal fashion. [024]

I would like to point out that the above cases are just a small representation of the total, and others of this ilk will appear in later chapters because they contain other properties that are even more important to our understanding. I would also like to bring to your attention as you read these cases, all of which have the sighting location, that it is not just crazy Americans who see these craft. Sightings occur worldwide .

1.5 b.   Entering Water

Here we enter the fight between ufologist and skeptic, as the "sighting" from the former's point of view is a strange or unconventional flying object versus the latter's meteor, flying fish, crashing aircraft, or Venus. However, many of these reports have additional qualities that put them into a class by themselves.

UNDATED—4   *Bismarck Sea off Pityilu Is., Admiralty Isls.*

"It was sort of weaving back and forth. And then, all of a sudden, it straightened out and of course the crowds gathered. More and more people—Look! Look! Look! Look! Then it went down the island, **it turned at the eastern end, and it landed in the water, which was all reef. It wasn't very deep.** And when it landed in the water, the men pulled their canoes into the water to go...to rescue anyone who needed it. But when they got there, there was absolutely no trace whatsoever. In fact, because it was a full moon, people were outdoors, and you could see the splash! You could see, it just...phussssh!" [025]

12-13-1956   *Off the coast of Venezuela*

A Swedish ship wired the harbor control at La Guaira saying that **a strange, cone-shaped object was falling vertically into the sea**, that it was very brilliant and gave off "strange glares." The time was 9:50 p.m., and the object was seen by the captain and several of the crew. When the object hit the water, an explosion was heard, **then the sea where the object fell became brilliantly colored. After the colors subsided, the sea became very disturbed with a "boiling motion," which continued for some time.** [026]

SM-??-1957   *Penobscot Bay off Camden, Maine, USA*

While enjoying the view of the sky and water of the Penobscot Bay area, which leads out to the Gulf of Maine, the witnesses become aware of a pinpoint in the sky that is growing and descending rapidly. Thinking at first that it is a plane, the reality of what it is becomes rapidly vivid. The craft is a silver, disc-shaped object and has what appears as a metallic dome. However, the shocking part is that **it crashes into the bay at a rapid speed approximated at about 100 mph with very little disturbance of the water. In the witness's own words: "It entered the water at a steep angle making no sound and barely a ripple or splash."** The distance to the craft from the shore is given as several hundred yards. [027]

01-08-1959 *Derwent River off Risdon, Tasmania, Australia*

When about 100 ft. from the surface of the river, its descent was arrested and it begxxxxxxxxxxxxxxxx [x's indicate text missing from document] horizontally in a southerly direction for about 100 ft. when it agxxx xxxcendxx vertically and **entered the water about 1800 ft. distant without any apparent disturbance.** [028]

FF-??-1966/7 Pacific Ocean off San Juan Capistrano, CA, USA

Suddenly the "craft" zipped to an extremely high elevation, almost disappearing. We thought the "event" was over. People were all talking to each other about what we were seeing. **Then the object sped at an angle towards the ocean and plunged directly into the water. It was gone.**

We spoke with several of the other campers and the state rangers, and we all agreed that this was something which was not of earthly origins. [029]

06-18-1974   Navegantes Beach, Santa Catarina, Brazil

"**A strange, metallic object** flew around over the sea off the State of Santa Catarina. It was at a distance of some thousand metres or so from the shore at Navegantes, a place 100 km distant from Florianópolis. **After hovering for some six minutes above the sea, it veered at a 90° angle towards the beach and then dropped into the water.**

"The **fisherman** Avelino Severiano Generosa, aged 46, at once put out in his motor-launch to the area but **found only foam on the sea where the object had disappeared.**" [030]

Keep this foam in mind for when you read about Shag Harbour in Chapter 18.

08-01-1975   Trinity Lake, New York, USA

Brad continued to watch as **the object slowly settled toward the lake's surface**. He said it was so slow that it was like watching the minute hand of a clock—you know it's moving, but can hardly perceive its progress. When the object was about midway between the tops of the trees and the surface of the lake, a nauseating odor "somewhat like rotten eggs" suddenly hit Brad, and he took one whiff and bent over and vomited. He retched a couple of times, then looked back up to the UFO which was by then just above the surface of the water, and Brad began to **hear a slight**

hissing sound. There was no splashdown; the object slowly settled into the water. Complete submersion took about two minutes, and ripples radiated out from the object's location. [031]

08-LL-1981   *Hamilton Harbour, Ontario, Canada*

Although the road had many curves, the craft was seen to move in a straight line towards Hamilton Harbour. As it approaches the harbour, **it appeared to slow down and then angle itself downward and gently entered the water.** At the point of entry (rough approximation), it can be seen on the nautical chart that the water depth at this point is 7 or more meters (23 feet) and gets deeper in the direction of Lake Ontario, so the UFO had at least 10 feet to spare. Therefore, a craft of 13 feet in height could be covered by water. There are many cases of UFOs submerging in shallow waters. [032]

11-03-1986 *Lake Trasimeno off Castiglione del Lago, Italy*

According to some sources, two people saw three objects arranged in the form of a triangle and, according to other sources, many spherical bodies, brightly lighted and blue in color. The formation descended to a low altitude and appeared to be about five meters in size. **Having reached the lake, the objects dived perpendicularly into the water.** [033]

04-14-2000   *Montauk, Long Island, New York, USA*

We were fishing off the pier at Montauk Point, and I see a blazing light flash in the sky. Then I see **3 disk-shaped objects hovering about 20 feet above the ocean. And about 5 minutes later, they plunged into the water.**
They were spinning in circles, but tight circles. It seemed there were windows in the craft, but I am not positive. When they plunged into the water, **there was really no splash; it**

was like there was an opening, and it just swallowed them up. [034]

06-27-2006   *A pond near Flagler, Oklahoma, USA*

On June 27, 2006, five of our grandchildren were fishing in our pond about 3:00 p.m., when **they saw a big splash on one side of the pond. Everything became still, and the pond became like glass as the object sank. Then they saw a whirlpool in the pond about 8 to 10 feet across that moved very fast across the pond towards them. When it hit the water, there was steam that went up from the object, then the object disappeared to reappear in another area.**

One of the children got on a boat that was at the bank, and she could see lights around the sides, and the object looked like metal. Two other children said they saw the lights, but two of them said it looked yellow under the water. Two were still watching it when **they saw the water and grass around the pond next to them start to move violently,** so they got scared and ran to the house. [035]

1.5 c.   *Hovering over Water*

Here the craft have been seen, not going in or out of the water, but remaining close to it without movement. Hovering over the water with no water disturbance or interaction is of little value as it duplicates what a UFO does over land as well. Since this is the situation, I will skip examples because almost all of the cases have this as a feature in the sighting at some point. In the next chapter, this same hovering has an unexpected effect on the water in several situations, and I will use case illustrations to explain them.

1.5 d.   *Floating on Water*

In this section, cases have the word "floating" in two senses. First is floating in the air, and second is floating on the surface of a body of water. Since the craft is described in many reports as having no apparent seams, it is hermetically sealed so that the craft can float. Also, there are other cases where the craft seems fixed in position, with waves crashing against it, which would indicate that the craft is under power and not in free movement.

06~08-??-1947 *Mediterranean Sea 20 miles south of Malta*

Fishermen on a boat 20 miles south of Malta were raising their nets with a catch of fish when **they saw an object floating on the water's surface** that looked like a black submarine. The fishermen were frightened because they thought it looked more like a monster than a submarine, so they quickly pulled in their nets and started the boat's engine. At that moment a bright light from the "submarine" lit up the whole area and "little men" began running over the deck of the object. The fishermen couldn't make out much detail from their boat, but whenever the light illuminated the "little men," they could see some sort of apparatus around their waist.

When the witness was asked how tall these men were, he replied, "About the size of a 10-year-old boy." After a few minutes, the "little men" entered the "submarine" which began to glow so brightly that the fishermen couldn't see the object. It then submerged. [036]

01-15-1956   *Pusan, South Korea*

Just a week later another incident occurred about 1,000 miles west of Komatsu. This one is unique in U.F.O. sightings for two reasons:

One: It was seen by a large number of witnesses including civilian and American personnel.

Two: The object was under direct and relatively close observation for about 90 minutes!

The incident began about 8 p.m. at Pusan, Korea, on January 15. The object was described to military authorities as being "about the size of a large washtub and emitting a blue-gray glow. **It was seen falling into the water about 50 yards offshore near Heunde** [sic-Haeundae]."

It was early enough in the evening to attract the attention of a large number of Korean townsfolk. They reported that **the glow continued for about an hour and a half before the object "apparently sank into the sea."**

By this time Korean National Police arrived at the scene and they, in turn, alerted U.S. Military Police. Cpl. Ben Elliot, an M.P. on patrol duty that night, was on the scene quickly enough to observe **the object floating in the water for almost an hour**. [037]

06-06-1965   *Fraser Island, Australia*

On June 6, 1965, private aircraft pilot, Mr. C. Adams, and a television cameraman, Mr. Les Hendy, reported seeing four or five **"mysterious objects" floating in the sea** 3 miles east of Fraser Island, 150 miles north of Brisbane, at about 11:30 a.m. Mr. Adams first noticed two of the objects from a distance of about 8 miles while flying over Fraser Island. The weather was clear and the objects appeared to resemble two big, dark-coloured logs. They were narrow and up to 100 feet long. As he steered toward them, two or three similar but smaller objects appeared near the other two. They did not appear to move, but seemed to **"sort of submerge"** when the plane was about one mile away from them. From the air they appeared to be lying just below the surface and, **when "submerging" from sight, seemed to do so without disturbing the surface**. [038]

09-15-1968   *Cornwallis River, Nova Scotia, Canada*

"We ran over and it was just floating in the air and appeared to be oscillating like a spinning top. We were 40 feet from it."

"We stood there and watched it for about 10 minutes—then it slowly started heading down for the Cornwallis River. **It went onto the water and moved with the current.**"

There was no splash when the object hit the water, and it made no noise in the air, the youth said in an interview.

"When we got there, it was nothing but a big, black shadow under the water. It broke surface twice." [039]

12-15/16-1968 *Pacific Ocean near Hawk Inlet, Alaska, USA*

Ken Marlowe, owner of the cargo boat *Teel,* and Ralph Kern reported seeing a "pure white light"—ball shaped and about 20 feet in diameter with two brightly glowing, 4-foot-diameter globes above it—while at Hawk Inlet at 3:30 p.m. on the 15th. When first seen, Marlowe passed it off as a reflection from an icy bluff or an aluminum boat but was soon observing it more closely when he noticed that it had begun to move slowly toward his boat. Using binoculars, Marlowe watched it but could not identify it. **By 7 p.m., the ball was floating on the water within a quarter mile from the *Teel*, then suddenly rose out of the water and slowly flew out of sight** over a nearby mountain ridge. [040]

07-26-1980 *Atlantic Ocean off the coast of Brazil*

New and surprising details are beginning to surface around this appearance of the flying saucer, seen by the crew of the tugboat *Caioba-Seahorse* in accordance with the reports given by the employees of the company that owns the ship. There were two, and not just one, unidentified flying objects seen by four of the crew of the tugboat, three of whom have already been interviewed by the Colonel-pilot Francisco Hennamann of the Barreira do Inferno.

One of the UFOs was floating close to sixty miles (almost 100 kilometers) off the coast, and it became necessary to make a sudden maneuver to keep the *Caioba* from colliding with it.

Next, a giant UFO, nearly 100 meters in diameter, came near and linked up with the floating object and they lifted together, remaining about ten minutes over the ship and later disappeared in the direction of the open seas.[041]

1.5 e.   *Out Of and Into or Into and Out Of Water*

Let us compare a UFO encountering a body of water to a sweating, exhausted, disheveled person who has traversed the fiery desert and stumbled upon an oasis with its "cool, clear water" (my musical favorite by Marty Robbins, Columbia Records, 1959). It is understandable in this respect that the individual drinks, splashes, swims, and frolics in the water. Similarly, but to a lesser extent, it would seem that the crews of these UFOs, although remaining in their craft, delight in the same joy of discovery as our fortunate friend above. For no apparent reason, they exit, reenter, and exit the water as if children playing in a pool.

03-29-1938   *Morecambe Bay, England, UK*

As with UFOs, our mysterious submarines have a history going back more than forty years. A group of fishermen in Morecambe Bay, England, reported one such oddity in March 1938. "I saw a sudden scurry of seabirds rise off the water," witness William Baxter told the Liverpool *Echo* (March 29, 1938), "and I looked at a spot nearly a mile away. **Out of the water there rose something large and black, like a big post. It was at least eight or nine feet high, and it rose and fell three times, then disappeared**. I've been all over the world, but I have never seen anything like this!" [042]

08-MM-1964 *Upper Nemahbin Lake off Delafield, WI, USA*

I am also enclosing a report given me by Mike Finello that I subsequebntly [sic] corroborated as true to the testimony of Gary R. He and his brother were out on a lake at four in the morning heading for a fishing spot when **the craft he described descended suddenly and entered the water very slowly, remained underwater for a while**

(he could see the glow in the depths of the lake), then emerged very slowly and, once it was free of the water, accelerated rapidly and had the appearance of a star once again in a matter of seconds. [043]

09-20-1964   *Golfo de San Jorge, Argentina*

An individual known and respected for his honesty, who requested that his name not be revealed, reported that on September 30 [date error], 1964, while he was driving at night to the city of Comodoro Rivadavia coming from Caleta Olivia, **he saw appear at the edge of the city some strange objects (3 or 4) that described parabolas and descended one after another into the sea.** The unusual spectacle caused the witness to stop his vehicle so as to carefully observe the flames which the objects were leaving as they passed.

After resuming his journey and having traveled some distance from the initial point of observation, **he noticed other luminous objects—or perhaps the same objects— emerge from the surface of the water and then circle around at a dizzy speed,** and, within an outline of shining light, again climb up and finally disappear into the sky. [044]

07-MM-1966 *Pacific Ocean between Seattle, WA and HI, USA*

The range was right on, as the craft had moved toward the general direction that we were headed. We watched repeatedly as **the strange craft reentered the water and then subsequently rose into the clouds over and over again** until finally we knew that it was gone for good. The episode lasted about 10 minutes. [045]

04-15-1974   *Between Ceuta and Algeciras, Spain*

"Round, intense, torch-like light.…Rose out of the water near a huge rock, traveled at low altitude, then fell into the water again. This happened twice." Witnesses were passengers on a ferry between the two cities.

(Note: Ceuta is a military station and seaport; Algeciras is also a seaport.) [046]

01-18-1975   *Puget Sound off Anderson Island, WA, USA*

[Transcription from a tape:]

RG   —   Appears we were disconnected.
WI   —   (Laughter) Yes, I guess we did...after all this program. Incidentally, I'll try to start this all over again. That it was coming from the Ketron Island area, along the far shore, and I could see everything clear beyond...Nisqually Flats. You know where that is?
RG   —   Yes.
WI   —   It came along and it probably was even with DuPont, but **it dove into the water.** You know where DuPont is?
RG   —   No . . . I don't think so.
WI   —   Well, DuPont is right across from where I live.
RG   —   OK.
WI   —   OK. And I live on Cole Point, Anderson Island.
RG   —   Right.
WI   —   OK.
RG   —   Right.
WI   —   All right. When it got just about to DuPont, **it dove into the water and left a big, foam, spray-like deal when it went into the water.**
RG   —   Yes.
WI   —   And it disappeared. And I kept looking and wondering what the heck it was, you know, and pretty soon, ah maybe...a...six seconds or so, **it surfaced again and skimmed . . . on top of the water**, oh...about a hundred feet above the water, and it continued on for quite a ways, and then when it got to approximately even with Nisqually Flats, **it dove into the water again and I lost...I couldn't see it again after that.** If it came up again, I don't know. [047]

07-20-1993   *Pacific Ocean off Oahu, Hawaii, USA*

Trail of light emerging from ocean and looping back into the water. Object must have been very large since sighting was from about a mile away. [048]

07-02-2007   *Okanagan Lake, Westbank, BC, Canada*

I don't know, but it didn't fit the usual description of a UFO round thing. It looked somewhat like a torpedo or something with wings, kind of like a fighter jet, screaming across the sky and downwards on fire. **It crashed into the lake, and it was a big splash**, almost like a large, large plane! The sound was incredibly loud and like a deep bass "womp womp womp" sound as it came down. I ducked for cover thinking of a meteor or asteroid or something. I started to freak out and yell, and as I was standing up, **something came literally shooting back out of the water, and I saw it start to hurtle towards a different looking object very high in the sky**, which kind of looked like a cone of some sort. [049]

1.6   Summary

Edward Gibbon, 18th century historian and author, once said, "I have but one lamp by which my feet are guided, and that is the lamp of experience. I know no way of judging of the future but by the past." What does this abbreviated history lesson teach us but that we are slow learners and not partial to changes in our thought processes. The UFO history lesson has its official start in 1947 and stretches, at this point, to 2015: sixty-eight years. Atomic energy came about slowly, with the earliest references to the concept of atoms dating back to ancient history. It comes to fruition in the 1900s and is etched in all our minds with a connotation of fear when we recall its devastating power as released by the bombs dropped on Hiroshima and Nagasaki during World War II. Perhaps UFOs do have some time left to mature.

In addition to those who have made progress in discovering ufology's mysteries, there are some who have focused their energies on the past, as it has become clearer to us that indeed there was a past. There are paintings from the Middle Ages and the Renaissance which have examples of what appear to be UFOs in them, and some researchers have also gone back even further to primitive cave drawings and cite what might be interpreted as UFOs. All of these avenues of historical research may uncover more as our "lamp of experience" grows.

With this compressed history as our background, I would like to advance to what has become more evident in our times and hopefully lead us to a greater understanding of what, back then, was a dark mystery and now has become an opening of the cosmic window: the true spaceship.

Chapter 2

Physical Influences of a UFO on Water

"The important thing in science is not so much to obtain new facts as to discover new ways of thinking about them."

—*Sir William Henry Bragg,*
Nobel Prize for Physics, 1915

This is the most important chapter of this book, so why introduce it at the beginning? Because the rest of the book depends on understanding how the field surrounding the UFO works.

First, I would like to give credit to those who initially noticed some relationship between the craft and water disturbance. Not being a UFO historian, I can only relate what I encountered in my research, and that came about from the retelling of case dated **09-LL-1954** in the May 1977 issue of *Official UFO Magazine*. The authors, Lucius Farish and Dale Titler, end the case with comments and questions that basically sum up their thoughts about this and other water cases. Mr. Farish has advised me that the technical aspects and thoughts on the UFO's influence on water were the work of Mr. Titler. Their text, in part, follows with the kind permission of the authors, and I have bolded the text which foreshadows what I had discovered:

> ...Did the UFO use the same—or a similar—force to melt the middle coils of the ship's nylon anchor rope? More importantly, **was an electronically generated and controlled field, caused by the UFO, responsible for the strange behavior of the water—the churning and thrashing and the plateau shape?**

Did this field, this force, push the water away from the submerged UFO's body as it rose out of the water? Did it also put the surrounding sea, at a radius of 300 to 400 feet, into a dead calm? **Is it possible that underwater UFOs, while moving submerged, never come into contact with the water surrounding them?**

If the UFO phenomenon is related to water that doesn't wet, then it has a direct effect on the water in contact—or nearly in contact—with it. Speculation on one characteristic of the UFO in flight is that the surrounding air moves along with it, accounting for the absence of sonic booms at high speeds and the rushing sound of a slipstream as it passes overhead at low altitudes. It is possible that water behaves similarly as the submerged UFO passes through the liquid medium, and this lack of friction explains why an air/space vehicle can perform equally well as a submarine. [This paragraph almost anticipates Chapter 8 in Paul Hill's book *Unconventional Flying Objects.*]

Can this also explain its sudden transition from air to water? The surface of any body of water is the physical boundary between air and water—a boundary not yet mastered by modern technology. And it is fatally hard[1], as shown by the shattered planes and men that have fallen on its surface from any appreciable height and speed. Although

---

1   Note from Dr. Bruce S. Maccabee, physicist and researcher, retired from the U.S. Naval Surface Warfare Center, author or coauthor of many UFO articles and books: The "hardness" of the water depends upon the size, shape and speed of the object entering the water. A pointed object can enter the water more easily than a blunt object at any speed. Boats and submarines move through the water at speeds determined by their shape and volume of water displaced and the power of the "pushing" mechanism.

submarines cruise its surface freely and move easily through its depths, while seaplanes use it as a platform on which to land and take off, we have no flying craft that can dive through the air-water interface, survive the shock, then move equally as freely underwater. When the UFO becomes a USO—Unidentified Submerged Object—it does just that, immediately, effortlessly, and oblivious to the medium through which it moves.

UFO investigators no longer doubt these objects are controlled by technically advanced intelligences. They have apparently mastered the electromagnetic state of Earth's waters with the quality and intensity of **their own self-generated control fields. Something about the UFO/USO influences the water around it. Some control source inside the object molds and shapes the water into mounds and water holes.**

Here may be the key to the sudden fogs and white mists that have enveloped surface craft. Several such cases have occurred in the "Bermuda Triangle" area of the Atlantic, according to author Charles Berlitz.

Water molecules are made of hydrogen and oxygen atoms held together by a covalent bond because they share a pair of electrons. There are two hydrogen atoms but only one oxygen atom in the water molecule.

The negative charges of the oxygen atoms and the positive charges of the hydrogen atoms are deployed unevenly, giving the water molecules positive and negative poles such as those found in a soft iron magnet. When enough water molecules line up in the same direction—as when a strong magnetic field is passed over or through them— they become polarized. That is, they take on light reflecting and refracting qualities.

Is this what happens when an undersea UFO comes near the water's surface or shoots up and breaks free into the atmosphere? **Is its powerful magnetic propulsion field** polarizing the water vapor on the area's surface, **causing it to form a fog or mist? Is this high intensity, electromagnetic force, intelligently aimed and directed at surface craft and airplanes, so strong that it can shape the surface water into mounds resembling boiling water?** Does the "boiling" appearance come from high-speed changes in the magnetic forces, which, in turn, cause high-speed changes in the water vapor molecules near the UFO?

Submarines can rise straight up and break the water's surface without "mounding" it. Polaris missiles make no lump in the water when fired from subsurface; no gradual up swelling of water occurs just before it breaks [the] surface. Yet the reports persist: "...the water was thrust away all around it..." There were places where the water **"boils in circles."** [Heat will be treated in Chapter 4.] Fishermen report "holes" in the Gulf of Mexico.

It seems the UFO/USO intelligence—when it changes from one fluid density to another (air to water or vice versa)—creates an area around it where gravity and normal magnetic attractions do not function in ways with which we are familiar[2].

Could there be another answer to the mysteries created by these alien denizens of the waters? [050] [End]

I ended this paragraph on my website by adding, "If only they had drawn the stages...," for it was when I drew

---

2   Dr. Maccabee's note: This may be true, but it is only speculation. We won't know until someone manages to measure the field strengths around a UFO/USO.

the phases of the UFOs' effects on water that my revelation came about and will be explained to you as we continue.

Eleven years before the above Farish and Titler article was published, Frank Edwards in his *Flying Saucers-Serious Business* talked about the magnetic field:

> As FDR used to say, "Let's look at the record." It has been known since 1947 that some sort of field associated with the unidentified flying objects causes observable disturbances in electromagnetic instruments[3]. It was first noted in connection with compasses. Later, pilots found that with a UFO in their vicinity the ignition systems of their internal combustion engines were adversely affected. Ships, automobiles, and tractors reported malfunction of their engines in the presence of UFOs. Motorists and airline pilots made many reports of radio interference which prevented transmitting or receiving signals when the UFOs came near. [051]

2.1 Physical Effects on Water—Tying It Together

The following proposes a possible explanation for the observed interaction between unidentified flying objects and water. Wrested from seemingly divergent eyewitness accounts, it speculates that the reaction of the water around an emerging craft implies the existence of a displacement field, one that exhibits electromagnetic tendencies. What at first seems like a series of unrelated accounts illustrates consistent behavior when the distance between the object and the body of water is considered.

What we will be dealing with here is what witnesses see a UFO doing with water. As important as that is, skeptics and debunkers keep saying that we have "no tangible proof." No,

---

3   Dr. Maccabee's note: There are many UFO cases that point toward the presence of strong magnetic fields often, but not always, around UFOs.

we do not have pieces that have fallen off a UFO, but we do periodically have pieces of *our* technology falling off cars and planes.

Before we start, I think a short review of some definitions is in order so we can keep our bearings straight.

Definition of **physical:**

> **1. Relating to the material universe or to the physical sciences**. 2. Pertaining to material things, as opposed to mental, moral, or spiritual; especially, relating to the human body apart from the mind or spirit; material; corporeal. **3. Of or pertaining to the phenomena treated of in physics. 4. Accessible to the senses**; external: physical characteristics of a mineral; physical changes.
>
> Synonyms: bodily, corporal, corporeal, material, natural, sensible, tangible, **visible**. Whatever is composed of or pertains to matter may be termed material: physical applies to material things considered as parts of a system or organic whole; hence, we speak of material substances, physical forces. [008]

Definition of **physics**:

> The science that treats of matter and energy and of the laws governing their **reciprocal interplay** under conditions susceptible to precise **observation**, control, and exact measurement. Physics generally includes the subjects of light and sound, electricity and magnetism and radiation, but not the phenomena peculiar to living matter or to chemical change. [008]

Scientists use repetitive results as proof of their experiments' conclusions. In land trace cases, researchers like Ted Phillips study the physical effects of UFOs on trees, grass, soil, and so forth. For those who investigate water cases, though, the situation is different because water, unfortunately, dissipates any physical evidence due to motion of the liquid or dilution of residual traces. However, could there be solid evidence in the form of *repetitive verbal* testimony? When I started to collect UFO cases, I had no concept of what I might find. I only wanted to gather all the cases I could to see if they fit any physical pattern.

One should read Paul Hill's book *Unconventional Flying Objects* for an understanding of his conception of the "field." Upon my initial reading, it seemed to confirm what I had mistakenly assumed the "field" was doing and how this craft could do what it does with water. What confused me was his statement: "By now the reader may have guessed the basic answer. The signal that the supersonic vehicle is coming will be given by a **force field** having a probable action **velocity** equal to the speed of light." [052]

Velocity, which is measured in feet per second or miles per hour, and so forth, implies movement of the field. Because of this, I had assumed that the field which surrounds the UFO moves the molecules of atmosphere it contacts to its rear instantly, thus avoiding a shockwave and its sonic boom. His conception is different but accomplishes the same end—removal of the shockwave. My idea of a rotational field stemmed from the correlation of physical evidence I noted in many of the cases.

An analogy to a light bulb can be drawn if we think of the filament as the functioning craft and the glass bulb as representing the UFO field. We can then see that this bulb protects the fragile filament from atmospheric and biological intrusion as well as other damage. In the majority of UFO cases, it is undoubtedly the function of the field to protect the UFO. The speeds that have been recorded, which would cause a tremendous heating of the surface of the vehicle due to atmospheric friction, as well as the possible collision with other aerial craft or life forms, are valid concerns.

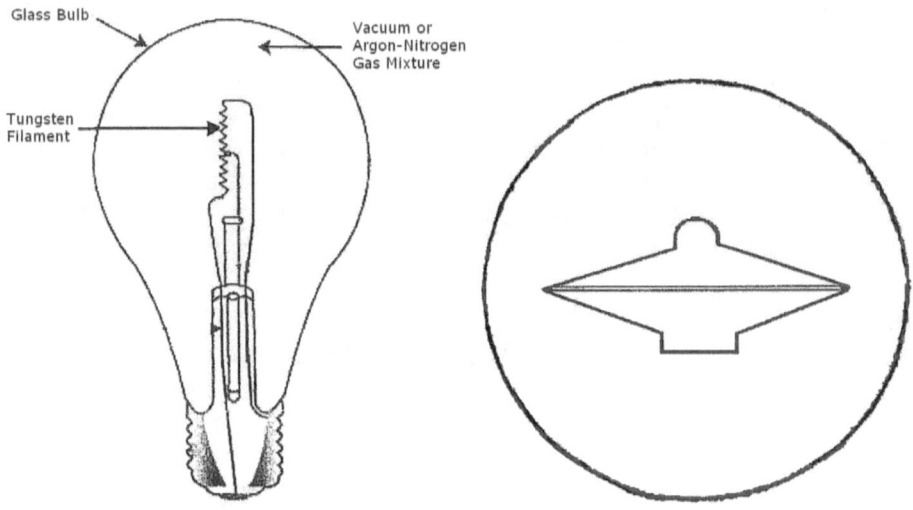

Sketch #1

As I continued my collection of reports, the major problem was not the lack of incidents, but the lack of text on *what the water was doing* during those incidents. For the most part, the cases report UFOs going into or coming out of water and then go into detail about the craft or its travel. This is understandable, as the craft is the unusual thing, not the water which we drink, bathe in, cook with, and sail upon. I eventually became frustrated that there was no clear correlation with what was occurring with the water in the different reports and began to think that perhaps all these disjointed groups of cases were leading me nowhere.

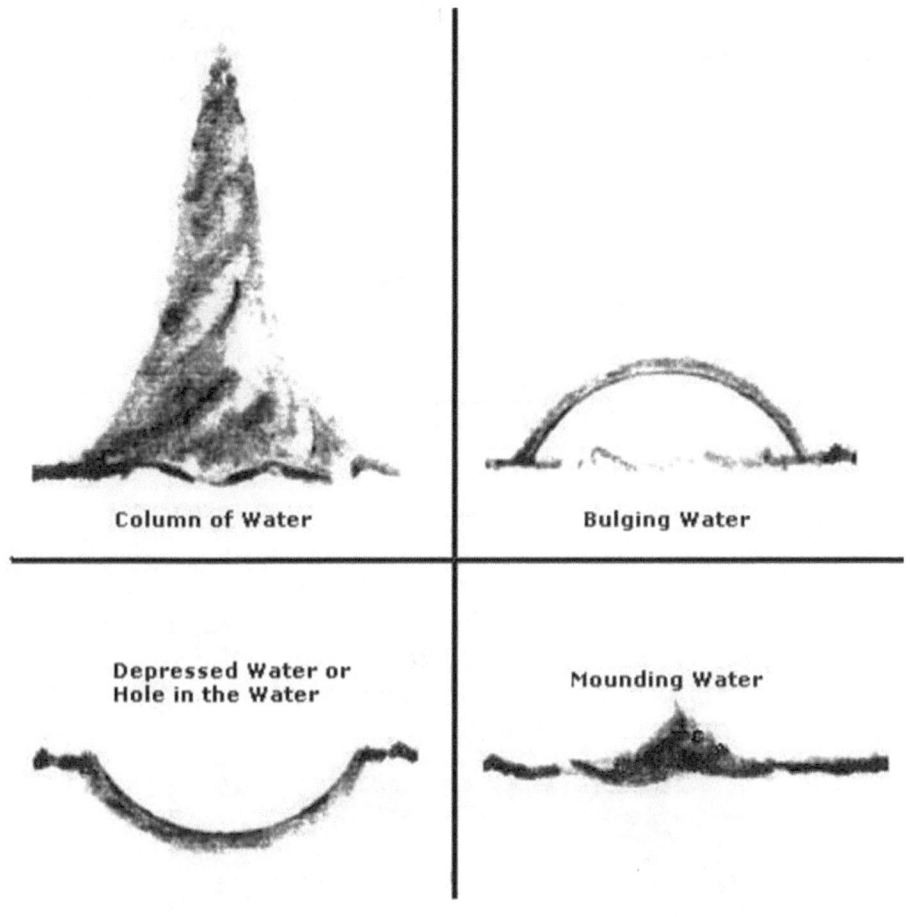

Column of Water

Bulging Water

Depressed Water or
Hole in the Water

Mounding Water

Sketch #2
These are the four groups I came across in
my research.

It should be noted that my hypothesis is based on what could be considered a very small percentage of the almost 1700 cases contained on my website and does not take into consideration any testimony concerning UFOs *entering* bodies of water. I purposely avoided this because of all the things that could possibly be identified as UFOs but could also be space junk, aircraft, meteors, or your neighbor's football.

In order to get an idea of what was happening in the above four sketches (sketch #2), I had to attach a UFO to each of these pictures. So I made one to my liking (sketch #3):

Sketch #3

Next step was to add a "field," which is commonly observed in ufology in the form of an ionized ball of light at night (sketch #4). I would like to explain here that I did not invent this field and that its existence has been a part of ufology for quite some time.

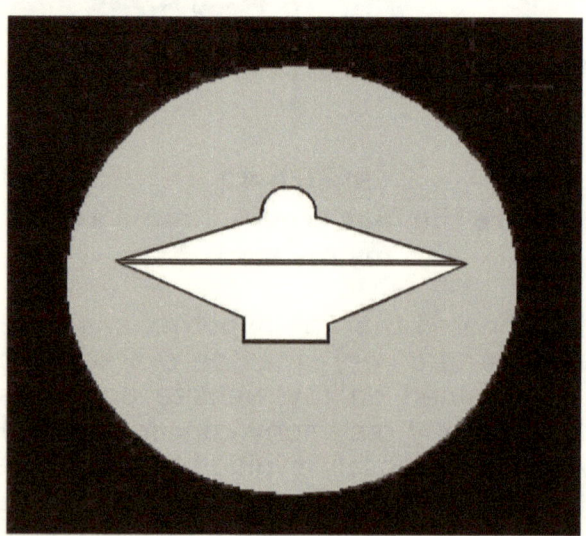

Sketch #4

It was very amusing one year when I used sketch #4 as part of a presentation I made at the Fund for UFO Research just prior to Halloween. In the sketch, black was black, the circle and all therein was orange (color of the field in many cases), and it resembled a Halloween pumpkin.

2.2   Ionization in General and As It Relates to UFOs

A UFO's *glow* as it is called is caused by our atmosphere in contact with the electrical field surrounding the craft, and this applies to water as well. It causes what is called ionization. The importance of understanding the whys and wherefores of what occurs will be discussed as we move into the different areas that are touched on in this book.

Encyclopædia Britannica explains what ionization is in general terms:

> **Ionization,** in chemistry and physics, [is] any process by which electrically neutral atoms or molecules are converted to electrically charged atoms or molecules (ions). Ionization is one of the principal ways that radiation, such as charged particles and X rays, transfers its energy to matter.
>
> In chemistry, ionization often occurs in a liquid solution. For example, neutral molecules of hydrogen chloride gas, HCl, react with similarly polar water molecules, H_2O, to produce positive hydronium ions, H_3O^+, and negative chloride ions, Cl^-; at the surface of a piece of metallic zinc in contact with an acidic solution, zinc atoms, Zn, lose electrons to hydrogen ions and become colourless zinc ions, Zn^{2+}.
>
> Ionization by collision occurs in gases at low pressures when an electric current is passed through them. If the electrons constituting the current have sufficient energy (the ionization energy is different

for each substance), they force other electrons out of the neutral gas molecules, producing ion pairs that individually consist of the resultant positive ion and detached negative electron. Negative ions are also formed as some of the electrons attach themselves to neutral gas molecules. Gases may also be ionized by intermolecular collisions at high temperatures.

Ionization, in general, occurs whenever sufficiently energetic charged particles or radiant energy travel through gases, liquids, or solids. Charged particles, such as alpha particles and electrons from radioactive materials, cause extensive ionization along their paths. Energetic neutral particles, such as neutrons and neutrinos, are more penetrating and cause almost no ionization. Pulses of radiant energy, such as X-ray and gamma-ray photons, can eject electrons from atoms by the photoelectric effect to cause ionization. The energetic electrons resulting from the absorption of radiant energy and the passage of charged particles in turn may cause further ionization, called secondary ionization. A certain minimal level of ionization is present in the Earth's atmosphere because of continuous absorption of cosmic rays from space and ultraviolet radiation from the Sun. [053]

Now let's see how Paul Hill, in his book *Unconventional Flying Objects*, relates ionization to UFOs:

The phenomenon of ionized and excited atmospheric molecules around a UFO also ties together a number of related mysteries about the UFO. It accounts for the general nighttime appearance of the UFO: the many observed colors, the fiery, neon-like look, the self-illuminating character, the fuzzy, indefinite or even indiscernible outline, yet an appearance of solidity behind the light. It also accounts for the general lack of heat radiation despite the fact that they sometimes look fiery or even like a flaming ball of fire, and even the ultraviolet burns sometimes received by close viewers of UFOs with a blue plasma. **In the daytime the same plasma is present, but usually invisible.** Early morning and evening, it is partly visible. Giant cigars and dirigibles are exceptions, for they can lay down a plasma wake or cloud visible in the daytime. The ion sheath also accounts for some daytime UFO characteristics such as a shimmering haze, nebulosity of the atmosphere or even smoke-like effects sometimes observed when high contaminant concentrations and chemical actions may be presumed to be present. [054]

2.3   The On-Off Switch

Given that the primary focus of this book is on water-related cases, there is a section about solid circular lights glowing underwater in Chapter 17. As stated previously, air and water are governed by the physical principle of fluid dynamics even though we think of them as separate entities. Unfortunately, seeing what UFOs can do underwater is next to impossible because of the density and coloration of the water. While the following cases are not on my website because they do not involve water, I have included them here to illustrate a trick that UFOs can perform to confuse us even further:

07-14-1952   *Chesapeake Bay, off Virginia, USA*

Immediately after the six objects redirected away from the plane, two additional objects passed under the wing of the DC-4. Nash stated, "They appeared to be higher than the others. They dived at the rear of the others on an apparent intercept heading." After the sixth and seventh objects joined formation, Nash exclaims, **"Suddenly the original six blinked out—just as though someone had**

turned off an electric switch [which would easily be accomplished by shutting down the field—switch off]. And almost immediately the second two blinked out." **Moments later, "They all blinked on again**—all together in an in-line formation."

In a letter dated "5 March 1970," from William Nash to Professor James E. McDonald, the pilot gave some further details of the incident. Nash explained, "The angle of elevation, relative to our visual horizon, was about 45 degrees up when they blinked out. **The blink-out was not in front to back order, but mixed order.** Not a disappearance by perspective diminishment." When the UFOs lit back up and glowed, they were in a single file formation. ⁰⁵⁵

07-19-1952   *Langley Air Force Base, Virginia, USA*

Later, Air Force officers would learn that as the fighters appeared over Washington, people in the area of Langley Air Force Base, Virginia, spotted weird lights in the sky. An F-94 in the area on a routine mission was diverted to search for the lights. The pilot saw one and turned toward it, but it disappeared **"like somebody turning off a light bulb."**

The pilot continued the intercept and **did get a radar lock on the now unlighted and unseen target**, but the lock was broken by the object as it sped away. The fighter continued the pursuit, obtaining two more radar locks on the object, but each time the locks were broken. ⁰⁵⁶

From the two cases above, we can see how the alien pilots play a game of illusion with us. Since the ionized field is invisible in daylight, this is mainly a night game. Here is a possible scenario: With the field turned on, the UFO is visible. A jet fighter pursues the UFO. The dogfight begins and during the chase, the jet turns toward the visible light, up, down, left, and right. At some point, the pilot of the UFO grows tired of the diversion, so he turns the field off, ending the ionization, and as if by magic, the UFO disappears. But as seen in the last case, even though the radar DOES detect the unidentified flying object, the UFO with its advantage of speed is able to lose the jet fighter.

This "switch" is also necessary in the change of field direction from north at the top to north at the bottom as a stop of one sequence and a start of the other. As an example, an electric stove with two burner sizes in one burner cannot be switched from one size to the other while one is active. The burner must be switched off first or else a short circuit will occur. This might also be true for the high-tech UFO.

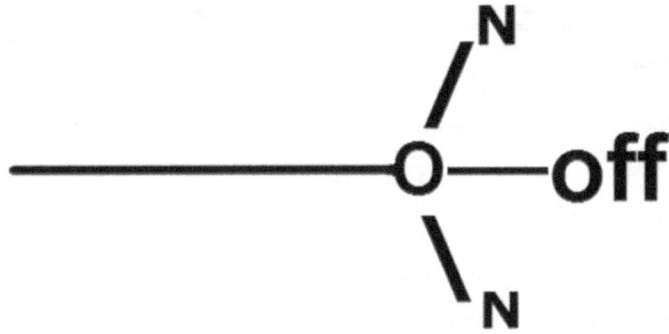

This ability to become invisible occurs in water as well, allowing the craft to remain inconspicuous. This is illustrated in Chapter 5 when an abductee, inside a UFO, goes underwater in the middle of New York City.

Furthermore, the off-switch plays an interesting part in making ice in Chapter 4, so keep this interplay of the field in mind with each case and its results.

Sketch #5

Finally, I needed some indication of how the field *might* be operating. So I borrowed this (sketch #5) from a book called *UFOs and Anti-Gravity* by Leonard G. Cramp [057], which introduced the idea of a magnetic field. This was interesting as the term "electromagnetic" is brought up constantly in ufology.

The only problem I can see regarding the force around the UFO in the sketch above is that it extends throughout the ship. The forces come out of the center, the midsection, and the tip, but if it is a high-energy field, it would probably reduce the occupants to ashes. I did not like this idea, so I made my own, shown in sketch #6.

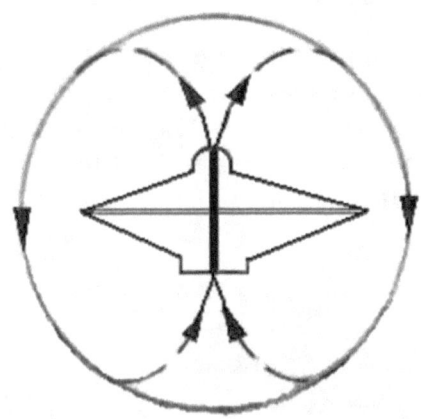

Sketch #6

This is a rather simplistic idea. Since it is based on the general principle of a magnet, I decided to use a simple bar magnet, with which we are all familiar from our early days in Physics 101, inserted into the center of the UFO. It was also simpler to draw. However, I do not consider this example to be a true representation of the UFO's field. Also, keep in mind that inserting a bar magnet into water will not produce any result other than a wet magnet. Another fact is that the field generated by a bar magnet does not rotate, whereas my assumption is that the field around the UFO does. From this idea, it is simply a matter of inserting this picture into each of the four pictures in sketch #2, shown previously, to see the resulting effect.

Since there was no continuity within the four groups, I decided to go from the lowest point which is under the water to the highest point above the water and see how this "field" fit the observations. So I started with the one still underwater.

Each drawing is followed by a description of that phase, along with demonstrative excerpts from published eyewitness case testimonies.

2.4   UFO Coming Up—Water Bulges

In sketch #7 below, I was forced to make a decision on how the field was rotating and thereby producing this up-flow or bulge in the water. The arrows showing the field moving up satisfy that concept in *this* instance. I had no idea at this point if it would work in other circumstances.

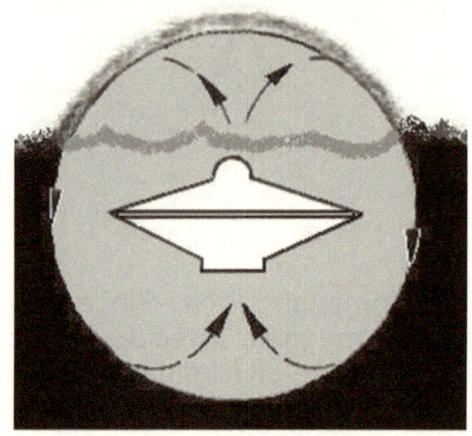

Sketch #7

In this section, we should visualize ourselves in a boat, leisurely enjoying the ocean view, when suddenly we notice a strange sight. A section of the water within our field of vision appears disturbed. At first it is like the water is boiling, with bubbles coming to the surface, but within seconds the surface becomes more agitated and appears to be rising and flowing violently. Suddenly, all this ceases as we are faced with a craft.

03-LL-1959   *The Baltic Sea off Kołobrzeg, Poland*

One day in March, while he was patrolling with some other soldiers not far from the coast near Kołobrzeg, he and his colleagues **suddenly noticed that at a certain point near the coast, the sea was becoming choppy,** swelling and spraying up. Immediately afterwards, a triangular object with sides about 4 metres long emerged from the sea and lifted into the air before the dumbfounded soldiers. Once in the air the spacecraft began to spin around above the area, more precisely above the unit's barracks. A few moments later it moved away at a dizzying speed and disappeared. [058]

06-03-1961   *Ligurian Sea off Savona, Italy*

While they were stopped offshore, four people who were taking a ride on a small motorboat noticed that suddenly the waves were getting bigger. Fearing that it was an oil tanker, they stood up and noticed that, about a kilometer away, **the sea was swelling like a big bubble, while large waves were widening in a circle. A few seconds later, the four witnesses saw that from the center of the bubble was emerging an object** whose lower part was similar to an upside-down plate, and the upper part was shaped like a cone. As it was emerging from the water, **the object seemed to push back the water as if there was a cushion of air**. After it had emerged completely, the object stopped for a few seconds in the air at about 10 meters above the surface. After some swaying, a halo formed around the base, and the object rapidly departed following a trajectory oblique to the sea surface, disappearing towards the northeast. [059]

09-24-1961   *Baltic Sea off Łeba, Poland*

"A sudden noise of rushing waters made me turn toward the sea again. And right in front of me, about 1000 feet from the shore, **the surface was rising in one spot. It looked like a round hill—pushed up from beneath.**" [060]

And from on high...

47

04-11-1963   *Atlantic Ocean near the Puerto Rico Trench*

The copilot was the first to see the underwater disturbance. It was 1:30 p.m., about 20 minutes after takeoff, when the jet was at its cruising altitude of 31,000 feet. The copilot saw, approximately five miles to the right, **a portion of the ocean's surface rising in what appeared to be a great round mound. He described it as "a big cauliflower" in the water, like an undersea atomic explosion that was about to burst through the surface.** [061]

08-08-1967   *Salina, Venezuela*

The disc rose out of the sea at about 500 meters (1650 feet) from shore, hovered for a few seconds, then, as in the Yepes incident, rose obliquely into the sky, disappearing within seconds. Unlike Yepes, Pastor Contreras heard what he described as a "buzzing" sound.

When he first noticed the phenomenon, Contreras was attracted by the fact that **the sea started to "wrinkle up" or "stir up" in a vast round area**. The color of the water in that particular area—Salina is six miles north of the resort city of Arrecifes—is much darker than elsewhere nearby, so the change was easily noticeable. A few moments after the stirring up was noticed, the water began to turn a lighter hue, then a light-blue shade which was very intense. The area continued to grow lighter, then turned whitish, then yellowish, and lastly a brilliant orange shade. Contreras then noticed the sound of the object, which he described as "intense and deafening"; **he also felt a tingling sensation in his feet**. At this point the huge "pancake"-shaped object emerged from the sea, hovered, and left toward Maiquetía in a slanting ascending pattern. [062]

Also important here are the yellowish tint, which is again a Shag Harbour foam color (see Chapter 18), as well as the "tingling sensation in his feet," which probably indicates an electrical relationship between the craft and the water.

09-??-1971   *Unnamed lake in Arroyo de la Miel, Spain*

A very responsible young lad of 17 years, an insurance agent, was camped out 40 m from a small lake, playing his guitar. The sky was clear, and there was a very relaxing, natural silence. Suddenly, he heard a great sound produced by a body which fell and plunged into the waters of the lake. Not giving it any importance, he took a cigarette and lit it. Then he heard a noise again and observed **a round object 1 m in diameter coming out of the lake, producing the normal rise of the water,** and it rose skyward at an angle of about 45°. The object was a luminous white, of blinding light. [063]

10-15-1979   *Atlantic coast north of Rio de Janeiro, Brazil*

About thirteen miles south of Saquarema, as they again neared Ponta Negra, they saw UFOs once more. This time, however, the UFOs were coming up out of the ocean just several hundred yards offshore.

"I saw seven to ten of them," said FX, then a university student.

Luli, sitting on the passenger side and looking past FX's head as he drove, thought there had been only five.

"But when they came out of the water, it was like a mushroom [cap] with the water spilling out as the UFOs came out," she said.

"Then we noticed a big black one just ahead of us. It must have been at least a hundred meters across with a dome on top. We could see it because it was blacker than the sky." [064]

As can be seen in sketch #7 on p. 46, the lines of force coming out of the top-center do push the water up and away from the craft. However, the next case is not consistent with the idea of a totally circular field:

09-LL-1954   *Off the coast of Georgia, USA*

ENCOUNTER IN THE GULF STREAM by Neil Deane as
told to George Earley [Snippets of the text follow.]

So in undershorts and barefoot, and the rest of me bare, I
came suddenly awake, went into the cockpit and discovered
she was dead on course, sails drawing. But she was not
making any progress while over her forward deck was
considerable depth of solid water and **a great thrashing
up ahead.**

Looking in the direction of the thrashing, I perceived a
great glow, like a big school of herring coming near the
surface and disturbing luminous bacteria...only it was
not the right color. It was more deeply orange. In color it
was like the annelids, which are common off Bermuda in
certain seasons, but this was one continuous glow and not
a collection of nickel-sized blobs of light. The glow seemed
to come from a **plateau of water,** which was causing solid
water to flow over my decks, hence impeding my forward
progress.

So, thinking I had gone ashore, even though nothing was
grating under the keel, I charged up her auxiliary engine,
slapped in reverse gear, and backed away. In a few yards
astern I was in dead calm water so I took in her sails and,
when I was squared away with power-plant running nicely,
I put her into forward gear and went back towards **this
plateau of glowing water and light**....

On seeing this "windjammer" apparently bearing down
on me, I had gone into reverse gear again, backing off
once more into calm water...perhaps a hundred yards from
the disturbance. Even as I realized my "windjammer" was
something else, **the splashing and sloshing of water
stopped, and this vessel, or whatever it was, rose out
of the water**, and the glow surrounding it subsided so that
her running lights were much more highly visible. She had
lifted off the water and was heading towards the now dimly

visible continent of North America in a long, upward slant and going at what seemed to be a tremendous speed. [065]

This bulge in the water can be reproduced in your home with a sink full of water and a rubber syringe (as used to clean the ears). Simply fill the syringe with water and place it below the water level. Point the nozzle towards the water's surface and squeeze the bulb. As can be seen, there is a bulge, but it does not go vertically up and out of the water. This is caused by the water's surface tension which has properties resembling those of a stretched elastic membrane.

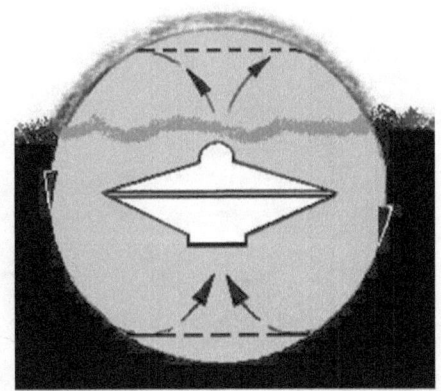

Sketch #8

It was not until I had put the sketch and text together that I came to realize that the point at which the water starts its downward motion is 360 degrees horizontally, thus making the bulge in the water of the above case appear flat.

The flatness, as represented in sketch #8, would be a plateau. In cases where the UFO is small (10 to 30 feet), the top might appear circular, but it too has the plateau. So you will note that we are starting out with a picture that is already in error. The field as represented here is *not* circular in a vertical direction, but donut- or apple-shaped (flat top and bottom in a side view).

I have a photo which was made from a videotape of a 1999 presentation by Ted Phillips, sold by Kentucky MUFON. [066] I was actually looking for something else at the time, but a moving UFO in the video caught my eye, and when I viewed it close-up, I instantly recognized the field layout. I was unable to reproduce it here in black and white due to fuzziness, but as in the prior sketch, there was a flat top and bottom as well as curved sides. The UFO was silvery in color, although there were streaks of other colors occasionally. This has been described by witnesses in various accounts.

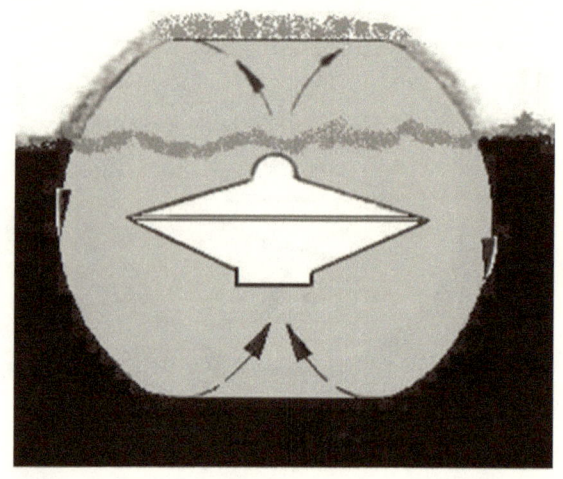

Sketch #9
(Sketch #8 as it should really appear)

2.5   UFO Just above the Surface—Water Depression

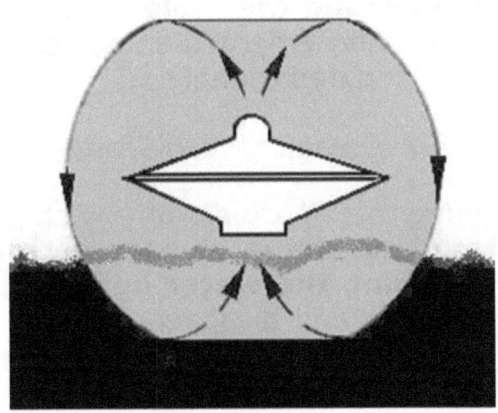

Sketch #10

Notice that in this position the force upon the water is pushing downward.

In these cases the UFO is seen *just above* the water, its field acting as a plow, while it invisibly moves water in the direction of the witness. This can be equated to the bow of a ship with its relatively sharp cutting edge slicing the water like a knife. However, if we rotate the knife 90 degrees, it is now pushing the water instead of cutting it.

In several cases where the UFO did not move in the direction of the witness but simply continued upward, we find the terms "depression" or "hole" used to describe what was observed below the UFO. A few examples follow:

06-12-1958   *Mediterranean Sea off Le Brusc, France*

"It grew bigger, and then it began to descend very fast towards the sea, and soon it was a great **big globe which was lying on the surface of the water....It caused a strong air displacement, for we could see the water being whipped up all around it.**" [067]

09-24-1961   *Baltic Sea off Łeba, Poland*

"Then, **splashes of water** gushed from the top and, like fountain jets, **fell around the 'hole' in the waves**. From this opening in the water emerged an object which at first I thought to be an elongated triangle...." [068]

07-MM-1966 *Pacific Ocean between Seattle, WA and HI, USA*

I was just lifting the binoculars off my chest when I saw it. **The giant saucer shape** plunged out of the clouds, tumbled, and, **pushing the water before it, opened up a hole in the ocean** and disappeared from view. It was incredible. [069]

LL-??-1972   *Unknown waters off Puerto Rico*

The most notable part of the observation was that the objects had a glowing greenish-blue field around them, and **they entered the water at considerable speed without any splash. The water just parted ahead of the visible field, and the objects flew into the hole, and it closed up behind them with only a flurry of ripples.** [070]

03-29-1974   *Sea off Togo, Africa*

The most interesting detail of his story concerns the surface of the sea under the object. **It was not flat but dug out in the form of a bowl-shaped depression.** A.W. estimates its depth at a few meters, 5 or 6 perhaps. Its diameter was comparable to the length of the object, on the order of 25-30 meters.

SEA SURFACE DISTORTION

Two elements in the account of A.W. are of great interest to anyone who tries to imagine the physical phenomena possibly accompanying UFO sightings: These are the **descriptions he makes, on the one hand, of**

the bowl-shaped depression dug out in the sea under it and, on the other hand, of the advancing waves on the beach up to the dune. It is not possible, given the conditions of the observation, to obtain a very precise description of the contour of the **bowl-shaped depression,** in particular of the bottom of it. [071]

02-EE-1977 *Gulf of Mexico off Ft. De Soto Park, Florida, USA*

"Suddenly, it seemed to become stationary just beyond the sandbar, and whatever it was seemed like it was hovering just a few feet above the water.

"**It seemed like there was a depression in the water out there.**" [072]

Again, this can be reproduced at home in your sink. Fill the sink with soapy water (representing white-capped waves). Take a clear glass dessert dish and press it into the water, keeping the dish's top edge just above the water. The result is that the depressed dish allows a view of the underlying clear water; the dish itself represents the UFO's invisible field.

Of interest here is a comparable case of depression on the ground with a UFO experience by a Mr. Sullivan:

> The night of Sullivan's experience had been a moonlit one, but three nights later on the 7th, the sky was overcast when young nineteen-year-old Gary Taylor was found dead in his car at the same spot. His automobile had run off the road. Police who checked the area after the accident found **a circular impression about two to five inches deep and about five feet in diameter in the freshly ploughed field**. [073]

In the above account, you have soil that has been loosened. Just as in gardening, when you step on this loose dirt, footprints remain. In this case, it is the imprint of the UFO's field.

2.6   Above the Surface but Close—Water Mounding

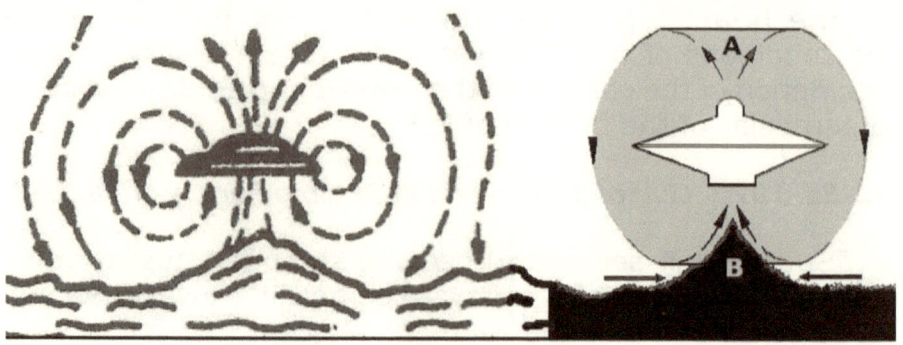

Sketch #11

Notice the water which assumes the shape of the upward flow of the field and appears to enter the UFO.

This is probably the single most important witness sighting in terms of understanding the total effect of the UFO upon water. There have been similar descriptions in other accounts, but they are usually tempered by terms like "upward swelling" or "raised water," language that does not clearly show a mounding or mountainous form. The more specific terminology gives us a fixed view of an effect as well as the beginning of what might occur when the craft moves farther up vertically. While this "group" consists of only one case, which I did not want to include because it is only a singular instance, it is necessary in making the transition to the group which follows. The witness happened to be at the right place, at the right time, so to speak. This was also the case that had the sketch of the magnetic field which I incorporated into my drawings:

04-??-1958   *Atlantic Ocean off Saúde, NE coast of Brazil*

The lower part was also a bowl, of the same size as the top, but dark in colour, and around the widest part where the two bowls met was a band with a number of square portholes from which came a reddish light. The portholes nearest to the onlookers were darkened "as though there were people looking out through them."

Beneath the machine, the water seemed to be boiling, or being *sucked up*, but without actually touching the under part of it; note this might well have been in part an aerodynamic effect. A faint humming could be heard at brief intervals, and from the under part of the machine, a number of things like leather thongs were *hanging* motionless. [074]

On the right-hand side of sketch #11, I have labeled two sections "A" and "B," which also have interesting properties: a funnel and inverted funnel shape, respectively. Inverted funnel (B) is important to the previous and following texts, while funnel (A) will show another physical trait in a later case.

This next case, from the Center for UFO Studies, also illustrates the upward movement at the bottom of the UFO. It deals with fog, which is water in a different form.

SM-??-1951/2   *Valley Forge, PA, USA*

During the summer of 1951 or 1952, John Van Roden, Jr. was employed by Freedoms Foundation, which at that time was using for headquarters a converted barn in Valley Forge, Pennsylvania. At about 8:30 on a very cloudy overcast morning, Van Roden, accompanied by another employee, an older man, left **the barn, over which hung a low fog layer, barely 50 feet above the roof**. Van Roden heard a sound like a train (which it might have been as railroad tracks ran nearby). He looked up—not, however, because the sound seemed to come from overhead—and saw, practically above the two men, a large disk-shaped object.

It had a surface like dull silver except at the top where a bright segment appeared as if it were reflecting the light. The disk did not appear to be spinning; if it was, the motion was so rapid that the bright reflection did not flicker. After about 30 seconds, the disk rose, steadily, without tilting, like an elevator. **The fog layer boiled turbulently (it did not spin all one way like a vortex[4] or a tornado cloud), and was sucked up in a funnel shape as the disk rose into the cloud and disappeared.** [075]

Sucked-up will be *the expression* in our next group's section.

2.7   Above the Surface—a Waterspout

The funny thing about this section is that I was going to leave it out. I had read the "Taking on Water" stories—your friendly alien knocks on the door asking to borrow a cup of water—and I dismissed these cases, as I could not see any value in them. However, my friend and fellow researcher Tony Rullán, who wrote *Blue Book UFO Reports by Ships at Sea*, which can be found on my website, kept insisting, "It's a part of ufology." So, I gave in. Thank goodness he persisted.

As I started collecting these types of reports, I began to notice that this category had two parts to it. One was the taking on water by obvious means: cups, buckets, tubes, and hoses, to which Chapter 11 is devoted. The second part gave the impression of taking on water with no visible means of collection other than the ship itself. It was here that I realized that this part is actually *not* the collection of water but some other influence of the craft itself. If you can

---

4   The lack of a vortex in this case may be due to the lack of rotation of a part of the craft, which is common in ufology. However, the suction is still there.

take it on this way, why bother taking it on with hoses, pipes or tubes?

This type of case involves a UFO hovering over water. The witness describes a column of fog, mist, haze, or water ascending from the surface of the water to the craft. The terminology generally used in the description is "sucking up water." This column, I believe, is caused by the high-speed, vertical rotation of the field, which creates a vacuum below the craft. As this field rotates, it causes a friction on the abutting atmosphere, creating an updraft at the bottom. This, in addition to an often-observed horizontal rotation of one section of the UFO, causes a swirling vortex to be formed below the UFO.

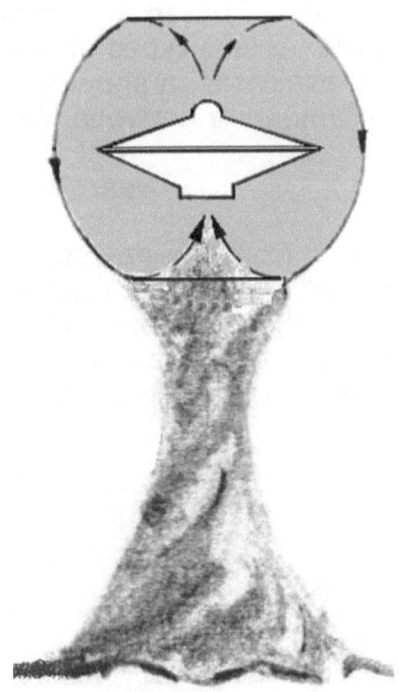

Sketch #12

This can be similar to a person rotating a finger through still waters in a pot at a rapid rate, eventually causing a whirlpool to occur. I once heard a radio program where a man was describing a waterspout underneath a UFO he had seen. He wondered where the water was going and said that he surmised it was going into another dimension because the UFO was constantly sucking up water, yet the craft was not that big. In fact, the water is not going into the craft. It is becoming more molecular[5] as it gets closer to the UFO, and it is just swirling at the bottom of the craft without going in. If this were a collection method, it would be a highly inefficient one, since the harvesting of molecular water would take much longer than the old-fashioned method of directly inserting a hose or tube into the water.

Next, I present several cases of this "sucking-up" scenario, but only with a limited amount of relevant text. Notice, if you will, the various types of bodies of water that are represented in these cases. It shows that in this situation the craft is not discriminating between fresh and salt water.

08-LL-1914   *Near Georgia **Bay**, Canada*

"The ship then rose from the bay surface **sucking with it a heavy upsurge of water** which sprayed the entire sea, leaving a mist above which did not settle for some time. As the ship continued to move straight up, it changed color from red to green and then made a tight left turn and flew off on its side." [076]

---

5   molecular: Of, pertaining to, or consisting of molecules.
     molecule: Physics. The smallest part of an element, substance, or compound that can exist freely in the solid, liquid or gaseous state and still retain its composition and properties. [008] e.g.: $H_2 + O = 1$ *molecule* of water.

08-27-1917   *Outside Grand **Harbour**, Malta*

"When he came out to look, he saw **a 'white cloud' coming out of the sea**. It was **as if the water was being drawn up to it** and, when the water fell back, there were splashes. The 'cloud' rose slowly, but as soon as it left the water completely, it shot up like lightning and we never saw it again."[077]

05-??-1953 *Loch Raven **Reservoir**, Towson, Maryland, USA*

The height of the object was given as 200 to 300 feet off the water. The passenger described the object as "BIG," and in dimension, the width of approximately 150 to 200 feet and 50 to 60 feet in height. The object hovered for a while, during which **a column of fog or haze or mist was seen by a light located in the center of the UFO between the object and the water's surface.** [078]

02-??-1955   ***Ocean** Beach near San Diego, CA, USA*

Around noon one day in February 1955, I was walking toward the beach at Ocean Beach, near San Diego, Calif., when suddenly something made me look up. I stopped, amazed and fascinated at what I saw—**a huge geyser of water with the rear end of what looked like a spaceship protruding from the top**. [079]

03-08-1957   ***Snow** in Baudette, Minnesota, USA*

At Baudette, the sighting had been made on the ground by a pilot with 2,000 flying hours. The object he saw was circular, from fifteen to eighteen feet in diameter. Moving upwind, its odd glow shone on the snow-covered ground. It was so low that **it seemed to suck the loose snow up under it as it passed.** [080]

??-??-1957   *Atlantic **Ocean** off São Sebastião, Brazil*

Suddenly he perceived that it was becoming clearer over the water, in the stretch between Bela Island and São Sebastião. **Next, a jet of water arose, like a waterspout**, which caused the thought of a whale to cross his mind. Shortly afterwards he made out that it was a pot-bellied machine moving in the direction of the beach. [081]

09-24-1961   *Baltic **Sea** off Łeba, Poland*

"**The object rose** a few yards," Kawecki continued, "**and hovered above this same spot, but there was now a whirlpool of water rushing inward with a loud sucking and gurgling noise**. The object itself was black and silent." [082]

??-??-1964   *Casitas **Reservoir** near Ventura, CA, USA*

The faster it turned (the more) it changed color, as though it were transparent and lights were going through it.

"When it came up out of the water, **it was like a funnel sucking up water. It brought the water up with it due to the fact that it was whirling so fast.**" [083]

08-??-1965   *The Red **Sea***

In August 1965, a crewman aboard the steamship *Raduga*, [which] was navigating in the Red Sea, observed an unusual phenomenon. At about two miles away, **a fiery sphere dashed out from under the water and hovered over the surface of the sea**, illuminating it. The sphere was sixty meters in diameter, and it hovered above the sea at an altitude of 150 meters. **A gigantic pillar of water rose as the sphere emerged from the sea and collapsed some moments later.** [084]

03-26-1966   *Atlantic **Ocean** off Florida Keys, USA*

When they were within a few hundred yards of it, they noticed "eerie pulsations of light around what appeared to be the nose section of the craft." Then, what looked like a greenish volume of light, **water or vapor, extended from the underside of the object down to the surface of the water**, which, they discovered, was strewn with dead fish. [085]

08-04-1971   *Tyrrhenian **Sea** off Milazzo, Italy*

A young man, who was with some friends, the director and two teachers of a summer camp, witnessed a somewhat unique event. About ten kilometers from them, the sea appeared to sink down; suddenly a ripple in the shape of an equilateral triangle formed. In place of the three vertices, **there were three marine vortices**. The vortices seemed to reverse the direction of their rotation and extend to reach the highest clouds. Only when **the three columns of water** disappeared was it possible to see the three lights in the clouds. [086]

07-??-1977   *Lake in Alaska near the Arctic Circle*

That was my last sight of the triangular thing, but in that last moment, I could see the **water of the lake surging upward—like a waterfall going upward, as if being sucked into the "machine!"** [087]

10-23-1978   *Adriatic **Sea** off Pedaso, Italy*

Some unusual marine events in the Adriatic Sea were quickly tied in to the UFO flap by the Italian press. Magazines like *L'Espresso* and *Gente* made mention of **a column of water rising from the sea** four miles out of Pedaso. Federicco Ricci, 35, and his son, Gabriele, 17, estimated the column to have hovered 100 feet high by 15 feet wide for a few seconds before falling fanlike back into the sea, leaving a large area of foam. [088]

07-17-1992   *Atlantic **Ocean** near Iguape, Brazil*

"They pulled their sail down to slow their boat to avoid a collision, but when the ship approached their boat, they realized it wasn't a ship. It was **a UFO** with a lot of lights on it. **It was sucking water up into it** and passed slowly within twenty meters of the boat, but the fishermen don't think the UFO saw them. It just went by them about six meters above the water, **sucking up a column of water almost as wide as the UFO**. The water didn't fall back into the ocean. The UFO went past them and disappeared in the distance." [089]

04-26-1994 *Santa Rosa **Sound**, Gulf Breeze, Florida, USA*

The green tree line along the distant shore silhouetted the UFO that appeared no farther away than the red buoy marking the center of the channel. Below the UFO the water began to churn. **Mist swirled into the air. A center plume of water shot up to the bottom of the UFO**. It was a waterspout. The UFO moved slowly to the left, and I aimed the camera and took Photo 49. Ten to fifteen seconds passed while I stared at this amazing scene. Suddenly the waterspout collapsed. [I will explain why in a bit.] The UFO angled off to the left, climbed at a sharp angle, and disappeared into the hazy sky. [090]

Every presentation or book must have its UFO picture. In the above case, the witness did take a photograph of the UFO, but, unfortunately, the UFO was a good distance away from him, so the details are very small and would not reproduce well in this black and white book. However, it is available in *UFOs Are Real: Here's the Proof* by Edward Walters and Dr. Bruce Maccabee, the same source as the above excerpt, or on my website under the case dated **04-26-1994.** Below I have included an artist's sketch of the photographed UFO.

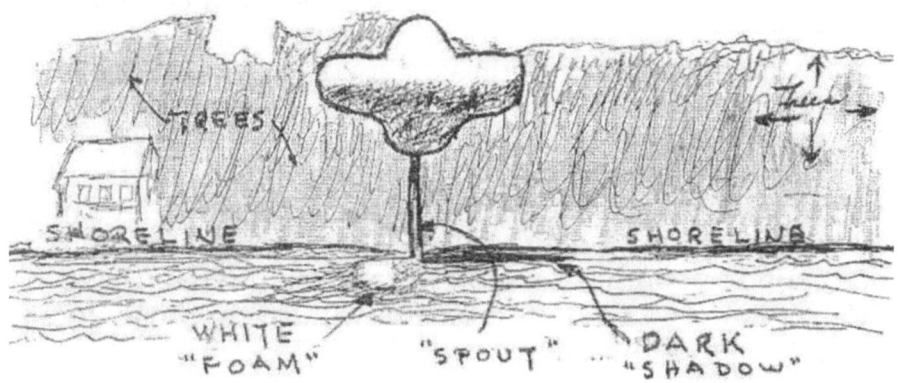

Sketch #13
Sketch of the photo of 04-26-1994

02-27-1995 *Mersey **River** near Fiddlers Ferry, England, UK*

A woman saw a **"bell-shaped object"** hover over the Mersey near Fiddlers Ferry power station. It appeared to be **"sucking up a column of water"** from the river. Six other passers-by also witnessed the event. [091]

LL-??-1996 *La Esperanza **Bay**, Vieques Island, Puerto Rico*

"What we were seeing," he continued to explain, "was what is called a flying saucer, a round apparatus, like a disc, but enormous, of gigantic proportions.

"The peculiar thing of all this was that **it was sucking out water from the sea**. The sea, the water, could be seen moving in circles, as if it was boiling and moving like in a blender, **like a whirlpool, and moving upwards toward the disc.** It was like a water jet from a hose, a great column of water." [092]

This "suction" on the surface of the water is exactly the same as if a tornado was to pass from land over a body of water. The bottom part would be viewed as a mounding of water. As we look vertically up the column, the water is more molecular and appears as fog, mist, or a continuation of the water itself. The vertical movement of this water or water vapor gives the impression of water being sucked into the UFO.

Separation of this column is almost predictable, as the height of the UFO above the water and the size of the UFO (which governs the width of the field) probably determine how much water weight the vortex can sustain. It would be interesting to compose a table with these figures. However, due to fright or awe on the part of witnesses, these details are lost most of the time.

2.8   What's It All About?

This is what you, the witness in each of the groups, see in the daytime. BUT something is missing. Can you tell me what it is?

The following sketch demonstrates the importance of the singular "mounding" case which shows the suction through the base of the UFO, opening the door to the explanation of the "waterspout" cases.

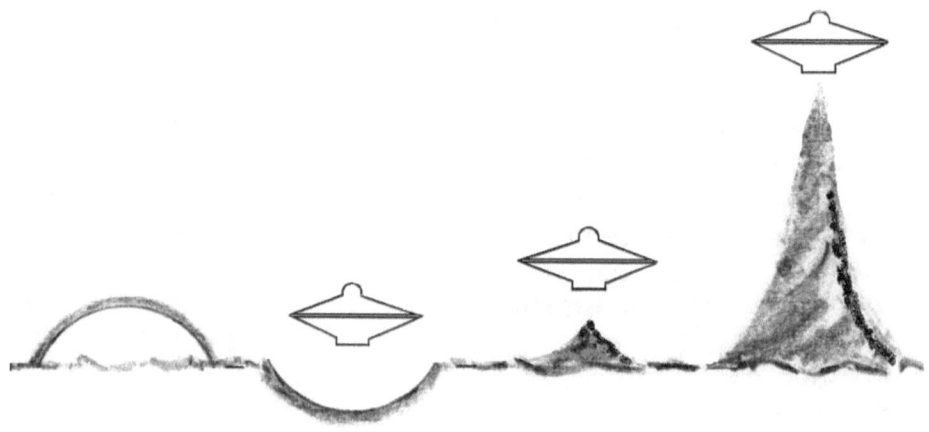

Sketch # 14
What the witnesses of each group actually see

2.9   A Function of Height

I finished sketch #15 on a night in February 2005. I had just finished putting up yet another case on my website involving the physical influence of a UFO on water and was frustrated that there was no relationship between the four groups shown in sketch #2 (p. 37). At 8 p.m. that evening, I decided to draw the groups together on one sheet of paper to see if anything would stand out visually.

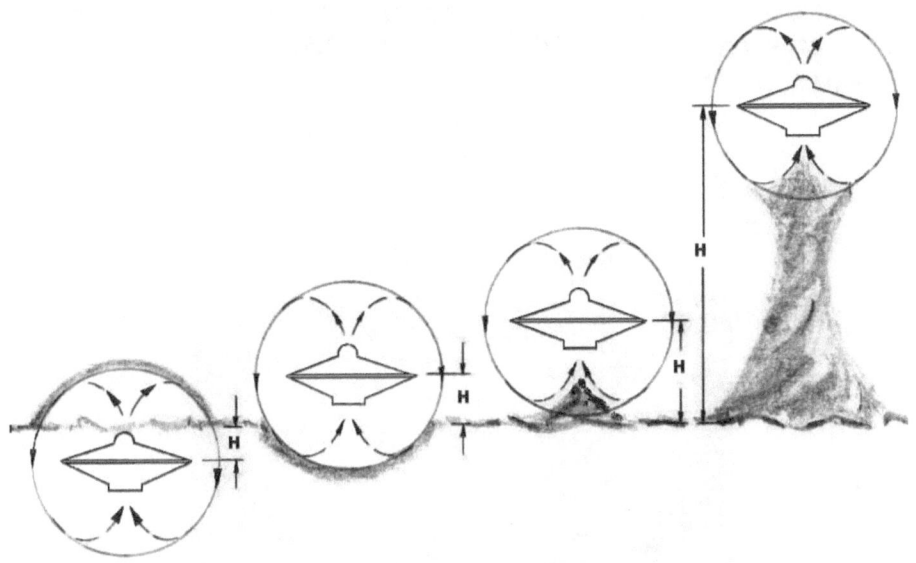

Sketch #15
My original drawing

By 10 p.m. I had finished the above drawing and was shocked by it. I went to bed but could not sleep. I just kept saying to myself, "They're real, they're real, they're real." What made me think so? All of the witnesses in the four groups are seeing one specific aspect of the UFO's movements. Yet, in the sketch, the field's vertical rotation does not change in any of these and complements what is occurring with the water. Therefore, although the aspects seen are different in each group, they *all* have the same field rotation in common. The reason the commonality is not seen is because the field is invisible in the daytime and therefore the witnesses only see the craft and the water—not the invisible field between the two, as illustrated in sketch #14.

It was only while preparing a paper for a PowerPoint presentation that I came across the same case twice, once in each of the first two groups. Upon rereading the case, I realized that it also included what would become the third phase of water alteration. I present the text here, in part, from this case dated **09-24-1961 (Baltic Sea off Łeba, Poland)** with illustrated examples:

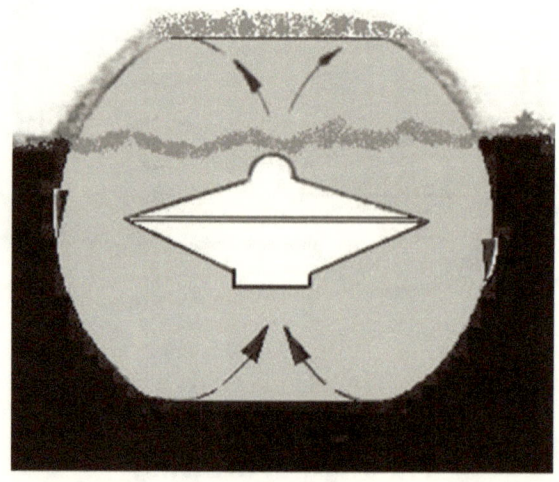

Sketch #16
Phase 1

Phase 1 – "A sudden noise of rushing waters made me turn toward the sea again. And right in front of me, about 1000 feet from the shore, **the surface was rising in one spot. It looked like a round hill—pushed up from beneath.**" [093]

Phase 2 – "Then, splashes of **water gushed from the top** and, like fountain jets, **fell around the 'hole' in the waves.**" [094]

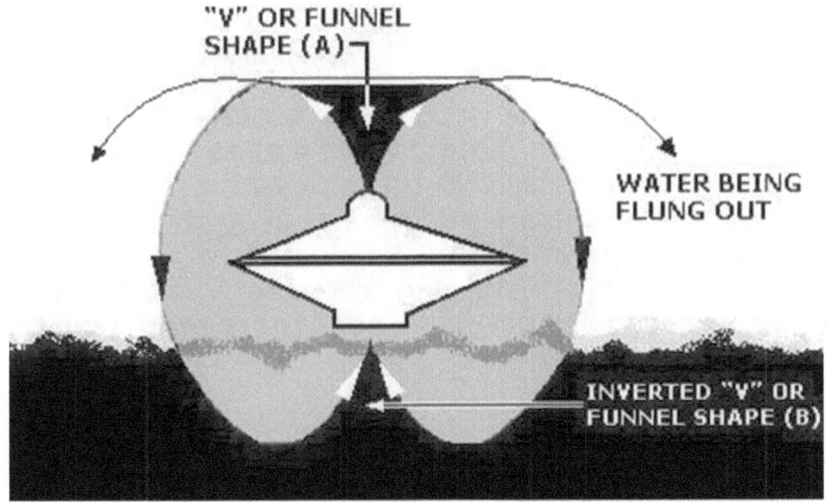

"V" OR FUNNEL SHAPE (A)

WATER BEING FLUNG OUT

INVERTED "V" OR FUNNEL SHAPE (B)

Sketch #17
Phase 2

This refers to the mounding in sketch #11 where we had concentrated on funnel (B). However, in sketch #17 above, we can investigate both (A) and (B)'s funnel. I start with funnel (A). This funnel is like the top of an apple, where the stem would be located. As the UFO rises from underwater, this area holds water until the UFO breaks the surface. Here, as in water pressure, air pressure is forcing the water against the field, which at this point is moving up and outwards, spraying the water enclosed in the funnel. One of Newton's laws is expressed as, "A body in motion tends to stay in motion unless acted on by an outside force." The force in this case is the field moving a molecule of water as it is pressed against the field, but the moment the field angles down, the field no longer influences the water molecule which is now affected only by its horizontal movement at the point of release and the force of gravity. There are a few cases where fishermen stated that the UFO dripped water on them, and this would be a result of the emptying funnel. This would not happen with a true circular field, as there would not be an area for water accumulation.

Focusing now on funnel (B) in sketch # 17, we notice the oft-repeated "hole" or "depression" in the water. The witness sees the hole, but not the field which prevents the water from touching the UFO, so the appearance that water is falling "*around* **the 'hole' in the waves**" is valid in that the water could not fall *into* the hole as the field prevents its entry. This was expressed differently in the Italian case **06-03-1961** Ligurian Sea off Savona, Italy (p. 47), where it stated, "As it was emerging from the water, the object seemed to push back the water **as if there was a cushion of air.**"

Sketch #18
Phase 3

Phase 3 - "**The object** rose a few yards," Kawecki continued, "and **hovered above this same spot, but there was now a whirlpool of water rushing inward with a loud sucking and gurgling noise.**" [095]

Both of these cases (09-24-1961 and 06-03-1961) which dealt with water alteration provide the positive proof that these separate groups were of *one* cause.

2.10 Water Nests (Crop Circles)

The vortex is a flow of hot air in a spiral most of the time, the result of the ionization of the air by the field. "Crop circles," the man in the third row screams. Yes, on land this vortex may produce crop circles, but we are not interested in them as artwork here. In many of the cases, crop circles are just that—circles, and the vortex produced by the field's rotation is, as I see it, the same as what causes the suction of water. As an example, I present a selection from a very famous case, found in its entirety on my website, which occurs in only six feet of water. I will only present the appropriate text from the case for review:

01-19-1966   *Tully Lagoon, Queensland, Australia*

George Pedley was driving a tractor....When he was approximately 25 yards from Horseshoe Lagoon, [he] heard, above the noise of the tractor, a loud hissing sound, "like air escaping from a tire."

"The tractor tires seemed O.K. to me, so I drove on," Pedley said. "Suddenly, an object rose out of the swamp. When I glanced at it, it was already 30 feet above the ground and at about treetop level. It was a large, gray, saucer-shaped object, convex on the top and bottom, and **measured some 25 feet across** and nine feet high. While I watched, it rose another 30 feet, spinning very fast, then it made a shallow dive and took off with tremendous speed...."

When Pedley drove around the bend of the track to the lagoon, there, at the spot beneath where the object had risen, was a huge, round, cleared area in the swamp grass. The water in this circular area was slowly rotating and appeared to be completely cleared of reeds....

71

Later in the day, apparently about noon, George returned along the track and stopped for another inspection. The cleared area of the lagoon surface was no longer visible. What was clearly evident was a floating mass of reeds, **approximately 30 feet in diameter** that had apparently come to the surface of the lagoon during the time Pedley was absent. The floating mass of reeds and grass was noticeably distributed in a radial pattern, in a clear clockwise manner. [096]

Note that the UFO is *estimated* at 25 feet and the reed circle at 30 feet, showing again that the field is larger than the craft. Also, this is one of the few cases where a UFO was *seen* leaving the site of a created crop circle[6]. There is no question of validity as this account was totally investigated and had good witnesses.

2.11 From Water to Land

Does this explanation of the field's direction and the corresponding vortex hold true for a UFO going from water to land? Does the action of the vortex also apply to what is seen when a UFO is close to the ground? To find the answers to these questions, let us fly inland with the UFO and see what effects it has on the environment and if this vortex still works.

First we should realize that there are no reports of the ground bulging and only one to my knowledge, which I presented earlier, of it being depressed by a UFO. The only other depressions are made by the weight of a UFO through its landing gear. This leaves only the "mounding" or "vortex" to be observed on land. The following cases are not on my website since they do not involve water.

---

6   Dr. Maccabee's note: I recall that a guy in Langenburg, Saskatchewan, Canada, Edwin Fuhr, saw 5 craft take off from a field in the early 1970s. After they left he found five circles in the crop.

08-13-1947 *Snake River Canyon at Twin Falls, Idaho, USA*

On each side there was a tubular-shaped, fiery glow, like some sort of exhaust. He said that when it went over trees, **they didn't sway back and forth but rather the treetops twisted around,** which suggests that **the air under the object was being swirled into a vortex.** [097]

As we continue through the trees we find...

10-10-1957   *Mariaville, New York, USA*

"Witness heard a whistling sound and saw a circular object descend to within 6 feet of the ground. It hovered 2 minutes, **sucking up leaves, grass, and dirt.** White material was found at the site, covering the grass in the disturbed area." [098]

05-20-1967   *Falcon Lake, Manitoba, Canada*

[The Stefan Michalak Incident—very famous case]

Paragraph 6. The attached photographs were taken at the site. No. 1. Taken from approximately 12 feet up in a tree facing in a southwesterly direction, **showing the outline of an approximate 15-foot-diameter circle on the rock surface where the moss and earth covering has been cleared to the rock surface by a force such as made by air at very high velocity.** For comparison, the prospector's ax and Beta counter were left in the approximate circle center. [099]

Let's get heavier:

EE-??-197.   *Furnace Creek, California, USA*

After four or five minutes of waiting, the ball backed away. Across the road, **it began creating a vortex** at the base of the mountain.

The rocks started rising into the air....They would shake from side to side. There were hundreds of them (the largest about the size of melons). They started going around in a circle, like it had complete control over them...Then the thing went way up in the air, and **it looked kind of like a tornado,** and it was all red.... [This coloration might have been due to particles of red soil in the mix.] The only noise you could hear was this clickety-clack, clickety-clack when the rocks hit together.

Then the light blinked out, and everything crashed down the mountainside and onto the road. The ball blinked back on again at the top of the mountain and meandered away [another example of the on-off switch in action].[100]

And heavier yet:

05-??-1991   *Unnamed location in Brazil*

This time Moises **believes he was held in the air for as long as fifteen minutes**. He remained conscious throughout but terrified. There was nothing he could do but pray.

Finally, the UFO moved away for good, going out of his range of vision. Until that moment it had made no noise at all, but as it drifted off, Moises could hear a low humming sound like a quiet turbine. **As it left, Moises felt the paralysis quickly end, and he fell to the ground like a lead weight.**

"They just dropped me hard," he said. In great pain and terribly afraid the UFO would come back...[101]

Of note here are two articles written by the eminent UFO trace ufologist Ted Phillips that appear in the December 2005 (#452) and January 2006 (#453) issues of the *MUFON UFO Journal*. In these articles, he gives 77 cases where the term "levitation" is used, not in the sense of the mystical rope trick or the magician's assistant rising into the air seemingly unaided, but in the true sense of a balance between weight and lift. Among these articles there might be a few that have a beam of light which is referred to in abduction cases as a cause of the levitation, but, in my opinion, the majority seem to be caused by the vortex.

2.12 How Tough Is This Field

I would like to bring to your attention something that any submariner would surely know about: the strength of the wall between him and the water, subject to high pressure on the other side of it, and the depth to which he can safely dive. But, there are stories of UFOs at incredible depths which would crush a normal submarine.

A very good look at humanity's reaction to the presence of a UFO can be read in Dr. Richard Haines' book entitled *CE-5: Close Encounters of the Fifth Kind*.[102] Within one chapter, readers will find 33 cases of aggression against UFOs involving everything from rocks that bounce off an unseen field to projectiles fired from pistols, shotguns and rifles. By the way, Dr. Haines, formerly with NASA, is a co-founder and chief scientist of NARCAP (National Aviation Reporting Center on Anomalous Phenomena).

2.12 a. Guess What, Dr. Haines, I've Found a Bigger Gun
Truth from Fiction?

The fictional elements in the following account are in the names, dates (case date 08-23-1988), and location of the ship (Atlantic Ocean N.E. of the Antilles) which were changed in order to protect the commander and his crew. The story stems from an interview between the ship's commander and Donald Todd, a highly respected researcher. In the following excerpt we get a view of how tough this high-energy field really is:

From *The Antilles Incident* by Donald R. Todd

> The hammer of the forward gun barely preceded the green-orange belch from the muzzle as a projectile hurled upward. Within seconds **a visible splatter of sparks, combined with a harsh crunch as the shell exploded, short of its target, and seemingly against some invisible field surrounding the craft.**
>
> Meadows stared. "What the hell?" He turned to Clough. "Try another!"
>
> Clough passed the order. "One round. Fire."
>
> Vomiting violent flame, the five-incher spewed another shell upward with **the same result; premature sparks and explosion; collision with the same invisible barrier**. [103]

Needless to say, the UFO was not damaged—shades of the movie *Independence Day*. The sparks mentioned above could be considered a result of the metal shell's contact with an electrical field or with the rapid rotating motion of the field itself, much like an ax being sharpened on a rotating grindstone.

But why the heavy-duty field? Well, the spaceship needs to defend itself against the gun-happy people on this planet.

Next, let us consider that out in space, there are a lot of very fast moving, variable-sized rocks, otherwise known as meteoroids. The combination of the craft's speed and these flying rocks would make one heck of a collision. Therefore a "collision field" is a necessity—don't leave home without it.

2.12 b. Our Technology versus Their Technology

In May 2007, I read an article that showed the dangers of maintaining an orbiting space station as well as extended space travel. The International Space Station, orbiting roughly 220 miles above Earth, was fitted with protective panels to shield against space debris such as discarded rocket parts, rocks, and planetary dust. While engineers on the ground use radar to keep track of objects larger than a softball and adjust the space station's position accordingly, the shielding protects against debris too small to be monitored. According to a February 2007 assessment by an independent safety task force, there was a nine-percent risk that the completed space station (in 2010) would be susceptible to debris that could cause the loss of the outpost or its crew members.

In addition to space debris, there is also the problem of micrometeoroids. To my understanding, if a micrometeoroid punctures the space station's multilayered wall, it pierces the outer surface and goes through some abrasive insulation (which could be thought of as steel wool) that continues to grind it down. A second wall, with additional insulation, is in place as a secondary safety to guard against further penetration by the micrometeoroid. The third and final wall is the one that faces the astronauts. If this wall is pierced, the crew and the space station have a major problem, the rapid exit of life supporting oxygen.

Astronauts in space also face other dangers posed by micrometeoroids. A couple years ago, my friend Kelly McKeown invited me to visit ILC Dover, a company based in my home state of Delaware that manufactures spacesuits for NASA astronauts and other countries that do space exploration. The tour of the plant was very interesting, but as I looked around, I saw nothing being produced except gloves. I inquired about this during a question-answer period and was told that while spacesuits are made of stock components (i.e., like taking a suit off the rack at a department store), the gloves must be custom-fit to the wearer so that they can stand up to the environment in which they will be used.

The International Space Station is equipped with hundreds of handrails which the astronauts hold onto as they maneuver around the station during spacewalks (well portrayed in the 2013 science-fiction movie *Gravity*, starring Sandra Bullock), but micrometeoroids sometimes impact these handrails and frequently make craters in them. These craters often have rough, jagged rims. When astronauts grip a surface with an exposed, razor-sharp edge, they are likely to rip their space gloves which could put the spacewalkers at risk of depressurization and even death. After several cut-glove incidents, the glove material was reinforced to provide extra protection. Additional measures have also been taken which have reduced the number of cut-glove occurrences.

Just like the International Space Station in the above text, I am sure the UFO is equipped with a means of detecting large projectiles in its path and avoiding them. However, it is the small pebble-sized particles that are not seen on "radar" that do the damage here. On the UFO, the field is rotating at a very high speed, causing the micrometeoroid or whatever dangerous matter impacts the *field* to be carried to the rear at the same time as it is pressing in against the field. Keep in mind that the field is a part of the UFO, so hitting it might also cause some slight movement of the craft. At the point where the field rotates back into the craft, the micrometeoroid's forward pressure against the field is finally released, and the object continues its trip past the forward-moving UFO. The time involved probably amounts to a fraction of a second, saving all onboard from an untimely end.

If we are to travel in space, it will not be just the propulsive, or more likely the attractive/repulsive power plant that we must think of, but mainly our survival in a strange and hostile environment. I personally do not see our current "layers" as a solution to micrometeoroid hits for long-term travel in the universe. However, our alien friends might be illustrating what can be done.

2.13 Tilting UFO—Loss of the Vacuum Cleaner— Vector Analysis

Definition of **vector**:

Math. A line representing a physical quantity that has magnitude and direction in space, as velocity, acceleration or force: distinguished from *scalar.* [008]

In many cases the UFO not only rises vertically, but also angles as it departs. We also have the "dead leaf" descent, where the UFO angles to the left or right and then back to the other side. When it does this, it is losing part of its vertical force component, which causes it to intentionally fall a short distance. When it rotates again to the other side, its vertical component is restored, and the UFO temporarily hovers until it rotates again in the opposite direction where it once again loses a portion of its vertical force and descends a bit more and moves in the direction of the tilt.

In several of the "waterspout" cases, it is noticed that as the UFO rotates for departure, the column of water drops away from the UFO and back into its source. This can be seen in the sketch below, with the loss of part of the vertical component as the cause. In its rotated position, the actual diagonal suction is drawing in more air and therefore its force on the water is depleted.

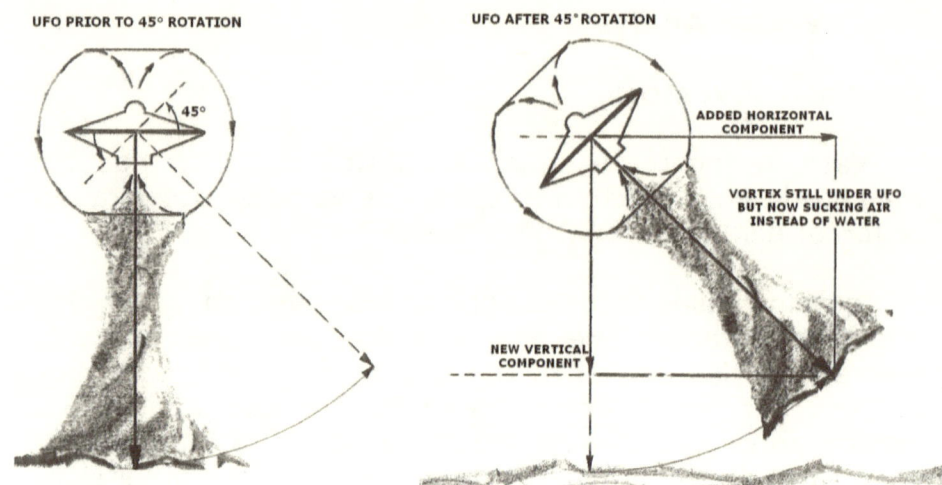

Sketch #19

This could be compared to vacuuming up sawdust. As the nozzle of the vacuum gets vertically closer to the sawdust, it eventually reaches a point where the sawdust is lifted by the vacuum force. If we were to measure the distance from where the force first lifts the sawdust and then repeat the experiment with the nozzle at an angle, we would find that the nozzle must get even closer to the sawdust in order to lift it.

This process is explained in a slightly different manner in *Unconventional Flying Objects* [104] on page 146 under the title of "Mechanics," although it does not correlate the application to water.

The vortex or suction below the craft is the cause of many physical events in UFO cases and is discussed often in this book. I am sorry to tell you, but this vortex that lifts water, dirt, people, and cars is a byproduct of the field, just as the turbulently flowing air behind an aircraft propeller, known as prop-wash, is a byproduct of the propeller's rotation.

My fascination with the mechanics of the vortex stems from how much it explains what was formerly "mysterious" in ufology but now turns out to be mere physics. What a disappointment. (I am, of course, being facetious.) As an example, I submit a very famous case that has mystified ufologists for a long time. It is called the Coyne Incident, and it occurred over Mansfield, Ohio, on October 18, 1973:

An Army Reserve helicopter crew of four men takes a helicopter to a hospital for their annual medical checkup. While returning to the base, the pilot sees a red light and calls the tower to check for traffic in their vicinity. The tower says "no traffic," despite the helicopter pilot's sustained visual confirmation of that red light. The light keeps getting brighter and coming on faster, so the pilot pushes the stick forward to dive and avoid collision with the unknown red light. However, as he looks up following his supposed descent, the UFO is now above them with its light filtering into the helicopter. So he looks at his altimeter thinking the UFO has come *down* with him, but instead of going *down*, he has gone *up...up* the funnel.

With its engine and propellers still functioning, providing the lift needed to counteract its own mass, the helicopter flies in that moment at the theoretical weight of zero. The UFO, with the suction created by the vortex below it, flies over the helicopter and pulls it up unintentionally. As the UFO realizes that the helicopter is continuously getting closer, it immediately takes action to avoid a collision and departs rapidly.

In the next chapter, we continue with the same culprit: the field—but in reverse.

Chapter 3

The Field in Reverse

A bridge between "Physical Influences..." and Heat

Perhaps I should clarify something about the field at the very beginning and that is I do not consider the field to be the propulsive or attractive-repulsive force used by the UFO for its travel. It is an extremely important force nonetheless in that it is *protective*. We have reviewed some aspects of this in the prior chapter and will even turn the field itself off in the next chapter.

Visualize a bar magnet with the north pole at the top and the south pole at the bottom with the field moving between them (granted, yet again, that a magnetic field is unmoving). This was the conception of the field's operation in all the sketches thus far. However, I will continue to use the example of a rotating field regardless of its composition.

3.1   Operation of the Field in Reverse

Now let us consider the UFO descending at some speed towards the water with the field as previously envisioned. Upon hitting the water, the force striking the water is not only the descending mass's speed, but also the speed of the field's rotation in a downward mode. This would cause a very large splash, which has occurred in some UFO cases. The fascinating cases are those where the UFO dives into the water silently and with *very little* surface disruption.

In this situation, as the UFO enters the water, we have the field reversed with the north pole on the bottom rather than on the top, pushing the water away 360 degrees horizontally from the UFO. At the same time, as the UFO descends further into the water, the water touching the field's midsection is being moved to the rear of its travel.

It would appear that the field is digging its own "hole" in the water as has been mentioned in several cases. The fact that there is very little movement of the water (ripples rather than waves) is a before-and-after effect of the field.

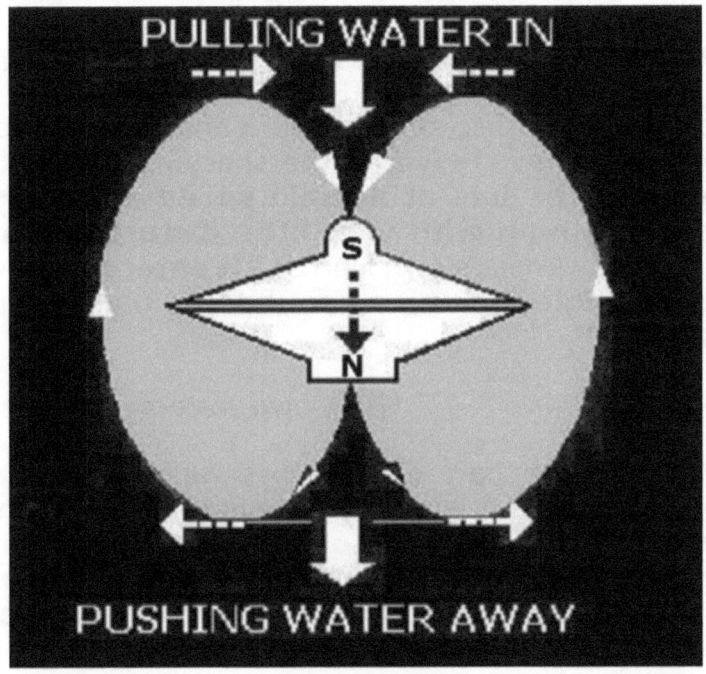

Sketch #20
In Chapter 2, the north pole is at the top,
whereas in the "reversed field," it is now at
the bottom.

So the force is not a continuous outward movement, as with a stone hurled into the water, but is now at the rear of the field as an inflowing force bringing the water back, thus eliminating any waves or splashes. In some cases this is similar to an aircraft rarely feeling turbulence in the vicinity of a UFO, as the turbulent air is pulled back by the field's rotation into the craft. Some examples of the water effects follow:

UNDATED—8   *Hartwell Lake near Anderson, SC, USA*

"A short time afterwards **the object** went straight up about 10 feet, **came down upon the water making no ripples**, and back up and out of sight." [105]

SM-??-1957   *Penobscot Bay off Camden, Maine, USA*

The craft is a silver, disc-shaped object and has what appears as a metallic dome. However, the shocking part is that **it crashes into the bay at a rapid speed approximated at about 100 mph with very little disturbance of the water.** In the witness's own words: **"It entered the water at a steep angle making no sound and barely a ripple or splash."** [106]

LL-??-1972   *Unknown waters off Puerto Rico*

The most notable part of the observation was that the objects had a glowing, greenish-blue field around them, and **they entered the water at considerable speed without any splash. The water just parted ahead of the visible field, and the objects flew into the hole, and it closed up behind them with only a flurry of ripples.** [107]

08-01-1975   *Trinity Lake, New York, USA*

There was no splashdown; the object slowly settled into the water. Complete submersion took about two minutes, and **ripples** [not waves] **radiated out from the object's location**. [108]

04-17-1999   *Lake near Invermere, BC, Canada*

"Three of us were looking out into the water; it was a quiet, peaceful afternoon when we all noticed a bright white light emerging from the water. **The object was breaking the water surface from the lake yet wasn't leaving**

any ripples in the water!—like a piece of wood would or a log could." [109]

06-27-2006   *A pond near Flagler, Oklahoma, USA*

A hissing sound was heard and a steam-like substance rose from the water. **There were no ripples from the entry of the object into the water**. [110]

09-03-2006   *White Fish Bay, Michigan, USA*

The object, visible only as a ball of light (which may have had a dark band across the center), lowered itself, and the object and its reflection merged at the surface of the lake, and the object was gone from view. **As I watched the object, it had descended into and below the surface of the lake. Not a ripple or other disturbance appeared on the surface.** [111]

3.2   Now the Important "One"

Like the "mounding water" case, the "reversed field" is an observation based on a single case that lends verification to the field's operation.

The following case is interesting in four aspects. First is the expulsion and suction of water, which were demonstrated in the last chapter. However, the unique feature of the expulsion and suction is that they occur while *totally underwater*, and the field movement is in reverse. Secondly, there is a reaction to the UFO by fish in the water. Thirdly, this case bears a connection to Chapter 5, "Abductees—Water and Bases." Fourthly is the feature of heat, the topic of the next chapter, where we see the inactive and cold craft heating up as it resumes its travel.

07-??-1989   *Aguadilla, Puerto Rico*

Inocencio Cataquet, another of the local fishermen, related an incredible episode he had experienced in July

1989. The professional skin diver and snorkeler had been one of the few people alive to have actually touched a sunken UFO.

"...I was alone, fishing in the deep, when around 5:30 in the afternoon, I became aware of an enormous, really shiny thing moving underwater...Ay, caramba! I swear to you that it was some two hundred feet in diameter," Cataquet told the assembled UFO investigators. "The thing came along and remained underwater some five hundred feet away from me, [and] **then it turned itself off**. You could only see a slightly dark form at the bottom. [This feature will become prominent in Chapter 5 in Budd Hopkins' story *Witnessed*.] It made me think, so I grabbed my flippers, my mask and snorkel and went in for a closer look. It was some 25 feet deep, and when I went down, I realized it was round. It was flat on the bottom and curved on top. I descended even more and got right over it. I touched it and felt that its surface was not metallic, but porous, like touching sand or cement, and it had an ashy-grey color. **When I touched it, I noticed that the thing was sort of absorbing water, sucking it from the top, which was full of tiny little holes. You could feel the suction on your hand.**

Cataquet's experience didn't end there. "I went up for some more air and went down again. This time, I swam under it, some three feet from the sand at the bottom, and **I felt it expelling water from underneath, as if expelling the water it was sucking in from the top.** [In from the top and out the bottom implies a motive force is at work.] **I could see fish swimming around me and it** when suddenly the thing lit up for a minute. I was scared, but I stayed. **Then, almost a minute later, it lit up** and made a nasty squealing sound, as if you were stuck inside a bell and you could hear the sharp vibration. **The fish took off**...I don't know why I thought they were telling me something, telling me to get out."

"The water started heating up," [indicating that the field is now on] Cataquet continued, "and I headed for the shore. After I'd gotten out of the water, I turned and saw the thing moving away really fast underwater. After a while, I looked toward the horizon and saw this tiny little point of light rise from the water and head straight up, vanishing in the sky." [112]

In the above story, we can also see another aspect of the craft itself. With the field off, as seen by the witness, we now have a submerged craft with no exterior protection other than its own outer skin. This would indicate that the craft is hermetically sealed and of a rational design, because if this craft were to enter any atmosphere without the field being active, it would still be immune to poisonous gases and any unwanted fluids.

However, a mystery persists. The field is off, which is confirmed by the fact that the exterior lighting is off and Cataquet does not feel heat. The question then becomes, what is making the water move in at the top and out at the bottom? The only solution that I can see is that there is a turbine of some sort in the center of the tube connecting the top and bottom of the craft that is still running and moving the water. Does this turbine also play a part in moving the electrons of the field? In addition, could it be that this turbine's operation is the reason that, in several reports, witnesses say the only sound they heard was a slight humming?

3.3   Yet More Proof of the Rotating Field in Reverse—Tiny Pyramid- or Inverted-Cone-Shaped Waves

More proof of the field's rotation comes to us, like a bridge between Chapter 2 and this one, in the form of another example of "function of height." This time it is of the uncommon variety and fits between water depression and mounding water. The field in this case, however, is barely touching the water, and, unlike the other examples in Chapter 2, the field is in reverse.

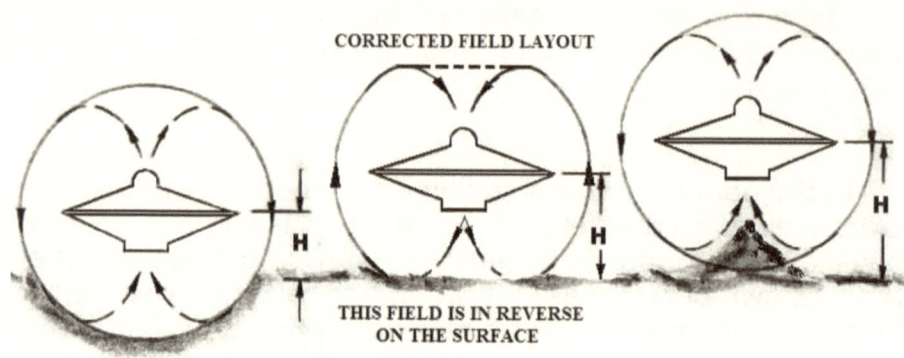

CORRECTED FIELD LAYOUT

H

H

H

THIS FIELD IS IN REVERSE
ON THE SURFACE

Sketch #21
Start of inverted-cone-shaped waves

Originally, I had only two cases of the instance described above, expressed by an author in one of his books, putting the idea of the usefulness of said cases in doubt. However, I have since found two other accounts involving sharp-pointed waves, as stated in the first case, and have had the good fortune to supplement it with an even stranger case dealing with snow and ice.

All five cases are trying to express what the witnesses are seeing, but it is so strange that there is no precedent for it. As a consequence, the experience becomes a matter of interpretation. I think I can explain.

The wave, which is mentioned in the four water cases, is not really a wave; the rising of the electrons of the circular field gives the *appearance* of a circular wave.

Inverted-cone and pyramid-shaped waves are references to the effects generated from thousands of electrons, which come through the surface of the water while pushing minuscule debris in the water ahead of them and have a capillary attraction to the rotating field. This is responsible for the formation of the perceived forms of inverted cones or pyramid shapes. (I would prefer to think of them as sharpened pencil tips, as the wood and lead would make a good example of what is seen.) An analogy can be seen in a heavy rainstorm as the "thousands of drops" hit the wet sidewalk and bounce upward.

If the field is just grazing the water's surface and the field is in a reversed mode, then hardly any depression would be seen. Both words, pyramid and cone, indicate triangles. In dealing with water, however, a cone shape would be a more reasonable assumption, and this shape could be similar to the candy known as a Hershey's Kiss. The rotation of the field would cause only a small surface wave or ripple to be created outbound. Also, it occurred to me that the field would have to be in reverse. If this were not the case, the field would suck up water, pulling it inward toward its center rather than creating waves outward. Again, this is proof of the field's rotation. Furthermore, this means that the field is in the reversed mode and that the craft may be preparing to submerge at some point. Since a specific height of the craft above the water would be required in order to make perceptible waves, accounts of this nature are not often reported. I highly doubt that the craft and their operators would be performing this maneuver for our entertainment, so the wave becomes just another physical manifestation of the field and its effect on water. To my joy, this is proof of the field's rotational movement. I would like to add that the sharp-pointed waves referred to do not occur naturally but are constructed by the field. As the field is circular, it *appears* that there is an outward wave. Other small waves may appear as the craft moves forward, causing a slight pushing motion upon the water. However, the waves with the pyramid/cone shapes as described above are not waves as such. Case examples follow:

??-??-1953   *Lake of the Ozarks, MO, USA*

They saw a shiny, disc-shaped thing. It was oscillating slowly, and both men noticed that directly beneath it, the water was dancing in **thousands of tiny, sharp-pointed waves**. [113]

06-??-1958   *Adriatic Sea off Casal Borsetti, Italy*

When he was a soldier, assigned to military antiaircraft, Mr. Giovanni Mantovani **saw the sea bubbling** about one thousand meters offshore at 1:30 p.m. Suddenly, a black object appeared, full of pipes, without windows or sails, and **covered by a whitish steam**. The depth was only 20 meters and the area was banned to navigation. The witness climbed on the aiming binoculars of the antiaircraft and **noticed that the object was suspended over the water and later disappeared. However, as the object moved, it caused small pointy waves to form beneath itself, as if a magnet was pulling the water.** [114]

??-??-1962   *Unnamed river, assumed in USA*

...they noticed that **the water beneath the UFO was dancing madly—a circular patch of tiny waves** that moved along with the UFO and was unquestionably caused by it. [115]

07-05-1997   *St. Margarets Bay, Nova Scotia, Canada*

The water which was flat calm had kicked up into tiny wavelets like little pyramids in a line towards the opposite shore, but he noticed no hint of a breeze. He offered that the water did not look like it normally would if there had been a breeze with little wavelets coming toward you but **choppy with little pyramids on it**. [116]

In the two sketches that follow, primarily made to illustrate case 01-03-1971, I have cut the saucer in half since we are only concerned here with what is happening below it. In all five cases, the action is on the water.

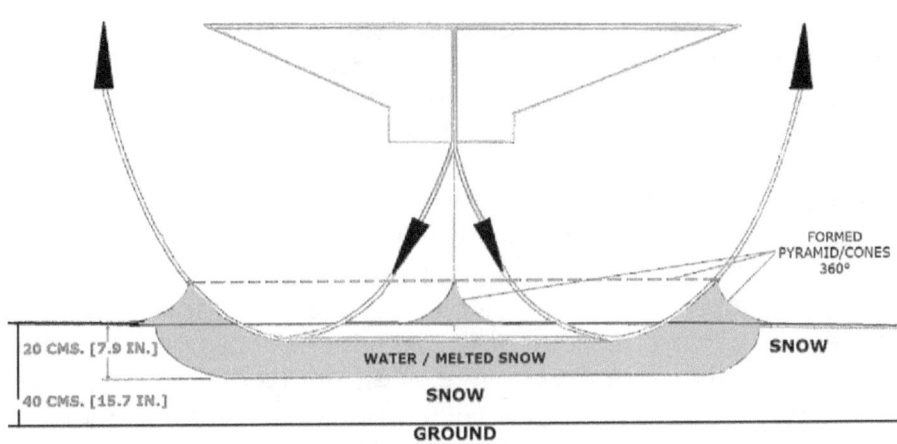

Sketch #22-A
The way it works

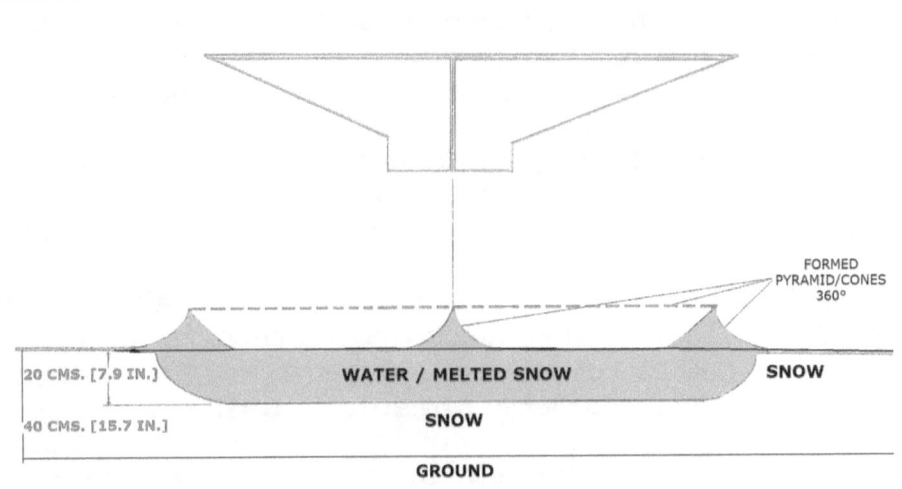

Sketch #22-B
What the witnesses see

Another manifestation of this same upward traveling of water is even more apparent in the next case. In this account, the same upward movement and the same minuscule submergence of the field accomplish the same movement of the *melted* snow. However, that water is quickly refrozen into an inverted-cone shape, and what is even more mysterious is its construction by the field: this upside-down icicle now has a hollow center. In short, while the field is scooping up the water, the electrons within the field are burning off the interior water in the main body of each created icicle, leaving in effect a hollow tube.

The following text comes from the last of three separate issues of *Flying Saucer Review* which contained articles regarding this case. Parts have been omitted since the case is quite lengthy and I only wish to include what is relevant to this topic. The account can be read in full on my website under the date indicated below.

01-03-1971   *Saapunki, Finland*

On the Sunday morning, January 3, between 5:58 and 6:15 a.m., a bright light phenomenon was seen in the village of Saapunki in Kuusamo....

The phenomenon was seen as a bright ball of light moving slowly along the length of the lake in a westerly direction. **The ball of light moved at a level of only about 8 metres** [26.25 ft.] **from the ice**, and it could be seen against the trees on the slope of the opposite bank. The first observations were made in the eastern part of the lake at about 6:00 a.m., and the phenomenon is then said to have moved at about walking speed, the watchers being at the side of the lake at a distance of 300 metres to 1 kilometre. In spite of what was **almost a 7 Beaufort storm** [32-38 mph] from the southwest, the phenomenon moved with a steady speed, slanting against the wind, keeping the same altitude.

The first observers were not able to tell the exact size of the ball of light since it was impossible to look straight at it because of its great intensity. There was **no sound to be heard**, but the snowstorm was so forceful it blotted out all weak sounds. **One of the first observations in connection with the phenomenon was the darkening**

of the electric lights as it was passing…. Mauno Talala estimates the size of the light to have been 10 metres in diameter, measured against the garden fence…. **The phenomenon remained at the spot about one minute** and the electricity was gone for about the same length of time after it had disappeared [again an indication of the depletion of electrical power in the vicinity of a UFO]…. **There was a temperature of about -20C** [-4°F or 36 degrees below freezing point on the Fahrenheit scale] and the wind was filled with snow. The sky was covered with clouds, but most of the snow carried by the wind was from the snow-covered ground and was not actually caused by the clouds…. The snow cover, about 40 cms. [15.7 in.] thick, had melted on the spot to a depth of about 20 cms. [7.9 in.], i.e. not all the way to the ground. In the middle of the spot there was a round area about 1.5 sq. metres in size [16.1458 sq. ft.] in which stood **needles of ice, thick as fingers, and empty inside like thin-walled tubes**. On top of each tube there was a ball-shaped formation the size of a fingertip and according to the children looked much like a candy that they collected for their games [caused by the melted interior water coming out of the top and falling back onto the icicle and refreezing at the top—lollipop]….

The UFO researchers in Oulu were told on the Tuesday, and got on their way to investigate the case by Wednesday, January 6. An investigation team of five men took the first samples. They also made radiation measurements on the spot, but could detect no deviation from the normal background radiation (four days had already passed since the incident). More check samples from the surroundings were taken on January 17, and vegetation samples taken later during the winter. **The melting of the snow and the forming of the ice needles could not be reproduced in the laboratory tests.**

The electrical distribution company for the Kuusamo area stated that there were no distribution disturbances at the time in question and therefore no breaks in the distribution. The electricity lines terminate at this area, so there is a possibility that the storm could have caused some local breaks [or energy depletion by the UFO]. [117]

3.4   The Roswell Gouge versus Underwater Gouges—More Evidence of the Field in Reverse

Earlier in this chapter was a case with one witness *underwater* with a UFO, then there were five cases that demonstrated the field in reverse near the *surface* of the water. Now I would like to introduce the Roswell gouge, which occurred on land, and follow it with five cases in which a gouge was left in the soil at the *bottom* of the water so that you may note the similarities. This case is not on my website.

07-EE-1947   *Roswell, New Mexico, USA*

"The debris field was oriented northwest to southeast. Marcel said that it (the debris field) was about three-quarters of a mile long and two to three-hundred feet across, with **a gouge at the top (NW) end of it that was about five-hundred feet long and ten feet wide.** Debris littered the ground, lying on top of it." [118]

As I understand it, there was an explosion onboard the Roswell craft, and not being in the craft, I can only surmise what might have occurred next. Since the craft travelled horizontally, the ship's landing gear, made for vertical ascent and descent, would have been useless. If there was a mechanical problem with the landing gear, the choice of which way the field should be rotating would have been a major concern. If the field was in what I consider to be the normal mode (on with the north pole at the top), the craft would have been bouncing on the field as it was rotating inward, almost pulling the craft down. If placed in the reversed mode, the field would be rotating downward and outward, making a cushion as it were, while still protecting the craft. Of course, this cushion, being a rotating field, is kicking any loose material out of its way in a 360-degree movement.

The gouge made at Roswell was shallow due to the type of soil and rock below it. Because the connection between water and UFOs has generally been overlooked in ufology up to this point, the similarity between the Roswell gouge and the gouges in the following water cases was not evaluated, so we will investigate the few cases that are available to us:

08-04-1888   *Highland Lake, Connecticut, USA*

The wind by this time was blowing with cyclonic force. Suddenly there came a roar, and far down the lake a huge flame of fire could be seen. **The water for yards ahead was parted, as though by a gigantic plow, and the billows seemed to rise at the sides of this furrow for fully twenty feet. The ball of fire appeared to force the water aside, and so deep did it go that the bottom of the lake could almost be seen as it passed through the narrows. The parted waters, with their singular propeller, advanced toward the head of the lake with great rapidity**. When within one hundred yards of the shore, there came another flash of lightning, and the fire disappeared as suddenly as it had come. [119]

08-??-1959   *Unnamed creek near Tulsa, Oklahoma, USA*

When it was 10 feet from the water, it sort of glided north, leaving the water badly disturbed. It moved slowly and soundlessly. **The water seemed to part until there was a ditch in the water about a foot deep**...as though there was a force blowing on the water. The machine finally rose above the treetops and went out of sight. [120]

02-??-1961 *Unnamed lake in the Karelia region of Russia*

Although the following day was a Sunday, it nonetheless saw the start of an enquiry by six specialists flown out from Leningrad who examined **the lengthy gash on the lake's edge**: a hole about 30 m long, 15 wide and 3 deep. **The crater was narrower towards the bottom**, which itself was smooth, though around the edges there were chaotic heaps of grass and soil. But **the bottom of the hole was as smooth as possible, as if someone had used a roller.** As for the soil excavated by the object, there was no trace of it.

...The ice on the lake nearby was largely broken, and at that battered spot there were ice floes, but elsewhere the ice was undamaged. Black and crumbling pellets were found on the bank which resembled buckwheat grains; these could be crushed between the fingers.

On the bottom of the lake, divers found **a portion of the soil displaced which was stretched out like a "tongue" across the base for about 100 m.**

The theory was that the object had first slid across the soil on landing and thence into the lake, yet there was no trace anywhere of any such object, and even when a metal detector was used, there was scarcely any reaction.

The divers dragged some of the ice floes on to the bank and noted that the submerged parts were green. Samples of the pellets, the soil, the ice, and some water were collected....

Meanwhile, the divers found roughly in the middle of the lake **another gully along the bottom about 100 m long, which almost seemed as if it had been excavated** for a piping system. At the end of the trench, there was a hillock about 1.5 m high which looked as if it had been piled up when **the trench was made**. All around the "tongue" and the gully, the ground was untouched and neither, once again, did the detector apparatus offer any clue. Who or **what had dug the hole** and where was the object lying?

It was gradually deduced that the object had skidded over the soil into the lake, **ploughed up the bed**, and then by some complicated means or other had managed to resume its flight. So this must have been a special sort of machine indeed, to say nothing of its pilot (if there was one) who, judged on our own criteria, should have been considered a candidate for a gold medal in sheer presence of mind. [121]

04-27-1961   *Lake Onega, Russia*

Among the sample of sightings made available to APRO, one appears different enough from the usual run of sightings elsewhere to merit describing in some detail. This concerns an object which came in low, at great speed, over Lake Onega, 200 miles northeast of Leningrad on April 27, 1961. **It hit the bank just above the ice and ploughed a furrow 27 meters long, 15 meters wide, and 3 meters deep through the frozen ground** before continuing its low trajectory and disappearing. The impact blasted the ice from near the shoreline, throwing out huge chunks to the bank with a strange intensive green color on their underside. Local residents, including a forest ranger, reported no sound except that of impact when the object crashed into the bank....

Scientists who interviewed residents of the nearby village reported agreement that the object was oval in shape, about the size of a large passenger plane, and bluish-green in color. Other scientists concluded that the object exhibited none of the characteristics of a meteorite and left none of the after-effects associated with meteorites. [122]

04-30-1976   *Lake Siljan, Dalarna Province, Sweden*

The object, estimated to be moving 90 km/hr, **cut a channel 1 km long and 3-4 meters wide through the 20-cm-thick ice**. Ice floes and water cascaded around it as it forced its way through the ice. The object was dark (2 witnesses) or gray (1 witness) and about 10 meters long. [123]

3.5   Energy Transfer between UFOs Using the Reversed Field

No, there are no pieces falling off UFOs, however, engine failures or slow starts...perhaps.

As I was writing the first edition of this book, I was also reading, during my time off, a book by Paul Potter titled *Gravitational Manipulation of Domed Craft: UFO Propulsion*

Dynamics. [124] While on pages 250-252, I suddenly realized that the text he was presenting (based on Raymond E. Fowler's book *The Watchers: The Secret Design behind UFO Abduction* [125]) revealed the operation and transfer of energy between two craft and was remarkably similar to a case that I had read many years ago and dismissed for two reasons: The account was not directly water related and also it was, yet again, a single case of its type. Now, though, with two cases having the same components, I view this not as a possible misidentification or hoax, but, indeed, as a parallelism of functions. The following texts were sent to me by Chester Grusinski, a researcher in a quest for answers to *his* own sighting (see Chapter 7, case dated 10-??-1958, page 241) which also took place on the USS *Franklin D. Roosevelt*. My thanks to Mr. Grusinski for sending them to me. The particular oddity described was seen several miles away whereas the land case, coming up in section 3.6, was seen by an abductee at very close range.

The first text is a memorandum dated March 1982, written by Leon Treadwell, possibly as a reminder to himself. It is titled:

U.F.O. SIGHTING RIO DE JANEIRO-BRAZIL
JULY 1956
ABOARD THE AIRCRAFT CARRIER U.S.S. *FDR*
CVA-42

These objects were seen by me as well as several dozen others including officers while crew members of the carrier *Franklin D. Roosevelt*. The time was in the early morning hours of July 26, 1956, while at anchor in the port of Rio de Janeiro—Brazil. Most of the crew were asleep and unaware of the sightings. As members of C.I.C. [Combat Information Center]—radar div.—these objects were watched through high power binoculars for several minutes before they finally vanished with tremendous speed.

When first sighted, the objects were suspended in mid-air over one another with several hundred feet between them. A radar fix placed them several miles from the *FDR*'s position at a height of 2000 ft. Two rows of bright, counter-rotating lights could be seen through the middle of each object that made them appear to be round in shape. For several minutes they never moved. **Then the upper one released a fireball object that dropped into the top of the lower one.** Within seconds they both vanished with tremendous speed. Radar contact was almost impossible [as they departed, I would imagine]. It was estimated the objects could have been between 75 and 100 feet in length.

All involved were told to keep a tight lip on what we saw. A report was filed, signed and sent to the Defense Dept. along with a color sketch by me of what was watched that night by many.

Since my discharge from the Navy in 1958, I told very few about this event and it was only now that I decided to sketch these objects again for future reference on a possible oil painting.

Leon Treadwell, March 1982 [126]
[END]

In April 1991, Leon Treadwell, who served on the *FDR* from 1956 until June 1958, wrote to Mr. Grusinski following Grusinski's request for information about a UFO sighting in October 1958 which he had witnessed, in hopes of finding others to confirm what he had seen. This letter follows:

Leon Treadwell
Address
April 1991

Mr. Grusinski:

I received your letter with much interest. I left
[the] *F.D.R.* in June 1958. I cannot help you on
the 1958 sighting [Grusinski's sighting in October
of that year] but I think you will find the enclosed
of interest.

I was part of the crew that put the *F.D.R.* back
into commission on April 6, 1956, at Bremerton,
Washington. She had the best radar in the fleet.
Before this I was on [the] U.S.S. *Midway* CVA-41.
I took advance Radar School at Treasure Island [in
San Francisco Bay] in 1955.

[The] *F.D.R.* was to report to Mayport—Florida.
We took her around the "Horn," [Cape Horn—the
bottom tip of South America] since she was too
large to fit the Panama Locks. Our last stop before
Mayport was Rio-de-Janeiro, Brazil.

At this time I was RD3 [RD=Radarman (Rate),
3=3rd Class Petty Officer (Rank), which no longer
exists] and on the night of *July 26, 1956,* I was duty
P.O. [Petty Officer] in the radar "shack"—C.I.C., on
anchor detail—0000-0400 hours. One other 3rd class
was with me. Four lookouts were topside. While
at anchor only surface radar was on line. Our VL
radar was down (height radar). One lookout, then
another called in that two strange objects were
in the air but did not know what. I put a VL into
operation—went topside to see for myself. What
you see in the sketch is what I saw that night 35
years ago. [A photocopy of his sketch is too dark to
reproduce here, but **it shows two disk-shaped
objects, one above the other, with a bright**

light in-between.] I called in the duty officer of the day and he in turn called others.

These "things" were watched by many for about 5-7 minutes. All possible reasons for these objects proved wrong other than the fact they had no reason to be there.

My original sketch went to the Defense Department, but it stuck in my mind all these years. I made the enclosed sketch in 1982 from a "rough one" that I had to tell no one to [sic-about], for 20 years.

The following names may help you if they are alive today: *F.D.R.* Commanding Officer: Captain John T. Hayward, 6 April 1956-1 October 1956; C.I.C. Operations Officer: Whitney Wright; Navigator: Commander James R. Thomson. [127]

[END]

As can be seen in sketch #23 below, both UFO fields are of the same rotation with the north pole on the bottom of each craft. This means that when one craft is positioned above the other one, the top UFO's north pole is facing the lower UFO's south pole, thus providing a magnetic attraction (through magnetic lines of force) between the two which could be interpreted as a corridor for the energy transfer.

This transfer of energy between UFOs is on a much greater scale than what we accomplish when jump-starting a car battery after it loses its charge due to some careless act, such as leaving the car lights on by mistake. This simply requires bringing in another vehicle and hooking up jumper cables between both of them. The nonfunctioning vehicle draws its charge from the functioning one, allowing it to start. From this point, the vehicle becomes fully operational and the cables are removed.

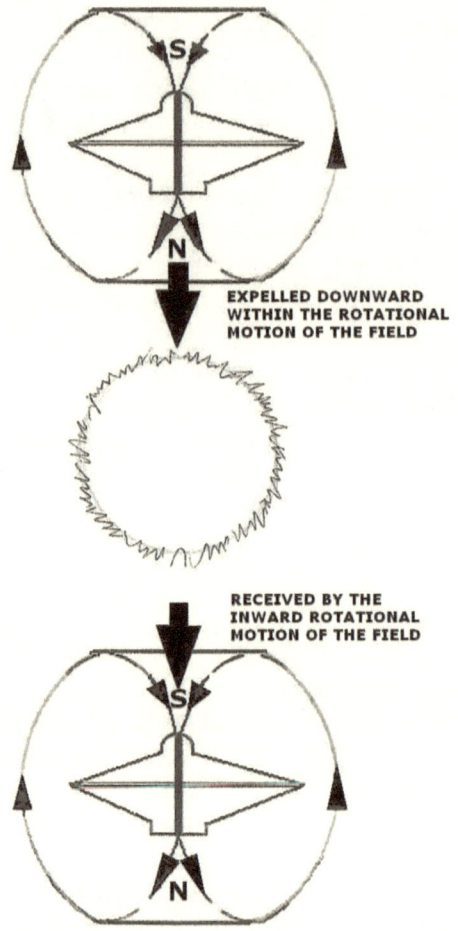

EXPELLED DOWNWARD
WITHIN THE ROTATIONAL
MOTION OF THE FIELD

RECEIVED BY THE
INWARD ROTATIONAL
MOTION OF THE FIELD

Sketch #23
was drawn by me (CF) as a replacement
for the darkened original sketch.

3.6 Similar Exchange As Seen by an Abductee in a Non-Water Landed Craft

Earlier in this chapter, I provided Sketch #20 to show how the reversed field affects water. But even with non-water landed craft, there is a situation when the reversed field can still be used.

In a famous investigation by Raymond E. Fowler, this account of an abductee's remembrances was acquired through regressive hypnosis. The abductee's name is Betty Andreasson Luca.

From his research into the text of this case, Paul Potter explains in his book that the aliens gave quite a bit of information to Betty concerning the UFO's engine, which he believes is a "quasi-black hole." Potter's book is quite technical, and his idea of a black-hole engine took me aback. I had heard of putting "a tiger in your tank," but my initial conception of a massive, consuming black hole inside the craft slowed my acceptance of the book. However, an artificial black hole was constructed on a small, controlled scale in a laboratory in China in 2009, so Potter's belief that the UFO's engine is a controlled, miniaturized, "quasi-black hole" seems plausible to me. As I continued reading, I found that I shared many of the author's ideas. Potter uses "**aa**UFO" to specifically identify the craft he is talking about as the "**a**ndreasson **a**ffair UFO," and the text for this section of his work is based on Fowler's book *The Watchers*.

From the text of *The Watchers* that takes place in 1967, we read about what happens when a second UFO comes to help an incapacitated craft:

> Betty: ...**And that huge craft is hovering silently right over that second craft on the ground**, and—(pauses) and, there's things like legs coming out, or arms coming out, from underneath the big craft. And they're coming down and they're clamping on to the edge, the edge, of the bottom craft. And it's like sitting there. And I'm standing there watching those other people (the lady and man [two other abductees]) sitting. And I'm trying to communicate

through my mind because I can't speak out as to what's going on to one of the beings by me. Oh!...

[NOTE: Due to the distance in the USS *FDR* case, these "grab-arms" might not have been seen. Of course, the distance between the two craft in that case was described as "several hundred feet between them" and the other craft was not "grounded."]

Betty: I'm wondering what it is. And, and the being says they are purging and lining the *cyclonetic trowel*. (phonetic rendering)

Bob: Is the *cyclonetic trowel* part of the craft?

Betty: I, I think it's that thing that is spinning counterclockwise, I don't know. It keeps spinning and spinning. Oh, and there's, there's—It, it looks just like water spinning around it, on it, on this smooth thing. It looks like water going around and round...

And it feels like it's starting to get warm here. Ah, it's getting warmer for some reason. I'm warmer. (sighs) The being says: "Balancing the oscillating telemeter wheels and leveling." Ah, I just can't understand some of that.

Bob: That's okay. Just, just try to repeat what you're hearing, even if you don't understand it.

Betty: But, I, I don't know the words to use.

Bob: Okay.

Betty tries her utmost to catch the phrases that the aliens are using to describe their ongoing operation but finds it almost impossible to do so.

Betty: Rotating series of semi-full swing back. Liquid line? Magnetic rings (pause) and the depolarized rim. And there was something else, I don't know. (sighs) And, oh! Is that something. Oh, this is beautiful, beautiful! Tch, there's bright light now right in that spinning part in the center? And the steam? It's like clouds all around. And it's causing rainbows. Ohhhh, this is beautiful. There's like bright, bright light coming from that spinning thing—and the middle—with water or something spinning around and round. And it's causing clouds all over the place and, and rainbows. And those balls of light hanging in the air over those jacks are turned, like, a blue color (pauses)—Real bright blue, blue. And, oh, it, that— Those rainbows are so beautiful! (sighs) Oh, this is so beautiful. But it is warm, so warm too. It's getting (pause)—You can hardly see anymore though because of the, (pause) the fog. That stuff is causing such a fog all over now. There's so much foggy fog. It's so warm, it's almost smothering. Whew!! And there's lightning coming all over the place! OOOOOOOh, that's scary!...

[In the next chapter we will be considering the heat generated, as above, by a UFO's field.]

Betty: Oooooh, there's lightning all over!

Bob: The lightning will not hurt you.

Betty: Yeah, but it's going all over the place! Oh, it's scary! (Betty is now exhibiting raw panic.)

Bob: Okay, I want you to step back and just observe. You've already been there and no harm has come to you.

Betty: (continues to talk excitedly) There's lightning! There's lightning coming out of that whirling thing. And it's going so fast. And it's coming out of the clouds. You can see streaks of lightning all over the place. Oh, oh, I'm so glad we're back this far (i.e., from the craft). Oh, there's lightning all over!

[In the dark photocopy of Treadwell's sketch, sent to me by Grusinski, the ball of energy seems to have streaks coming out of it, which could be the lightning as described by Betty. (See my sketch #23.)]

Bob: It's pretty to look at though—isn't it?

Betty: It's scary. It's too much lightning. I don't like it! The lightning. It's too strange—All over the place.

Bob: Okay.

Betty: It's even hitting those balls, those blue balls. Those blue balls of light. And, and going all over it. Oh!

Bob: Okay. Now, just relax. Let's go—

Betty: (interrupts) Oh, it's scary! The lightning.

Bob: Let's go further ahead—

Betty: Ohhhh!

Bob: —after the lightning stops. Let's go to where the lightning stops. We're moving ahead in time now. The lightning has stopped. The lightning is gone....

Betty: It's raining. We're getting drenched. Oh, I'm so drenched. It's just heavy, heavy rain—like buckets of rain coming down on us. (sighs) But it's cooling us off because it was so warm. Oh, it's raining, it's just raining. It's starting to slow down a little but I feel drenched.

It seems as if the quizzical operation performed by the aliens may have produced a localized artificial cloudburst.

Bob: Are the beings out in the rain also?

Betty: Yes. They're just standing there and the rain is just coming down. It's starting to go real soft now. Oh, and it's washed off all that shiny stuff off of me [an electro- insulating gel the aliens put on the abductees to

107

protect them]. My hair is soaking wet from it. It's starting to let up. And that bigger craft is—the spin is starting to slow down. The water is going round and round and it's going slowly. And the rain is just very lightly now. I can see on the big, big craft—There's like huge indentations, like windows. They were spinning—Big silvery-colored craft. It's slowing. And it's just slowing down. And it's stopped. Oh, I feel prickly. All over my body feels prickly. (static electricity?) Oh, oh, that feels just like my whole body is fast asleep. And it's, oh, it's all prickly, prickly, prickly feeling—All over my—(Bob interrupts)

Bob: Are you still outside?

Betty: Yeah, next to that being. The lady's there. (sighs) All her shine is gone too—Just sitting there and they're wet. And the center thing stopped. And now it's starting clockwise very, very slowly. (pause) Very slowly. (pause) Very slowly. (pause) Still going clockwise. (pause) And it stops. (pause) And swinging back again to counterclockwise. (pause) Slowly. (pause) Stops. (pause)

Bob: Do the beings indicate what the purpose of this is? Why it's stopping and going back and forth?

Betty: No. I don't know. It's going forward... [128]

We will meet Betty Andreasson Luca again in Chapter 5 as an abductee who has also been underwater. Additionally, in another section of Fowler's text, Betty mentioned that the aliens had attached hoses to a UFO that went into a lake. At the end of a power boost these hoses were removed. This is an indication that water is an essential part of a UFO's operation in some way. More cases of taking on water with hoses, buckets, and so forth will be discussed in Chapter 11.

3.7   Fuel

Though one of my main concerns in this book is the field surrounding the UFO, I would like to briefly touch on possible sources of energy. Other than conversations that have occurred between aliens and witnesses in some accounts, we have no documentation regarding an engine in a UFO. In several of these cases, though, electrical energy is mentioned, and in numerous other reports, electromagnetic effects are seen or felt by witnesses.

In both of the cases previously presented that included the transfer of electrical energy, the question of where this energy is obtained comes to the fore. If, as observed, UFOs have no propellers, no jets or rockets or other propulsive-force engines, the only source left for us to consider is electricity. While fossil fuels and nuclear power provide much of our electricity here on Earth, abductees usually report a sterile environment within the UFO, which would generally rule out the burning of fossil fuels within the craft. What could heat the water into steam to operate the generators? Is there another means of gathering energy while traveling the depths of space? We must consider alternatives for power generation.

There has been speculation that the source of energy for a UFO is drained from a black hole. In the preceding section of this chapter, I made reference to this. Developed from a proposal in a paper published in 2009 by Evgenii Narimanov and Alexander Kildishev of Purdue University in West Lafayette, Indiana, USA, a black hole was constructed in a laboratory in China. [129] Normally, we would associate the term black hole with its nemesis, the gigantic galactic black hole which is a planet eater. However, this laboratory model is much smaller and evidently controllable or else we would have heard about its destructive tendencies here on Earth by now.

The advantage of having a black hole as a fuel source is that it is extracting energy in the form of electrons from all around it, wherever it is. I am sure that micrometeoroids, aside from being an energy snack, would also make the UFO a vacuum cleaner of space as the craft travels.

Now that all this electricity is available, what can be done with it? Well, it seems humans have discovered a use for it: electromagnetic trains. Could aliens use that magnetism for attraction/repulsion of their craft? Perhaps, and maybe we have already encountered its "in use" effects. If a UFO is extracting energy from its surroundings, would it not be logical that lamps would dim and any engine with an electrical system would have power drained from its wires? The same could be said, of course, for cell phones and video cameras as well.

But there is a problem with this fuel source. Black holes increase in size as they gain more energy. As a result, a relief valve of some sort would be necessary to contain the monster within, as it would threaten the craft itself by its growth. Some energy would have to be dumped before it started devouring the craft. The transfer of energy to the lower craft in section 3.5, similar to car jumper cables, made me realize that in other cases where this drop occurs and there is no other UFO below, it is for the purpose of reducing excess energy. This could be an automatic function within the craft that causes a predetermined amount of energy to be dumped when the power level reaches a certain limit. As these balls drop, they explode and energy is distributed into the surrounding atmosphere, over a wide area. The following cases may demonstrate this action. Do not confuse this dumping of energy with balls of light or fire that travel like guided UAVs (Unmanned Aerial Vehicles). Examples of this dumping of energy follow:

02-23~24-1740 *Mediterranean Sea off Toulon, France*

During the night of 23 to 24 February, people saw a purple "globe of fire" that rose gradually and then appeared to plunge into the sea, where it rebounded. Reaching a certain height, **it blew up and spread several balls of fire over the sea and the mountains. It made a sound like that of a violent thunderclap or a bomb as it burst.** [130]

10-31-1926 *Indian Ocean southwest of Sri Lanka*

A little more than a year later, on Oct. 31, 1926, the S.S. *Somersetshire* again encountered strange sea luminescence. The ship was in the same general area as that crossed by the *Omar* [on my website, case dated 10-14-1923] and the *Preussen* [case dated 08-23-1925 on my website]. P. H. Potter, the second officer, described this occurrence as **"balls of brilliant light [that] seemed to shoot from the depth, burst on nearing the surface, irradiate and cover an area, seemingly of a couple of hundred square yards."** [131]

07-29-1957 *Oldsmar, Florida, USA*

In Oldsmar, Florida at 11:45 a.m. [sic-2345 or 11:45 **p.**m. per UFOCAT] witness E. E. Henkins **observed a pale yellow fireball glide into the water and explode**. It was viewed for approximately one minute. [132]

11-13-1978 *Adriatic Sea off Giulianova, Italy*

The crew of the *Trozza* saw the water bubble and the boat, whose engine was running at low speed, began to move fast, to the point that the stern was pointing toward the land. The radar had stopped functioning. At that point an object emerged from the sea. The object was also described by other fishermen as an opaque, pear-shaped ball. **The object, which was red, disintegrated in the air with bright light effects.** [133]

08~09-??-2004 *Pacific Ocean 200 miles off China's coast*

Then I saw a large ball of fire in the distance ahead of the ship. I can only guess as to how big it was, or how far away. No one else seemed to notice it. I continued to watch it thinking at first that an aircraft was on fire. But what got me was that it was moving very slowly, **and dripping what looked to me like slag that was still burning**. [134]

3.8   Let's Play Roswell Again

If the abovementioned quasi-black hole is to be considered the "accumulator" of electrical energy, consider the consequences of having such a device aboard a craft under conditions such as those described during the Roswell event: a stormy evening with a great deal of lightning. If the accumulator was in operation in the craft at the time of a lightning strike, it could have, in an instant, overpowered the energy-dumping mechanism and possibly caused an explosion on board the craft (a secondary safety device?).

So much for the intangible fuel and engine. I will leave that for better minds.

In section 3.2, case dated 07-??-1989, witness Inocencio Cataquet stated, "The water started heating up." This brings us to a more in-depth discussion of the UFO's field and the physical characteristic of *heat*.

Chapter 4

Heat and Water-Related UFOs

Heat [is] energy transferred from one body to another as the result of a difference in temperature. If two bodies at different temperatures are brought together, energy is transferred—i.e., heat flows— from the hotter body to the colder. The effect of this transfer of energy usually, but not always, is an increase in the temperature of the colder body and a decrease in the temperature of the hotter body. A substance may absorb heat without an increase in temperature by changing from one physical state (or phase) to another, as from a solid to a liquid (melting), from a solid to a vapor (sublimation), from a liquid to a vapor (boiling), or from one solid form to another (usually called a crystalline transition). The important distinction between heat and temperature (heat being a form of energy and temperature a measure of the amount of that energy present in a body) was clarified during the 18th and 19th centuries. [135]

In this chapter I will attempt to show that the "field" that encompasses the UFO is *hot*, although we have no evidence as gathered from instruments to support this. Descriptions by eyewitnesses of UFO interaction with water, and even in cases on land, suggest that heat is common to most UFO sightings.

But what produces this heat? Is it the friction between the rotating field and the atmosphere which is being pressed against it? Or is it the energy of the field itself? Ufologists are already certain that this friction accounts for the ionization of the atmosphere, producing a colored ball of light around the craft. Could this be a new form of radiation? At this point all we have is speculation, as there have been no means of measuring the attributes of the "field." However, the fact that heat is there gives us a crack-in-the-doorway to understanding the physics that govern a UFO's operation.

4.1   How Hot Is UFO Radiation?

In his excellent book *Unconventional Flying Objects*, Paul Hill, who was a NASA scientist, tackles this aspect of UFOs in Section IV, pages 70-82, and presents several examples:

The Radiation Questions

From ionization, heating, and vibration data, we have gotten some pretty definite indications that UFOs are radiating energy. The following questions are now being asked. Of what energy level is the radiation? What is its intensity? From the observed data, what can be judged regarding its basic type? That is, is the UFO radiating high energy waves, or high energy, ionizing particles? Do these facts, whatever they are, give us useful information about the propulsion system? Can we answer these questions now, or must we await field investigations with expensive scientific equipment combined with lots of good luck?

The answers to these questions can be roughly given now; more refined answers will require more data. An important connection between these answers and the power plant system will be discussed in Section V. [136]

The fact that this field is, in the very least, "dangerous" is shown by the examples where, in several land cases, the craft has backed away or risen vertically when an overly curious witness approached the craft. Is the UFO terrified of the witness? I highly doubt that. Are the crew members of the UFO concerned that the humans might injure themselves on this invisible force? This would be the equivalent of backing up into the high speed rotation of a turboprop, which from the front of the aircraft cannot be easily heard.

An example of a witness's curiosity, as well as the field as an electrical force, comes in the following text from a case in France (not a water-related case, so it is not on my website). Keep in mind that this craft is only 3 feet long and would probably not require a very powerful field:

10-27-1954   *Plougasnou, Finistère, France*

"A young girl age 12, Anne Jegou of Plouezoc'h, boarding at a school in Plougasnou, affirms having seen an ovoid machine, about one meter long, of a creamy white color. [It was] resting on the road to Morlaix. **She approached within one meter of the object and immediately felt something like an electrical disturbance which paralyzed her for some seconds.** The 'egg' then went away smoothly, then suddenly gained altitude." [137]

Heat in the vicinity of a UFO is not a new discovery. Although it is not water-related, one very famous case, investigated and written about in *The Cash-Landrum UFO Incident* [138] by John Schuessler (retired aerospace engineer and former International Director of MUFON), shows this most vividly.

It is a winter night in Texas, December 29, 1980. Three people are in a car with the heater on, wearing winter coats. This scene changes with the arrival of a UFO in that due to the heat of the craft, they remove their coats and change from the heater to the air conditioner. Sometime after the event, inspection of the old road surface reveals that the asphalt has been melted into a new surface, which could only have been caused by heat. Another deadly aspect of this encounter is that all three occupants of the car that night came down with radiation poisoning. Betty Cash, who had ventured outside the car, endured the worst of this radiation poisoning since she had gone closer to the object for a better view. On January 21, 1999, she died of cancer. Investigations into the types of radiation that are known to exist have been unable to explain the effects on the human body as experienced by the witnesses. This leaves only the possibility of either an unknown form of radiation or some other combination of unknown causes.

As in the above case, where heat makes itself evident, the number of cases reviewed later illustrates that this occurrence is a rather common scenario for UFO witnesses. It is not used intentionally as a weapon, so there must be another reason for its presence. Since car engines also produce heat, we could think of its generation in a similar manner—a propulsion byproduct—if we knew that the field surrounding the UFO contributed to its propulsion. The only thing that could be said with any certainty is that water in contact with this field boils as we will show later.

Arthur Conan Doyle's fictional detective, Sherlock Holmes, is quoted as saying: "It is a capital mistake to theorize before you have all the evidence. Insensibly, one begins to twist the facts to suit theories, instead of theories to suit facts. It biases the judgment." [139] To this end I will present circumstances that demonstrate that heat is an integral part of the field. Whether it performs a function or is a byproduct is not clear at this point. In many situations, I have used a sketch of the field, as in Chapter 2, to demonstrate the relationship of the water to the field.

4.2 Explanation of Terms

There are eyewitness accounts where there are several observed atmospheric conditions that appear similar and in some cases, they are mentioned as if they were equivalent in meaning. In many instances these cases use at least one of the following terms, so let us begin with their **dictionary definitions**:

cloud — A mass of visible vapor or an aggregation of watery or icy particles, floating in the atmosphere at various heights and exhibiting a large variety of shapes, which partly aid in the determination of weather conditions. [008]

fog — Condensed watery vapor suspended in the atmosphere at or near the earth's surface. [008]

haze — 1. Very fine suspended particles in the air, often with little or no moisture. 2. A thin vapor of fog, smoke, dust, etc. in the air that reduces visibility. [008]

mist — An aggregation of fine drops of water in the atmosphere at or near the earth's surface floating or falling slowly: used either synonymously with fog or distinguished from it as being less dense, or as consisting of drops large enough to fall perceptibly but slowly. [008]

smoke — The volatilized products of the combustion of an organic compound, as coal, wood, etc., charged with fine particles of carbon or soot; less properly, fumes, **steam**, etc. [008]

steam — 1. Water in the form of vapor. 2. The gas or vapor into which water is changed by **boiling**. [This is of interest.] [008]

vapor — 1. Moisture in the air: especially, visible floating moisture, as light mist. 2. Any light, cloudy substance in the air, as smoke or fumes. [008]

Meteorological terms in witnesses' accounts in this chapter can be quite subjective, since they are sometimes dependent on the distance that an object can be seen without doubt. As UFO events are unexpected, witnesses are not usually prepared and would not, in all probability, have in their possession tools to measure distances, etc.

Cloud density would serve as the best example of obscuration since objects, such as aircraft, mountains, and our stealthy UFOs, to name a few, are often hidden behind them.

Fog, being an effect at ground level, would greatly impact visibility since fog can be very dense and would require ships to signal their position by sounding a foghorn or using radar/sonar to avoid collision.

Another consideration is the time of day in which the sighting occurs, as sunrise and sunset affect the viewing of the field in a varying manner. At night, the object glows brightly, and as sunrise comes, the object begins to lose that brightness due to the increasing sunlight that diminishes the field's visibility more and more. At sunset, the opposite is true, as fewer rays of light make the field appear brighter.

Words such as haze, mist, steam, and vapor could be used interchangeably in the texts of different accounts, consolidated into one group as they are similar in "fuzziness," and I think that this is where witnesses cross lines in their meanings and may refer to them as if they were the same when they describe conditions that they have seen.

4.3   On or Close to the Water's Surface

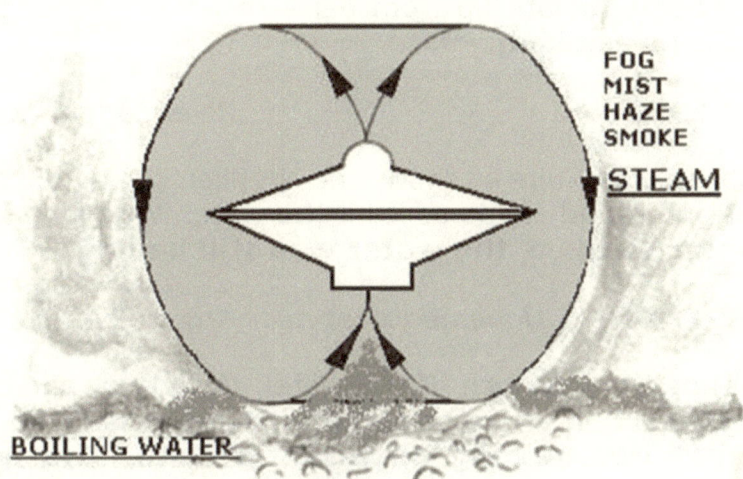

Sketch #24
UFO's field touching surface of water,
causing steam (fog/mist, etc.) and boiling.

Here I have so many cases that it is impracticable to use large amounts of text, so I will only present a relevant sentence or two from each case. Also note the countries where these events occurred, yet again showing that these are worldwide events.

02-21-1893 *Atlantic Ocean off coast of the United States*

...they flew through the air with the appearance of red hot chunks of iron, **struck the water with hissing sounds and disappeared only to send up great masses of steam where they had gone down**. [140]

??-??-1927   *Po River near Corbola, Italy*

...she saw a **shiny, round object** come down on the river close to her. **It submerged then came out again causing the water to boil**; it emitted a hissing sound... [141]

03-??-1931   *Indian Ocean north of the Maldive Islands*

Laser beams radiate up from beneath ocean, water glows and roils [is agitated]. [142]

10-??-1952   *River Lågen, Norway*

The object was white as snow, and Nordlien clearly observed its flat and round shape as it hit the water. When his fellow workers showed up, **the water was still boiling**. [143]

11-12-1954   *Unnamed river near Arquà Polesine, Italy*

It skimmed the surface of the water, nosed-up, and sank, raising a column of water and **making a noise like that of a hot iron resting on a damp cloth. The water bubbled** for several minutes and gave off a **dense cloud of steam**. [144]

12-13-1956   *Off the coast of Venezuela*

When the object hit the water, an explosion was heard, then the sea where the object fell became brilliantly colored. After the colors subsided, **the sea became very disturbed with a "boiling motion,"** which continued for some time. [145]

04-19-1957   *Pacific Ocean south of Yokohama, Japan*

...two silver saucer-shaped metal craft plunge into the water nearby. Immediately afterward, the sea where they fell became furiously disturbed, **boiling and churning violently**. [146]

04-??-1958   *Atlantic Ocean off Saúde, Brazil*

Beneath the machine, **the water seemed to be boiling**, or being sucked up, but without actually touching the under part of it... [147]

06-??-1958   *Adriatic Sea off Casal Borsetti, Italy*

...saw, at 1:30 p.m., **the sea bubbling** about one thousand meters offshore. Suddenly**, a black object appeared**, full of pipes, without windows or sails, and **covered by a whitish steam**. [148]

07-20-1958   *Private lake in Glennie, Michigan, USA*

RE: Alleged object seen beneath surface of water, **sizzling and leaving foam in wake three feet high**, observed by three widely separated observers. [149]

??-??-1959   *Off the coast of Havana, Cuba*

He noticed that **the sea "began to bubble like soap suds."** An enormous metallic disk suddenly emerged from the water... [150]

07-19-1962   *Unnamed creek, Asheboro, N.C., USA*

...it apparently settled down in a nearby creek. The women heard **the sound of bubbles** as the thing, whatever it was, settled to the bottom of the creek. [151]

Now that the bubble has been introduced, I think it is necessary to insert the fact that not all bubbles are created by heat. So, before we continue on:

4.4   Bubbles and Boiling—Separate Causes?

Definition of **bubble**:

(n.) 1. A vesicle of cohesive liquid, filled with air or other gas. 2. A globule of air or other gas in any confined space, as in a liquid or solid substance. 4. The process or sound of bubbling. (v.) 1. To form bubbles in, as a liquid. [008]

Definition of **boil**:

1. To be agitated, as a liquid, by gaseous bubbles rising to the surface; also said of the container in reference to the contents: The kettle *boils.* 2. To reach the boiling point. 3. To undergo the action of a boiling liquid, as meat. 4. To be agitated; to seethe: The water *boiled* with sharks. [008]

Naturally, in our search for answers, the agitated condition of the water, along with the appearance of bubbles, must lead us to search for causes other than heat as a possible explanation for what is seen.

One explanation offered by fellow ufologist Tony Rullán is the release of gases, likened to the venting of tanks on a submarine to become less buoyant in order to submerge. He added, "but this will not explain the heat felt. I recall the Shag Harbor case had bubbles that formed foam with a sulfur odor. However, the sulfur odor is not reported in the cases you list above." [152] Don Ledger, prominent Canadian UFO investigator, replied:

> Some claimed there was the odor of sulfur and additionally there was some evidence of bubbling from below and "roiling" of the water. It's not claimed that the bubbling caused the foam. If the Shag Harbour object was hot, it might have caused the water to release or gel agents in the water such as plankton. If there is an electrical field around the object, then perhaps that could cause plankton-rich water [Nova Scotia's waters are plankton rich] to react with the production of some foaming agent. [153]

Regarding gases as a reason for the bubbling, I had many times in the past thought of electrolysis as the culprit. I was fascinated in chemistry class when the teacher produced bubbles of hydrogen from one submerged wire and bubbles of oxygen from the other submerged wire. This evidently assumes that the UFO's electromagnetic field is capable of producing this separation of oxygen and hydrogen.

The rotation of the field, as previously explained, might be yet another cause. Again Don Ledger came to the rescue:

> Addressing boiling water vs. bubbles caused by churned water, such as the propeller of a submarine, the right word is cavitation.[7] The propeller literally sucks oxygen out of the water—this causes the water to be less dense in its immediacy which makes the propeller inefficient. [Note: Aircraft propellers can cavitate as well but cause vacuous pockets (low pressure areas) behind the propeller which create drag, making the propeller inefficient.] A submerged body moving very fast through the water can cause cavitation, e.g., a trail of bubbles. [154]

In some cases there is no mention of the excessive heat generated by the UFO. This might be due to the distance between the witness and the craft. This could mean that bubbles are caused by heat which is produced by the field itself due to the lack of any other external heat-generating influence.

With all this bubbling, it should be easy to track a UFO underwater, right? Not necessarily. The heat generated by a deeply submerged UFO is dissipated as it rises through the cooler layers above it. However, as the UFO rises and the depth decreases, the heat has less water through which to dissipate, and therefore the water above the craft begins to boil.

We continue on with more cases which, besides bubbles, contain the words boiling, steam, hissing, and so on, all indications of the presence of heat.

---

7   cavitation: The formation of vapor cavities in the water flowing around the blades of a propeller, due to excessive speed of rotation, and resulting in structural damage or loss of efficiency. [008]

04-11-1963   *Atlantic Ocean near Puerto Rico Trench*

As they left the **area of the great boiling mound of water**, they noticed it was beginning to subside. [155]

10-31-1963   *Peropava River, Iguape, Brazil*

...the object was seen by more than ten persons and that "from the bank in the river, mother and daughter, frightened, **watched the water boil and become dirtier than ever before.**" [156]

11-13-1965   *Off Rugged Islands, New Zealand*

"There was a **great surging of water, like a tide boil**..." [157]

05-29-1967   *Trout Brook Lake, New Brunswick, Canada*

...he saw **a big cloud of steam** and heard **a gurgling sound.** [158]

08-04-1967   *Atlantic Ocean off Arrecife, Venezuela*

"Suddenly—he says—I felt a vibration of the boat that bothered me quite a bit. I thought that it was an earthquake, but in a few seconds the water of the **sea began to boil** in a well-defined circle of some six meters in diameter. The vibration and **the bubbling intensified**." [159]

12-??-1967   *Gulf of Mexico, unknown location*

They told of "holes" in the Gulf of Mexico and places where the **water "boils in circles."** [160]

05-20-1968   *Moore Lake, Littleton, NH, USA*

...**sounding like the bubbling** of an aqualung underwater. [161]

07-12-1973   *Stanley Draper Lake, Oklahoma, USA*

...saw **steam rising from the lake** immediately under the object. [162]

08-01-1975   *Trinity Lake, New York, USA*

...the UFO...was by then just above the surface of the water and Brad began to hear **a slight hissing sound**. [163]

06-??-1976   *St. Johns River, Jacksonville, Florida, USA*

A hundred feet or so ahead and just off starboard, **a patch of water exploded with churning bubbles**. [164]

12-28-1978   *Ionian Sea off Santa Tecla, Italy*

...**a stretch of the sea suddenly started bubbling**. From the water, a big dark spot arose...and rapidly headed inland. [165]

06-??-1979   *Ligurian Sea off Cinque Terre, Italy*

...into the sea at high speed; shortly after the same object reemerged amid **a violent boiling of waters**... [166]

05-16-1981   *Thompson River at Kamloops, BC, Canada*

...a noise **"like water being poured into a hot frying pan"** is heard. The water in front of the fisherman, about 100 to 150 yards away, **suddenly bubbles**. [167]

07-15-1981   *Newberry, South Carolina, USA*

He said it slowly **settled on the surface of the water causing steam to rise around it**. [168]

10-12-1985   *White Oak River near Swansboro, NC, USA*

Immediately I heard the noise of masses of minnows jumping in the water. My sister and I walked down onto the dock to investigate the seemingly amplified noise. What we discovered was **a large area of foam. In fact the water seemed to be boiling.** All around us the minnows were jumping. [169]

02-20~28-1986   *Tyrrhenian Sea off Milazzo, Italy*

Suddenly **the sea seemed to boil** and a strange object with an elliptical shape emerged from the water... [170]

06-??-1986   *Atlantic Ocean off coast of New Jersey, USA*

...as I walked along the water's edge, I noticed **a line of bubbles breaking the surface of the bay**. Approximately **thirty yards distant, the bubbles became larger and larger**, and suddenly an object emerged from the water. [171]

07-06-1990 *Off the shore of Palominos Island, Puerto Rico*

...both men became aware that **the sea off the sailboat's stern was bubbling and heaving**. [172]

01-31-2005   *Farm pond in Columbia City, Indiana, USA*

It moved over the frozen water and lowered slowly. It sat motionless without a sound for a minute or two. **As it rose silently, "steam" rose from the lake....The area of the pond where the object hovered was melted into about 2 inches.** [173]

05-04-2006   *Caspian Sea off Baku, Azerbaijan*

Representative of...project's operator diplomatically evaded the question on the reason for **boiling water near the platform**. [174]

06-27-2006   *A pond near Flagler, Oklahoma, USA*

A hissing sound was heard and a steam-like substance rose from the water....The object never was observed to come out of the pond. At 10:00 p.m. when they returned, **there was a mist or fog over only the pond**. [175]

4.5   Close Encounters above the Surface

The reports which follow deal with UFOs near water but not interacting with it. Still, heat is felt and is described:

08-15-1663   *Ozero (Lake) Zarobozero, Sëmkino, Russia*

"At noon, a large ball of fire came down over Robozero, arriving from the clearest part of the cloudless heavens. It came from the direction whence winter comes, and it moved toward midday (south) along the lake passing over water surface. The ball of fire measured some 140 ft. from one edge to the other and over the same distance, ahead of it, two ardent rays extended....[T]he big fire and two smaller ones disappeared.

"Less than an hour later...the same fire suddenly reappeared over the same lake, from the same place where it first disappeared. [It again disappeared and then reappeared a third time.]

"As the fireball was coming over water, peasants who were in their boat on the lake followed it, and **the fire burned them by the heat**, not allowing them to get closer." [176]

06-18-1845   *Mediterranean Sea near Malta*

Two hours it blew very hard from the east, and whilst all hands were aloft reefing topsails, it suddenly fell calm again, and **they felt an overpowering heat and stench of sulphur**. At this moment three luminous bodies issued from the sea, about half a mile from the vessel, and remained visible for ten minutes... [177]

06-12-1958   *Mediterranean Sea off Le Brusc, France*

"The fantastic wheel didn't in fact go over us, but it passed very close by, making such big waves that we nearly capsized. **When it was close to us, we now felt very powerful heat from the thing and a strong blast of air.**" [178]

01-19-1966   *Tully Lagoon, Queensland, Australia*

"One of the group observed a dazzling and powerful light in the area where **burnt cane scarred by great heat was found**. When they reached the driver and fireman, they found them stupefied with one man repeatedly pointing to the sky. Both were taken to a hospital." [179]

05-04-1968   *Seal Island, Nova Scotia, Canada*

Suddenly it burst into a blood red light and appeared to be about 50 to 75 yards away, coming towards the boat. **As he watched through the window, it became so hot he had to move away.** The light floated overead [sic – overhead] for about five minutes then lowered and seemed to float towards Brown's Bank....Capt. Atwood said **the heat was intense, and he expected the boat to be burned before it passed**. [180]

03-29-1974   *Sea off Togo, Africa*

While before the appearance of the object the temperature was mild and agreeable, **there was now a heat in the beam of the spotlights** that A.W. describes as frightening. **The woman and he were sweating profusely.** [181]

12-26-1980   *Atlantic Ocean off Paco de Arcos, Portugal*

When directly overhead, a sensation of heat was felt. Remember this was on the Atlantic coast in late December. The night itself was on the cool side, so both men were adamant about this heat being given off by the object. That same night they both suffered sleeping problems and headaches. [182]

10-24-1981   *Mediterranean Sea off Licata, Italy*

The flashlight illuminated a hairy being two meters high standing in the water up to its knees. **Immediately the two fishermen felt a sensation of heat, so much so that the wet sand on which they were barefoot dried quickly.** [183]

09-17-1985   *The Bosphorus off Istanbul, Turkey*

It hovered over them, and everywhere was like daylight. They said that **the object was emitting a great deal of heat**. The yachtsmen tried to talk to each other on the radio, but even though they tried 16 different channels, the radios just wouldn't work. [184]

07-24-1990   *Tyrrhenian Sea off Amantea, Italy*

Two witnesses who were on a terrace heard a noise, like a buzzing, which was repeated in successive waves, together with **emissions of heat** and, at the same time, observed a light about one and one half kilometers from the coast, which was reflected on the sea below. [185]

4.6   In the Rain

Definition of **rain**:

rain   — The condensed vapor of the atmosphere falling in drops. [008]

   While it would seem that rainwater would be insignificant in a study of UFOs going into, floating on, and coming out of bodies of water, the cases that follow are highly significant in trying to determine what the field is and how it works. There have been instances where skeptics have attempted to debunk such cases by substituting kites for UFOs; however, kites would not fly in the rain for long because their wood, fabric, and paper would absorb the water and become heavier.
   As with other cases involving bodies of water, mist is seen. This equates to the oft-reported fog, haze, and steam seen in relation to a horizontal body of water. The fact that these are seen at the top of the UFO rather than the bottom indicates where the water is at the time, falling vertically to the top of the vehicle; the water hits a hot, ionized field, causing it to turn to steam, which is interpreted as mist.

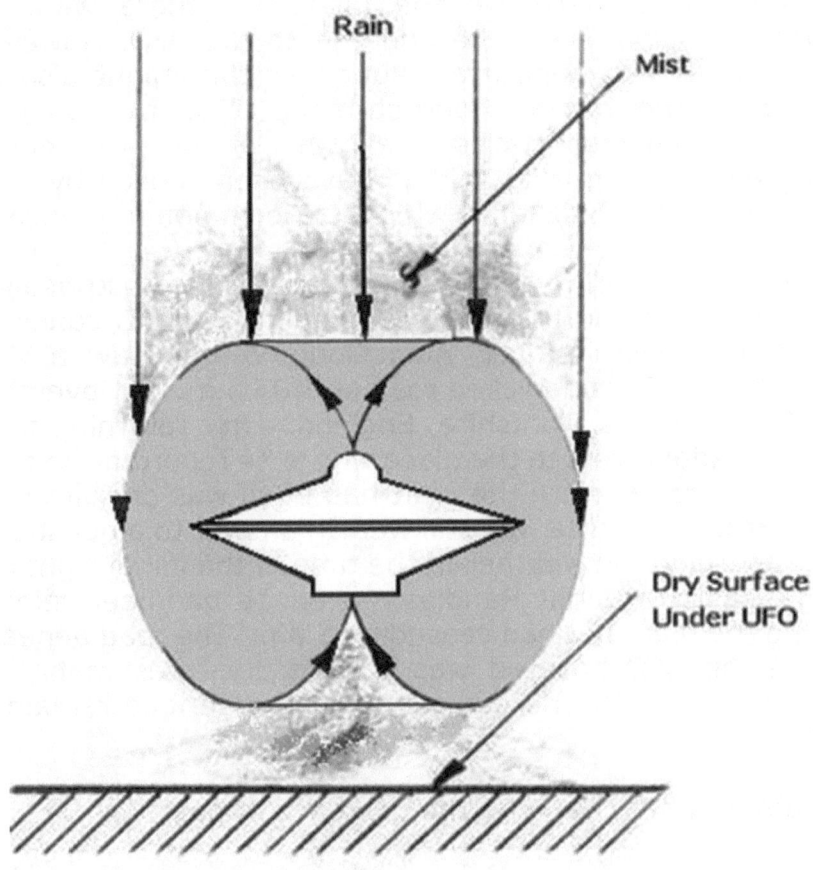

Rain

Mist

Dry Surface
Under UFO

Sketch #25

09-30-1977 *Day's Corner, Prince Edward Island, Canada*

At the time of the sighting, it was **raining quite heavily**. Rosemary and her eldest daughter, Kim Greencorn, age twenty-one, noticed a lighted object ahead of their vehicle, which came close enough that they were able to get a good look at it. Rosemary said the object appeared to be similar to two lampshades inside a wide V, with the light from the lampshades brighter than that from the V. **Encircling this object was a sort of glow and, over the top, a mist.** [186]

Conversely, there are the instances where water is evaporated from a surface underneath the UFO. However, here I must add a note of caution as this might also be caused by the vortex underneath the UFO, mentioned in Chapter 2. In crop circles, the cause of the plant nodes being expanded is also said to have been caused by heat stemming from ionization, which is a common influence of the field itself.

Though not on my website, an interesting case dealing with rainwater and the field surrounding the UFO concerns an English police officer, Alan Godfrey, who saw a UFO hovering over a rain-slicked road at 5:05 a.m. on November 28, 1980, in West Yorkshire, England. After returning from police headquarters to the place where he reported the craft to be, the road where the craft had been was dry although the surrounding area was still wet. In answer to a question I had regarding the weather at the time of the initial sighting, UFO researcher Jenny Randles replied, "It had been raining during the night but had ceased by 5 a.m. The road beneath where the UFO hovered was *swirled* dry—presumably by the vortex—whilst the surrounding road surface remained wet." [187]

And another case with the presence of heat:

10-20-1954   *Lusigny Forest near Troyes, France*

This heat sensation was reported again on that same afternoon in the department of Aube, 125 miles west of Turquenstein. **It was raining hard that evening** in the Lusigny Forest, not far from Troyes. Mr. Roger Réveillé, a lumber dealer, was walking along a wood[ed] road when his attention was attracted by a loud rustling sound such as would be made by a flight of pigeons. Looking up, he noted, at treetop height, an oval-shaped object perhaps 20 feet long. At the same time, **he felt an ever more intense heat**. In a few seconds the machine disappeared upward.

In the woods **the heat was now intolerable and was producing a thick fog**. It was almost a quarter of an hour before Mr. Réveillé was able to approach the site. He then found that, **in spite of the rain, the ground and trees at that spot were as dry as if they had been exposed to full sunshine**. [188]

These remaining cases do not mention any observations of mist or heat but do show us that the skeptics' misidentified "kites" do fly in the rain. Certainly, if it is raining and a UFO is seen below the low cloud cover, this would indicate that the craft is closer to the witness.

02-15-1963   *Moe, Victoria, Australia*

Question: You didn't actually see anybody in it through what appeared to be the glass portion on the top, on the dome?
Answer: No, on a clear day you may have but, as I said, it was **raining heavy** and no, I can't honestly say I did see anybody although I was lookin' hard enough. [189]

09-15-1965   *Sudbury, Massachusetts, USA*

When his car engine skipped and its lights dimmed, the witness stopped to check his engine **in the midst of a rainstorm**. He got out of the car and saw a disc-shaped object with a glowing rim which sparked in sequence about 200 feet above him. [190]

11-16-1965   *Salem, Massachusetts, USA*

A large red light and a smaller white light were seen hovering over the N.E. Power Plant **during a rainstorm**. They sped off three times as fast as a jet aircraft. [191]

04-07-1966   *Lincoln, Nebraska, USA*

A group of doctors, lawyers, and pilots saw a constellation of lights, including a large white "beacon" with 4-5 bright red lights, hovering at low altitude (estimated to be about 400 feet) for about 7 minutes **during heavy rain**. The object(s) suddenly sped away at "fantastic speed." [192]

08-30-1980   *Dumfries, Scotland, United Kingdom*

Through the swirling clouds and **drizzling rain**, the "thing" suddenly appeared—a magnificent cluster of coloured lights.
    Late-night revellers on their way home stared skyward in amazement as it hovered just below the low cloud base.
    It was large, very large. For a full 20 minutes its brightness lit up the sky near the town of Dumfries in southern Scotland. Then it vanished over the hills. [193]

4.7   Through the Clouds

Speaking of "the low cloud base" from the case above, (08-30-1980), there is an article written in the *International UFO Reporter (IUR)* about a UFO that was below the cloud layer but decided to go up through it. It is reprinted here with the permission of the authors:

A UAP[8] AND ITS SAFETY IMPLICATIONS:
O'HARE INTERNATIONAL AIRPORT, Nov. 7, 2006
By Richard F. Haines, Chief Scientist
Used by permission of NARCAP, all rights reserved 2008.

Based on eyewitness testimony, the UAP would have ranged in size from about 22 to 88 feet in diameter. It accelerated at a steeply inclined angle through the 1,900-foot cloud base, leaving a round

---

8   Unidentified Aerial Phenomenon

hole [hole in cloud similar to hole in ice, which follows later] approximately its own size that lasted for as long as 14 minutes. This is suggestive of a **super-heated object or otherwise radiated (microwave?) heat energy on the order of 9.4 kJ/m³**. Ref. 194, Footnote 9

Continuation of "A UAP and Its Safety Implications":

HOLE-IN-CLOUD CONSIDERATIONS
By NARCAP members Kim Efishoff and Larry Lemke

In this section we concentrate on one particularly striking assertion occurring in the reports; namely, that the apparent oblate spheroid-shaped object or phenomenon produced a sharp-edged **"hole in the clouds" (HIC)**. If we assume only that the witnesses are not mistaken or dissembling, then the HIC must be considered to be a physical trace capable, in principle, of providing some information about the nature of the object or phenomenon.

What can the HIC tell us about whatever caused it? Ultimately, there is insufficient data in the reports to uniquely and definitively identify the presumptive "object" that caused the HIC. Indeed, we cannot even determine whether the object was solid or as ephemeral as, for example, self-organized plasma. However, we may hope to eliminate from

---

9   kJ = kilojoule = **Kilo-** *prefix* One thousand (times a given unit) used chiefly in the metric system of weights and measures.

Joule *n. Physics* The meter-kilogram–second system of units of work or energy: equivalent to the work done, or heat generated, in one second by an electric current of one ampere against a resistance of one ohm, or in raising the potential of one coulomb by one volt.

m = mass per unit time of fluid (in kg/sec) [008]

consideration classes of explanations that do not make physical sense. Thus even if we cannot identify the ultimate cause of the sighting reports, we can make some reasonable inferences about what it was not. As usual, in attempting this we should apply Occam's razor[10] and avoid needlessly invoking any unconventional physics.

The phenomenon we are attempting to explain is described in the words of one of the witnesses: At around 4:30 in the afternoon of November 7 several employees of United Airlines witnessed a "disc-shaped object" that was seen "hovering over Gate C 17 at the C concourse" of the Chicago O'Hare International Airport. The object, which could not be identified by witnesses as any known aircraft, was said to be "holding very steady and appeared to be trying to stay close to the cloud cover."

According to testimony given to NARCAP by one witness, after looking away for a short while, the witness "noticed that the craft [was] no longer there but there was an almost perfect circle in the cloud layer where the craft had been. The hole disappeared a few minutes later." One highly qualified witness (B) confirmed that he and witness (C) saw the hole at about 4:20 p.m. He said, "I guess it had just left." Estimates given for the time of departure of the UAP and a fairly definite time when the hole in the cloud was still visible range from five to ten (mean seven and one-half) minutes [5+10=15/2=7.5].

Although the multiple-eyewitness accounts of this HIC may be unusual, they are not unprecedented. Reports of this odd manifestation have been associated with UAP sightings as far

---

10 Principle which calls for the least number of assumptions when making an explanation; the simplest explanation may be the best one. [195]

back as 1947, and as far afield as Newfoundland, England, and Scotland.

Because the freezing level above Chicago O'Hare on November 7 was at 10,000 feet, we know that the clouds in which the hole appeared consisted of water droplets only.

Classically, there are three ways to make a cloud of water droplets disappear. One way is to evaporate them (turning them back into invisible vapor); another way is to freeze them into ice particles (causing them to drop out of the sky); the third way is to aggregate them into large rain drops (which also fall out of the sky). We know it was not raining on November 7, 2006, or in any of the other three cases, so we need only consider evaporation and freezing. In fact, we may generalize this conclusion to eliminate any form of freezing as the removal method because falling ice crystals were not present at the time and location in question, and so concentrate instead on evaporation.

Only about 10% of the total drag force of a blunt body moving at subsonic velocities could show up immediately as heat. This means that the presumed oblate spheroid that ascended nearly vertically above Chicago O'Hare on November 7, 2006, could only have provided about 2.95 kJ/m^3 of heating to the surrounding air due to its kinetic energy. This calculation yields an effect more than three times too small to have produced cloud-droplet evaporation and provides a semiquantitative basis for the common-sense observation that solid objects such as aircraft, rockets, artillery rounds, and the like do not punch sharp-edged holes in clouds. We must look for an evaporation mechanism not limited to the energy transfer mechanism of aerodynamics.

In order for the size and shape to be preserved, the cloud water droplets must have been evaporated out to a distance that is of the same order as the body

137

radius. Moreover, all the energy required for this evaporation must have been transferred from the object to the cloud droplets during the brief period they were within this range (about 10 milliseconds). Consideration of this fact yields a startling realization regarding the motion of the object—the relatively high power associated with its passage.

We have eliminated aerodynamics, or motion of the object, as the cause of the evaporation energy source, but still require some sort of energy-transfer mechanism that is attached to and centered on the phenomenon.

We postulate that sudden *in situ* evaporation of the water droplets constituting the cloud represents the least extraordinary physical process capable of explaining the observations. We estimate the minimum volumetric energy density required to cause such *in situ* evaporation as approximately 9.4 kJ/m^3, in the form of heat. We consider the remote possibility that a blunt body moving at high subsonic velocities through the air may dissipate sufficient amounts of aerodynamic energy through viscous friction to cause this amount of heating, but find that the likely heat production rate is too low.

We postulate that the instantaneous nature of the HIC formation, the circular shape, and its sharp edges all point to the direct emission of, for example, electromagnetic radiation from the surface of the oblate spheroid as the proximate cause of the HIC. We cannot identify the object or phenomenon lying inside the oblate spheroid surface, but two conclusions seem inescapable: (1) the object or phenomenon observed would have to have been something objectively and externally real to create the HIC effect; and, (2) the HIC phenomenon associated with this object cannot be explained by either conventional weather phenomena or conventional aerospace

craft, whether acknowledged or unacknowledged.
[END] [196]

Following this well-investigated case, I would like to present two cases from my website that occurred over or near water.

The first is from a woman who had UFO abduction experiences and therefore had an interest in the UFO phenomenon. While she was not an actual witness to this particular sighting, she saw it on a local webcam video the day it occurred and was able to take photos from it. On my website, there are five colored photos of the event she captured from that webcam video.

10-07-2008   *St. Pete Beach, Florida, USA*

I live on St. Pete Beach in Florida. I caught this from a webcam video posted at a site for the TradeWinds Resort near my home. **This little disc-shaped UFO came out of the clouds, left a little hole in the cloud it came from**, and bounced on the water, and took off. I tried to get the best pics I could and this is one of them. I have several more and I am very excited about this because I am always searching the skies....

[H]ere you can see **the little hole it made in the clouds above it**, and this is right before it splashes on the water. **It appeared to be ejecting some steam** just before it hit the surface; that is not a splash because it had not touched the water yet. [197]

Note that in the above text, the person who sent me the e-mail described seeing the UFO getting close to the water and also seeing steam, an indication of heat again.

02-10-2010   *Bayonne, New Jersey, USA*

UFOs occasionally appear when weather conditions are poor. During a heavy snowfall about 8 p.m. on February 10, a man entering a Dunkin' Donuts in Bayonne, New Jersey,

was surprised to see a domed object drop through a hole in the clouds. The witness said that the domed disk resembled a bowl atop a plate. It was dark gray with a series of windows or vertical lines at the junction between the dome and "plate." He reported the lines extended perhaps 10 percent of the distance to the apex of the dome. **Air around the object seemed to be distorted or wavering as though from intense heat.** [198]

4.8   In the Snow

Definition of **snow:**

Precipitation taking the form of minute ice crystals formed from an aqueous vapor in the air when the temperature is below 32° F., and usually falling in irregular masses or flakes. [008]

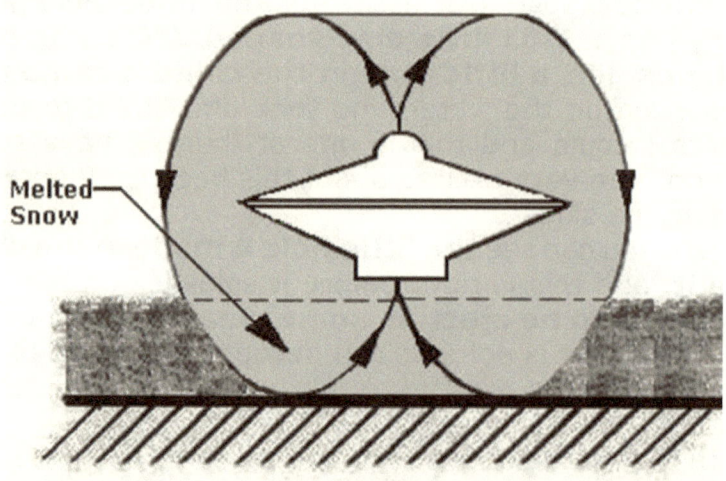

Melted Snow

Sketch #26

Here we have a UFO in close proximity to snow. The field, operating at a high temperature, changes the snow to water and, in some cases, dries or scorches the ground below it. It has also been pointed out to me that there is the possibility that, rather than heat, it is compression that could cause the melting. However, in the cases I have read, when a UFO rests on the ground, it uses its landing gear, which might leave traces of the gear imprint by compression, as when a car's tires melt the snow, not to mention a depression resulting from the weight of the craft. However, in the case of a melted 30-foot circle, I must lean to the field surrounding the UFO as the culprit of the snow melting.

To show what we can learn from visual observations of snow, I have also listed cases that can be found on my website that do not concern heat yet show effects in the snow in other ways:

Traces

02-22-1922   *Hubbell, Nebraska, USA*

...landed like an aircraft and left traces in the snow. [199]

11-23-1954   *Torpo, Norway*

...descended rapidly...knocked snow from the treetops... [200]

01-??-1968   *Baraga, Michigan, USA*

I could see it snowing but it wasn't snowing on me or the truck....The next thing I recall was standing about 20 feet away from the truck and I don't know how I got there. It was still snowing heavily and, although the snow was at least 5 to 7 inches deep, there was not one trace that showed that I had walked that distance [possible abduction, which is discussed in Chapter 5]. [201]

Vortex

03-08-1957   *Baudette, Minnesota, USA*

...it seemed to suck the loose snow up under it as it passed. [202]

11-29-1965   *Springhill, Nova Scotia, Canada*

Snow was blown away and bushes were flattened. [203]

02-23-1967   *Severna Park, Maryland, USA*

...snow in the backyard was kicked up in swirling eddys [sic - eddies]... [204]

02-10-1969   *Dartmouth, Massachusetts, USA*

...causing snow on the treetops to swirl. [205]

03-13-1975   *Near Mellen, Wisconsin, USA*

...round area on the road where the snow was "fluffed up."[206]

01-12-1976   *Brantford, Ontario, Canada*

"It didn't touch the ground, but it did kick up snow." [207]

The following snips of text deal primarily with the aspect of heat, which is our focus right now:

01-15-1924   *Neillsville, Wisconsin, USA*

While homeward bound in a sleigh from a wolf-hunting expedition west of here, Leo Williamson, editor, and Will Dangers, merchant, witnessed **a strange phenomenon when the snow in the road ahead of them for several hundred feet was melted by a nebulous green fire, leaving the highway covered with several inches of mud**....[They] noticed a faint, green light in the sky before them....The light hovered a moment and suddenly descended to the road. **A hissing noise followed and a huge cloud of steam swept about them**....

Williamson said he and Dangers **examined the road and found it was too hot to touch**... [208]

01-??-1958   *Niagara Falls/Depew, NY, USA*

"My lights suddenly came on. I started the car and it was all right. I pulled up to where that place was, got out with a flashlight, and walked over to where it was sitting. **A large hole was melted in the snow about a foot across and grass was showing. The grass was warm but nothing was dug up around there.**" [209]

WW-??-196.   *Swift Current, Saskatchewan, Canada*

In ultra-cold weather (c. minus 30 to minus 40° F.), a couple, driving home at 2:00 a.m., came across **a dense but extremely localized area of fog**. Stopping the car, **they observed that all the snow in the area was melted and the ground was steaming. All snow outside the area was intact.** The trace was not perfectly circular but gave witnesses the feeling that something hot had been hovering overtop the area. [210]

01-12-1965   *Near Custer, Washington, USA*

According to *Spaceview*, published at Vidor, Texas, **a strange, circular craft was seen to land on a snow-covered field** near Custer, Wash., in the winter of 1965. **Next day a circular pattern was found in the snow at the landing site**. [211]

[Another account of the same sighting:]

On the same night, according to reports NICAP is still checking, a similar glowing UFO (perhaps the same one) touched down on a farm near Blaine [about 7 miles from Custer], a few miles from Lynden. Reportedly, the farmer phoned the AF radar station at Blaine, but before investigators arrived, the UFO took off, at high speed. **Where the machine had touched down, the snow was said to be melted and the ground scorched.** [212]

01-29-1967   *Knox City, Missouri, USA*

While it hovered nearby, two smaller white, luminous spheres approached the object and entered through an opening, after which the central ring jetted out red fire. Then the object ascended vertically and disappeared in the distance. **The snow on the ground under the spot where it hovered was melted,** and the area over which the object approached the house failed to produce plants, while plants elsewhere grew normally. The sighting began at 2200 local time. [213]

02-21-1967   *Sheboygan, Wisconsin, USA*

Only one of the five officers responding went with the witness into the field to look for the object with a flashlight, as **by this time it had clouded over and started snowing.** Winds at the time of sighting were out of the southwest with gusts up to 35 mph. **There had been several inches of new snowfall and considerable amount of drifting since the time of the sighting.**

I went out to talk with the witness on the 23rd and also to check the area. The witness and I went into the field, trying to see if we could find the object in daylight. **We noticed an area about 15 feet in sort of a circular shape which had a crust of ice ¾ to 1 inch thick.** Using a compass, the needle spun around and was very arratic [sic] at one spot on the ice area. Moving a few feet away, the compass went back to normal operation. We did not find anything other than the magnetic fluctuation. We went to his house and he filled in the report form. [214]

12-26-1968   *Berlin, New Hampshire, USA*

George Hamanne of 70 Hinchey St. said he saw a "bright light which lit up half the mountain" around 9:30 p.m. He said the light was on Mt. Forist—a solid mass of mountain that spills down on the west side of the city.

Hamanne said he and his whole family viewed the burning spectacle which he described to be "about three feet in diameter." **He said the snow melted in the area of this object** and the glow lasted for about five minutes. [215]

WW-??-1973?   *Stratham, New Hampshire, USA*

It was not enormous. "**It stayed on the ground for maybe 10 seconds, went up slow, then phoom, it was gone**," he says. It ascended at a 45-degree angle and went from 0 to 10,000 feet in 10 seconds—"that quick." The next day a UFO investigator interviewed the young Perrys and viewed the field. Tim recalls that he and his sister were interviewed separately.

"I was in the family kitchen," he says. "I remember his dark blue pants and light blue shirt and being apprehensive as a boy would be if answering questions to a policeman."

Perry says **the craft had melted a circular formation in the snowy field, about 50 feet in diameter and marked with weird patterns**. [216]

02-21-2003   *Wellington, Prince Edward Island, Canada*

It even **left a melted area in the snow of a triangle with circles on each tip and a strange sign. Like it melted the drift of snow in 25 minutes.** [217]

4.9   Through the Ice

Definition of **ice**:

Congealed or frozen water; the solid condition assumed by water at or below 32° F., or 0° C. [008]

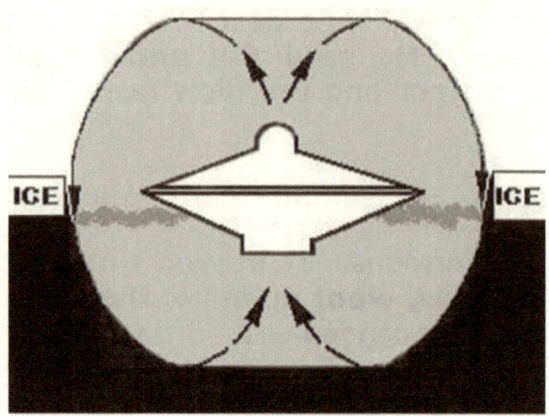

Sketch #27

Ice is yet another form of water, but the hardest one. The idea of one of our aircraft entering water without damage is hard to conceive, but for one to penetrate ice and then water is completely incomprehensible. The fact that UFOs can penetrate this medium without damage brings us to focus on why these craft remain untouched by the force required to accomplish this feat. Applying the concept of the surrounding field might explain this. Examples of the heat from this field (or shield) are numerous, so it is conceivable that heat alone is the solution. There are also examples, however, where blocks of ice have been hurled through the air, indicating instant kinetic force rather than a slow melting process. In most instances of blocks being hurled, it seems that the craft is hitting the water horizontally from the shore. The following stories describe such events, not caused by our aircraft, of course.

04-25-1946   *Anima-Nipissing Lake, Ontario, Canada*

Don Cameron, his wife, and their children had their first brush with flying saucers on April 25, 1946. On the same lake, cutting a channel for their boat, they saw between 12 to 14 small, disk-shaped objects come down at a 45-degree angle a mere 75 feet in front of them.

Appearing to be about twice the size of a dish, the objects jumped on and off ground of height of two feet, **spinning in the ice**. [The previous sentence is incomplete in thought.] The family all watched. As Don Cameron walked toward them, they [multiple UFOs] rose at the same angle and shot off through trees on a nearby ridge without touching the branches.

The objects left **black marks on the snow and ice**. [218]

02-??-1961   *Unnamed lake in the Karelia region of Russia*

...lengthy gash on the lake's edge...**The ice on the lake nearby was largely broken**, and at that battered spot, **there were ice floes, but elsewhere the ice was undamaged**....

On the bottom of the lake, divers found a portion of the soil displaced which was stretched out like a "tongue" across the base for about 100 m.

The theory was that the object had first slid across the soil on landing and thence into the lake, yet there was no trace anywhere of any such object, and even when a metal detector was used, there was scarcely any reaction. [horizontal entry] [219]

04-27-1961   *Lake Onega, Russia*

It hit the bank just above the ice and ploughed a furrow 27 meters long, 15 meters wide, and 3 meters deep through the frozen ground before continuing its low trajectory and disappearing. **The impact blasted the ice from near the shoreline, throwing out huge chunks to the bank with a strange intensive green color on their underside.** Local residents, including a forest ranger, reported no sound except that of impact when the object crashed into the bank. [horizontal entry] [220]

12-29-1972   *Saranac Lake, New York, USA*

As she walked around the **lake,** which **was covered with ice estimated to be several inches thick** in this particular cove, she heard a muffled, rumbling sound. She said it sounded like it was coming from under the surface of the lake. As she stared in amazement, **she observed the ice breaking up, as if something under the ice was pushing it up, and the ice disintegrated as a shiny, aluminum-colored, submarine-shaped object without any protrusions on its surface came streaking out of the water and rose rapidly into the sky**. No sound was emitted by the object outside of the noise from the ice breaking up. [221]

04-30-1976   *Lake Siljan, Dalarna Province, Sweden*

In Lake Siljan, Dalarna Province (central Sweden), April 30, 1976, three witnesses saw a pointed object tearing a channel through the ice between 5:15 and 5:30 p.m. **The object**, estimated to be moving 90 km/hr, **cut a channel 1 km long and 3-4 meters wide through the 20-cm-thick ice.** Ice floes and water cascaded around it as it forced its way through the ice. The object was dark (2 witnesses) or gray (1 witness) and about 10 meters long. [222]

01-23-1981   *Grand Forks, North Dakota, USA*

But Lynn also saw "a ball of light" land in the Grand Forks Country Club parking lot.

"I was facing the window when the sky lit up. I then saw this light, about six feet in diameter, sail over the portico and hit the parking lot. It was orange but was blue around the edges."

Her husband, the country club manager, had his back to the window "but the light lit up the entire interior of the building. It was like a camera flash blub [sic], only it lasted a second or two. It wasn't bright enough to hurt the eyes, though."

Thinking at first someone was playing with fireworks, they ran outside.

The only evidence they found was a 20-foot-diameter circle of dry pavement. The rest of the lot had a slight ice coating.

"It was bare and completely dry," Duval said, "but there were no burn marks, no crater and no object." [223]

03-04-1988   *Lake Erie, several cities, USA*

In their incident report sent later by teletype to Coast Guard headquarters in Detroit, MI, the men were quoted as saying that **"the ice was cracking and moving abnormal amounts as the object came closer to it."** [224]

05-23-2001   *Ottawa River, Ontario, Canada*

J. said he watched the lights slowly move down river till they disappeared. He didn't follow them; he says now he wishes he had. J. said he went back to the same spot on the bridge in the morning light and noticed that **the ice was broken up directly down the middle of the river, and as if it had been cut by a knife. Not usual in ice-breakup with chunks floating willy-nilly about. But a very exact, straight cut through the ice**, straight down the course of the river, probably the same path the lights had taken.

My friend J. and I both surmise we witnessed the same craft, J. from the bridge, me an hour or so later at the water's edge. [225]

A hole in the ice is made when a craft hovers vertically above the ice. Examples follow:

04-EE-1968   *Upprämen Lake, Sweden*

The hole, discovered yesterday near Malung by two villagers, is 700 square yards in extent and was **made through ice almost 3 ft. thick**.

The Defence Ministry was notified of the discovery by the police and sent experts to examine the hole. Colonel Curt Hermansson, who is leading the investigation, said today that an aircraft crash was out of the question.

"There are no traces round the hole, only big blocks of ice which have been thrown up, indicating that whatever went into the lake was incredibly powerful," he said.

Some experts think that the hole is too big to have been made by a meteorite. [226]

01-07-1971   *Scargo Lake, Dennis, Massachusetts, USA*

It traveled horizontally, then on an oblique path toward the ground, and disappeared behind some trees in the direction of a small body of water called Scargo Lake. One of the boys, unaware anyone else had seen the object, ran to a dock at the lake's edge and discovered **a large hole in the ice covering the lake. Steam was rising from the hole, and the exposed water appeared agitated.** Except for the hole, the lake and shore were normal, and the object was not in sight. [227]

01-10-1977   *Wakefield, New Hampshire, USA*

Horse farmer William McCarthy was looking out his window at the falling snow when he was surprised to see a hole in his pond. The pond, 105 by 75 feet, had been frozen solid just the day before, when McCarthy's horses had played over the surface. He put on his coat and went outside for a closer look.

The hole was perfectly round and cut smoothly through 14 inches of ice. Eight inches of slush surrounded the hole, suggesting that something had melted through. [228]

In the above case, it is stated that "slush surrounded the hole." If the UFO intended to go underwater, its field would have to be in reverse. In that scenario, the field would be pushing out and up, causing the slush to fall outside the field and onto the pond's ice.

12-25-1987   *Lake Elmo, Minnesota, USA*

An object bearing three bright orange lights which was observed by several people hovered about **40 feet above the frozen lake**, then extinguished the lights, and **apparently left a partially melted area about 100 feet long and a straight line of holes in the ice, 3 to 4 feet apart.** Those closest to shore were about the size of a small dinner plate and got smaller as one followed them away from shore to the southeast until they disappeared. **The largest holes were funnel-shaped and look[ed] as if they had been melted by a beam of energy that was strongest in the center.** I think it is likely that the support system of the UFO caused all of this ice evidence. [229]

01-21-2007   *Unnamed pond near Anson, Maine, USA*

After about three minutes Olivia's vehicle started rolling. A small stand of evergreens blocked our view for about ten seconds, and when we could see the sky again, the thing was gone.

The next day, I drove past the little pond and noticed **a round hole in the ice where the thing was hovering**. I also noticed a rash on the side of my body that burned and itched for several days; it also hurt, like I had been kicked. [230]

02-15-2007   *Unnamed pond near Karki, Latvia*

Riga—**A hole in a frozen pond in northern Latvia** achieved unexpected notoriety on Wednesday as rumours began circulating that it was created by an object falling from outer space.

"We've been collecting information on the story all day, and we'll definitely report it in the morning," Inguna Plume, editor of local newspaper *Ziemellatvija*, told reporters.

The **"unusually large hole"** appeared in the ice of a frozen pond near the northern Latvian village of Karki in early February, the Leta news agency wrote. **Despite sub-zero temperatures, the hole reportedly did not freeze over for two days.**

Locals initially reported seeing "strange things" in the area.

One girl said that she had seen **"a small bright object with a silver ring around it,"** while other witnesses reported seeing **up to six symmetrical beams of light emerging from the pond.** [231]

4.10 Engraved Ice Circles

One of the mysteries that has ufologists scratching their heads is a circle in the ice that appears as if it has been engraved with a tool, such as a milling machine, to manufacture a groove in the ice in the shape of a perfect circle. Examples follow:

01-21-1976   *Lake Vikern near Gyttorp, Sweden*

We have no proof, but we have one person who saw the object, and **we have the traces in the ice on Vikern that are very interesting and cannot be explained by natural explanations**, says Thorvald Berthelsen, chairman of the UFO-association, which has been in the town of Gyttorp and examined the case....

At 3.00 a.m. the object had gone again and then Allan thought he just should forget about it entirely. But when he was on his way to work **a couple of days later, he looked out on the ice and discovered a round ring on the ice where he had seen the object**. [232]

12-28-2003   *Mud Lake, Liberty Township, Michigan, USA*

It all started when lake resident Vaughn Hobe saw the light. An unidentifiable, brilliant white light illuminated the area over the lake on a bitterly cold and windy night at about one thirty a.m. on December 18. He stared in disbelief for a while, then told himself it was simply the best moonrise ever and went to bed. Ten days later, on the twenty-eighth, he and his son were collecting firewood from a bluff near the lake when **they looked over and saw the anomaly: "a three- to four-foot-wide ring around forty feet across with white snow in the center,"** according to a report on michiganufos.com. **Outside the ring was an additional pattern that radiated as far as the shore.**

Something had melted the ice in a perfect circle while leaving the center of the circle intact. [233]

12-09-2005   *Sudbury, Ontario, Canada*

...another "ice circle" was found on December 9, near Sudbury, Ontario, in snow-covered creek ice. Approximately 12-18 metres (40-60 feet) [in] diameter, with a 2.5-5 centimetre-(1-2 inch) deep **"V"-shaped groove in the ice defining the circle's perimeter**. There is another possible groove a few inches inside of the outer groove. [234]

Based on the following text, however, I believe I might have an explanation:

01-07-1990   *Mzha River, Ukraine*

SOVIET ICE RING
Vladimir V. Rubtsov, Ph.D.

On January 7, 1990, about 8:40 a.m., Mr. A. E. Vorontsov, a resident of the little town of Merefa, about 30 kilometers SSW of Kharkov, went by bicycle to the bank of the river Mzha to inspect his baited fishing tackle. When he was not far from the bank, he noticed a strange luminescence over the river. As the witness came nearer, he saw a **"saucer,"** **or rather a big top-shaped object** with the diameter of its base approximately 25 meters, and the height, including its spire, about 5–6 meters. **The object was situated on the ice or slightly above it in a small bay**. The spire and the base of the object were greyish-blue, and its body—orange or rose-colored. The witness said it was "something like the color of the clouds in the sky at sunset." The base was pulsating "as if some balls were rolling around there."

The witness observed the object in astonishment for, as he believes, about 10 minutes. Then the UFO suddenly took off vertically, to an altitude of some 30 meters, hovered for several seconds and then flew eastward. **A big round piece of ice on the landing site sank into the water, then it rose again back to the surface.**

Thanks to a lucky chance, a member of the Kharkov regional UFO study group, Dr. Pyotr I. Kutnyuk, was advised about this case only one day later. He promptly took photos of the landing site, when the trace was clearly visible (photos 1–4, taken on January 8) and informed the group about the event. I was able to visit the site on January 13, when the ring was less distinct, but still quite visible (photos 5–8). A few days later the members of the group took samples of the ice from the site. A month ago we obtained, at last, the results of its chemical analysis. No peculiarities were found, except that in one sample there proved to be a slightly increased concentration of platinum (2.5×10^{-7} grammes per liter, versus some 0.5×10^{-7} in other samples).

The diameter of the circle was 20.7 meters and the **ring was one meter wide** [1 meter = 3.28 ft.]. It is worthy to note that on January 7 **the ice was still rather thin due to a comparatively warm winter**, and it was not until January 13 that it became thick enough to safely bear the weight of a man walking on ice. Thus, the traces could hardly have been hoaxed. The depth of water under the circle is about eight meters.

The ring or rather the rings, as one will note on photos 2 and 3, were very regular and clearly outlined, as if made by a giant milling machine cutter. Three weeks after the date of the event the rings were still visible, but when the ice melted, this very interesting trace obviously vanished. (Date of report filing: June 5, 1991) [235]

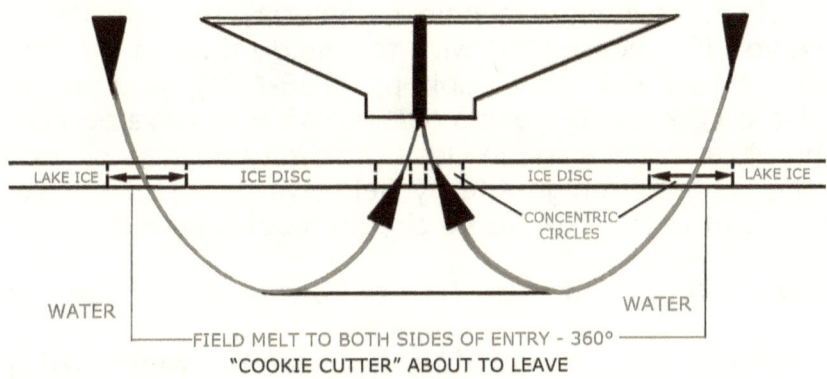

LAKE ICE   ICE DISC   ICE DISC   LAKE ICE

CONCENTRIC CIRCLES

WATER   WATER

FIELD MELT TO BOTH SIDES OF ENTRY - 360°

"COOKIE CUTTER" ABOUT TO LEAVE

Sketch #28
Ice melted in 360-degree circular groove

Note that in sketch #28 (above), the field returns into the craft and, therefore, in the last case, had the witness been able to walk across, he might have seen a small hole in the center. But, as was mentioned before, "the ice was rather thin," thus preventing a closer examination of the ring. Nevertheless, the following case gives credence to that idea:

01-15-1991 Charles River, Waltham, Massachusetts, USA

Alongside the cemetery lies the Charles River. At about a mile and a half into my run, I noticed on the river off to my right an unusual circular formation in the ice. The circle was located about 40 feet out from the river bank, and I estimated the impression to be around 20 feet in diameter.

The impression had a roughly circular mark in the center, a ring going around that, and a concentric ring on the outside of that. [236]

My conception of what is happening is based on what I have learned thus far about the UFO's protective field. Had the field been on as the UFO descended to the frozen river, it would have melted a hole completely through the ice. However, if it descended with the field off as in upcoming case 02-18-1968 (this book pp. 158-159), and perhaps landed on the ice, the pilots of the craft may have perceived a threat, and while still in close proximity to the ice, turned on the protective high-energy field, which would act like a cookie cutter—a 360-degree circular cookie cutter.

12-28-2003   *Mud Lake, Liberty Township, Michigan, USA*

"**It looked like something had come down with a cookie cutter** and whoosh," Hobe said raising his hands to explain. (...) [237]

Digressing for a moment, in the previous case dated 01-07-1990, the text says, "A big round piece of ice on the landing site sank into the water, and then it rose again back to the surface." In Australia, a similar sinking and rising occurred, but this time, instead of ice, it involved reeds:

01-19-1966   *Tully Lagoon, Queensland, Australia*

When Pedley drove around the bend of the track to the lagoon, there, at the spot beneath where the object had risen, was a huge, round, cleared area in the swamp grass. The water in this circular area was **slowly rotating** and **appeared to be completely cleared of reeds**....
Later in the day, apparently about noon, George returned along the track and stopped for another inspection. **The cleared area of the lagoon surface was no longer visible.** What was clearly **evident was a floating mass of reeds**, approximately 30 feet in diameter **that had**

apparently come to the surface of the lagoon during the time Pedley was absent. [238]

Returning to our discussion of ice, the high-energy field must have some width, and the fact that it is very hot would guarantee a melting between the lake ice and the ice circle that comes back up to reoccupy the hole it has originally covered. However, there is a loss of ice that is now replaced by the water below the ice and fills part of the gap between the two bodies of ice.

LAKE ICE   360° "COOKIE CUT" ICE   LAKE ICE

WEIGHT OF ICE KEEPS CIRCLE BELOW LAKE ICE

Sketch #29

Due to the weight of the circular ice cookie, the vortex causes it to sink. After this, the ice rises only a fraction of its original height above the water. An example of this can be seen by dropping an ice cube horizontally into a glass of water, and just like an iceberg, the greatest mass is below the waterline.

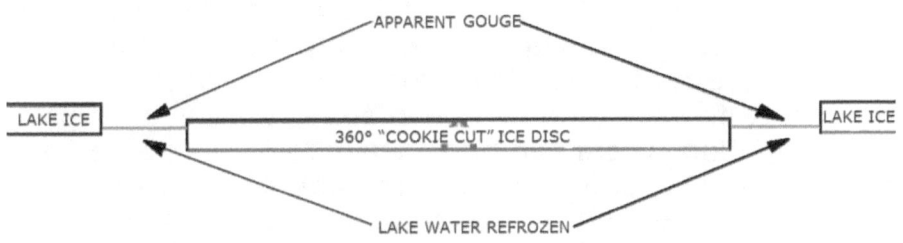

APPARENT GOUGE

LAKE ICE   360° "COOKIE CUT" ICE DISC   LAKE ICE

LAKE WATER REFROZEN

Sketch #30

The water between the lake ice and the ice disc refreezes the two bodies back together again but with a groove effect. See sketch #30.

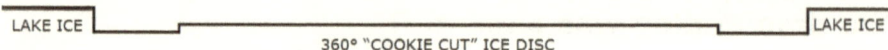

LAKE ICE   LAKE ICE

360° "COOKIE CUT" ICE DISC

Sketch #31
What you see

4.11 Flipping the Field Switch to Off—UFO Makes Ice

In contrast to the preceding cases where UFOs melted ice, the following account has water which has been rendered frozen by the craft. The field in almost all situations is invisible except at night, so we have no reference to use at any other time to determine whether it is active or inactive. However, here with ice as a medium, we can see the off or on quality of the field. The next case concerns a pond freezing beneath a UFO with the air temperature above freezing.

02-18-1968   *Pond on Vashon Island, Washington, USA*

When more people crowded out to the gravel pit, the object was gone. But something very strange had happened—the one-hundred-foot pond located in the gravel pit was completely frozen over! The discovery was reported to Sheriff Don Holke at 2 a.m., and he went out to investigate the situation immediately. He was startled to find that the report was true, for temperature in the area was above freezing and had been so for several days prior to the sighting. There had been heavy rain in the area for a couple of weeks, and besides, there was no other ice in the vicinity. Investigation showed that small puddles bordering the pond, which should have frozen first if this were a normal freeze, had no ice whatsoever, and the mud was not frozen either. Another unusual feature of the pond was the fact that the ice was very dry, and it was raining during the whole episode.

Other details uncovered during subsequent investigation by Mr. Akers were that the ice on the pond measured as much as three inches thick in spots and that it was formed in layers ranging from two to five in quantity. The ice was also riddled with large numbers of bubbles containing air and dirt. It was concluded that the possibility of a hoax was unlikely because of the size of the affected area of the pond. [239]

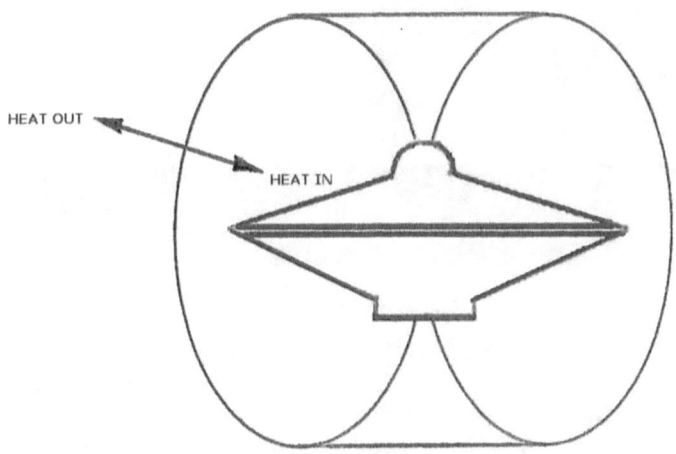

Sketch #32
Heat would act both in and out.

How and why a UFO makes ice is a mystery because this is a feature that is not normally observed. Indeed, if the field is radiating heat out, it is also radiating it inward upon itself. That would mean that the craft itself would be subjected to intense heat. That would necessitate a system to cool the craft, such as a high-powered air-conditioning unit of some type.

Perhaps the field generator is the answer. In order to produce a powerful magnetic field, a material used for the travel of the electric current must be extremely conductive. This feature is the basis of a superconductive medium, absolute zero (-459.67°F), in which all molecular and atomic motion ceases and electricity flows freely.

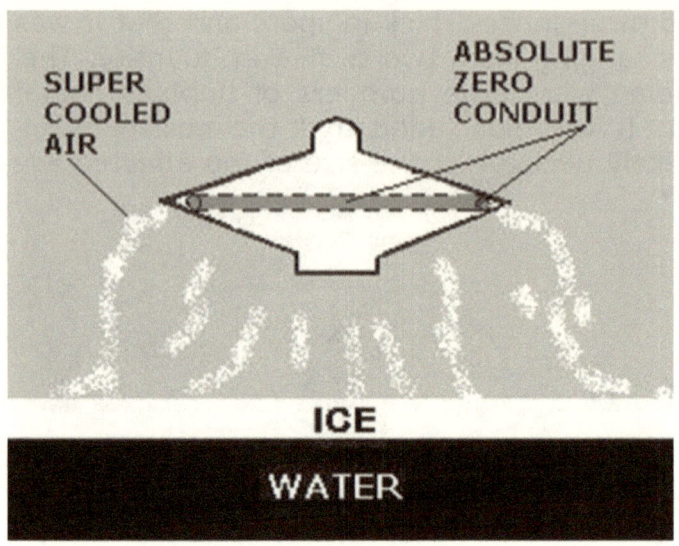

Sketch #33
Making ice below the craft

The absolute zero conduit produces the power for either the field or the propulsion mechanism. At the same time, it produces the cold air used to counteract the heat produced by the UFO's field and so protects the craft from the invading heat by venting the escaping cold air to the exterior of the craft or to the skin that faces the exterior. There may be a certain amount of seepage of cold from the conduit's exterior surface to the interior of the UFO's occupied quarters. It always amazed me that in abduction stories, the abductee would mention that the interior of the craft was cool. Naturally, our first thoughts go to the idea that the temperature is what the aliens are used to on their home planet, but now there is another reason—a system used to offset the craft's heat.

If the field is off and the same amount of coolant is still circulating to the outside of the craft, what is it cooling? Since cold air is heavy and sinks, the combination of this craft over water and an extreme temperature differential amounts to a "rain" of extreme cold to the surface below the craft, resulting in the freezing of any water below, and, in this case, the pond on Vashon Island.

The following cases (which are not on my website since they do not involve water) do not deal with ice but do show that, with the field off, the craft's temperature is cold.

SM-??-1933   *Cherryville, Pennsylvania, USA*

2:30 a.m. While repairing a flat tire on a clear night with no moon, an 18-year-old motorist noticed a violet-purple light in a nearby field and went to investigate. On the ground was a bell-shaped object about 10 feet in diameter and 6 feet tall. The light was emanating from a one-foot-diameter circular door opening into the object. The witness stuck his head through the door but the light was blinding. He could make out some sort of tubing and dials and a kind of console. There was an odor like ammonia and **he felt cold air**. The witness then walked around the object to inspect it. The outside surface was smooth, metallic in texture and **cold to the touch**. There was no sign of life at any time. After about 10 minutes, he walked back to the car, finished fixing the flat tire, and drove home. (Allentown *Sunday Call-Chronicle*, Pennsylvania, Feb. 16, 1964). [240]

The Roswell Crash is still muddled in controversy, as is the witness. Nevertheless, not much has been discussed in regards to the particular aspect of the case which deals with the temperature of the craft. The following interview of the witness **(GA)** relating to the crashed saucer in Roswell, New Mexico, was done by Stanton Friedman **(SF)**, nuclear physicist, ufologist, and the original civilian investigator of the Roswell incident. The witness was 5 years old at the time of the Roswell crash. He and his family had just moved to the Albuquerque area on July 04, 1947. Since the interview was lengthy, what follows is my transcription of the portion of the DVD covering the description of what the witness "felt" at the crash scene:

07-EE-1947   *Roswell, New Mexico, USA*

SF – OK, so you're there; you take all of this in; everybody's mystified. What were the circumstances outside? Hot, cold?

GA – Very, very hot. Incred... well to me, being my first time in New Mexico and coming from back east, that dry heat was just like being inside an oven. It was unbelievable to me, and that was the odd part about this thing, **the closer you got to it, the cooler it was. And standing under it in the shade there, next to these creatures' bodies, it was like refrigerated air conditioning**. And...

SF – Did you feel air coming out of this thing?

GA – No. It was just like it was ambient. And I remember reaching up and putting my hand on the side of it, that I think I was afraid I was going to hit my head, because there was enough room for me as a small child... you know...I was approximately the same size as these creatures...to walk up under there and stand there... but I kind of...did like that...put my hand up against this thing.

SF – What did it feel like?

GA – **It was ice cold. It felt like it just came out of a freezer.**

SF – Was it smooth, was it rough?

GA – It was very smooth. It had a very smooth texture to it, it was obviously made out of metal, it was very solid, and **it was very cold, ice cold**. [241]

09-19-1961   *New Hampshire, USA*

When Hynek inquired about the temperature inside the interior of the craft, **Barney stated that it was different from the night air. It was cooler.** Later, when Barney was "switched off" and Betty was asked the same question, her reply was consistent with Barney's. This detailed information cannot be found in Betty's dream sequence, and their stories correlate point by point in minute detail. [242]

The following is another case with a similar impression by the witness. Because it is not a water case, it will not be found on my website:

??-??-1958   *Las Vegas, Nevada, USA*

Paul Wilson and Joe Brown conducted the first interview in August 1968 with a female abductee, who had been a cocktail waitress at a Las Vegas casino in 1958 and who lived in a trailer not far away from the bus stop. She was walking home after work at night and had to traverse a few hundred yards of desert. She encountered a man on the way. He encouraged her to walk with him farther out into the desert, and as she stood there, in front of them a small light ball at the bottom of a hollow puffed out a smoke-like ring, which seemed to absorb the surrounding volume, and then the ball grew larger. It did this several times, and soon she was standing in front of a craft. She said that as she crawled up the ladder into the craft, **it was very cold**. When inside, she walked around a spiral ramp and saw something like liquid mercury in a pipe. [243]

An interesting thought came out of this conduit idea. Instead of the vertical bar magnet scenario constructed in Chapter 2, the conduit would fulfill the same purpose in producing the field in a saucer-shaped UFO as well as a triangle- or rectangle-shaped craft. To visualize this for a triangular craft, you could substitute three bar magnets with all three north poles connected to the other magnets' south poles in a series to make one complete triangular circuit. The same would be done with four bar magnets similarly arranged for a rectangular craft.

And what is in the conduit? We do not know, but in the following case, we find a possible answer:

SP-??-1951 *Paarl, South Africa*

The[n] he asked if H. M. had any questions about the craft. H. M. asked where the engines where [sic-were]; he replied, "We don't have any engines." Pointing at the levers, he said, "**We nullify gravity**; that is how we rise." Asked how this was done, **he answered that they used "a very heavy fluid" which circulated in a tube at near light velocity, creating a magnet.** H. M. said such a speed was impossible. "No, it is simple," he replied. "When the fluid is leaving the tube, it is already entering at the other end, thus its relative speed is infinite." [244]

So what is this "very heavy fluid which circulated in a tube" [or conduit]? Sorry, but the aliens did not divulge that information.

4.12 Eminent Physicists Discuss the Reality of UFOs

Before moving on to the next chapter, as a comparison to Chapters 2, 3 and 4, I would like to present a "contactee" whom I found very interesting. I came across this person while reading the book *Earth: An Alien Enterprise: The Shocking Truth Behind the Greatest Cover-Up in Human History* by Timothy Good, [245] published in 2013. I would like to thank the author for granting me permission to use some of his text here.

This contactee is Carl David Anderson, born of Swedish parents in New York City on September 3, 1905. His credentials are impressive. He received his B.S. in 1927 and his Ph.D. in 1930, both from the California Institute of Technology (Caltech). For his doctoral thesis, he studied the space distribution of photoelectrons ejected from various gases by X-rays. In 1930, he began his cosmic-ray studies which led in 1932 to the discovery of the positron [or positive electron], an accomplishment for which he received the **Nobel Prize in Physics.** [Additionally, in 1936 he discovered the muon (an elementary particle similar to the electron)]. Starting in 1930 as a Research Fellow at Caltech, he went on to become a Professor of Physics there a few years later. During World War Two, he was also active on projects for the National Defense Research Committee and the Office of Scientific Research and Development. During his lifetime, he was given many awards for his work. Anderson died January 11, 1991 [246] in San Marino, California.

In my quest for the physical reality of UFOs, I discovered that much of what I found out from witnesses' water-related sightings, which was initially included in my first edition of *UFOs and Water* in 2010, correlated with much of the contactee's text. In Chapters 2, 3, and 4, I attempted to show what I had learned from the contact of water with the field, writing about it fifty years after Dr. Anderson had this explained to him. As a Nobel Prize-winning physicist, he definitely had an advantage since he got his information firsthand from the aliens themselves.

In the quoted material that follows, I have italicized my personal comments and enclosed them in braces {} to distinguish them from those of Mr. Good, Dr. Anderson, and Hermann Oberth, who is also mentioned in the text. I bolded some parts for emphasis:

CONTACT
Further encounters ensued, including a contact in February 1960. I have condensed the following report from a lecture given by Anderson in 1966:
"On the night—or I should say morning—of February 14, 1960, I was ushered *{military escort?}* personally aboard a 200-foot craft on the Mojave Desert, *{northeast of Los Angeles}* some ten to

twenty miles north of the town of Yucca Valley. **I remained on board this craft for two hours and twenty minutes, during which time I was given some very important information.** And I was told that I was to take this information to some great scientists. And when I questioned this, the man who told me this informed me that the way would be made clear, that all expenses would be taken care of, and that I would go to Germany and talk with some noted scientists and physicists in that country.

"I asked why I had to go to another country [and] why I couldn't see physicists in my own country, because it so happens that in 1932 or 1933 I was employed by none other than Dr. Vannevar Bush [a pivotal pioneer of nuclear weapon technology]. At that time, he was with the Massachusetts Institute of Technology. I lived with him in his house in Belmont, Massachusetts, for two years. I was very well acquainted with him—he's one of our noted physicists and scientists. And I asked why I had to go to a foreign country to impart this knowledge [and] I was told that scientists here in my own country would turn a deaf ear to what I had to say, therefore I would have to go where people were capable of listening and understanding, and would accept the truth." [3] [Note 3: Lecture at a conference in Reno, Nevada, July 10, 1966.]

Anderson flew to Germany on October 17, 1960, and participated in a UFO conference at Wiesbaden. Another participant was Dr. Hermann Oberth, one of the true pioneers of astronautics (whom I met in 1972 at his home in Germany). In 1955 Oberth had been invited by Dr. Wernher von Braun (his former assistant) to go the United States, where he worked on rockets at the Army Ballistic Missile Agency, and later with NASA at the George C. Marshall

Space Flight Center in Huntsville, Alabama. Oberth returned to Germany in 1958.

PROPULSION TECHNOLOGY

The information imparted by Anderson at the 1966 Reno conference included much data relating to alien propulsion technology, based on the information imparted to him by his alien friends in February 1960:

"Most everything we read about the craft—the way they're propelled—**we read it in terms of an electromagnetic field. This is not entirely true. There are two forces involved. There *is* an electromagnetic force field; however, there is also a very, very high-voltage static charge involved....And it is this static charge that has caused the scorched bushes and the scorched grass, where these craft have landed and made contact** {*not to mention melted snow and ice as presented in Chapter 4*}. **This is from the static charge, which contains billions [and] billions of volts. But as such, it is not actually a danger to human life because there is no amperage involved.**

"Now many people wonder why these craft have been seen going through the sky **at terrific rates of speed that have been clocked on radar and by other means**, and being able to negotiate the seemingly impossible maneuvers and turns that they make—instantaneous stops and starts—traveling near the speed of light and making 90- and 45-degree turns. {*This foreshadows my Chapter 6, "Radar and Sonar," where there are many accounts of UFOs being detected by military-operated radar/ sonar. And we continue:*} And they say these people cannot be human people inside these things, they must be machines, they must be robots, they could never stand the pressures that are being brought about during these quick turns and quick stops and starts. **This [would be] very, very true, provided they use Earthmen's means and methods. But we know they are much more highly evolved—** not only spiritually, morally, physically, but also mentally and electronically, and in every other thing that you can name they are much, much further advanced than we are.

"**This is what they have done: they have made a vehicle and they have harnessed the forces of nature; they have copied a real planet.** {*In aviation we imitate birds; aliens imitate planets.*} **Now, the planet Earth that we know spins on its axis, which is nothing more in itself than a static generator—a huge static generator.**" [04] [Note 4: Lecture at a conference in Reno, Nevada, July 10, 1966.]

During a Chamber of Commerce luncheon in Pueblo, Colorado, on July 22, 1952, Joe Rohrer of Pikes Peak Radio Co. recounted an interesting story relating to early knowledge of alien propulsion technology which endorses some of Anderson's claims: "A Californian air pilot told me that, in 1942, he had been right inside a giant saucer and **seen giant flywheels sheathed in metal skins, and found that the motive force came from electrostatic turbines, whose flywheels create an electro-magnetic field of force, creating tremendous speeds....**" [5] [Note 5: Harold T. Wilkins, *Flying Saucers on the Attack*, Ace Books, New York, p. 261.]

Anderson learned that part of the propulsion system of the spacecraft he went on involved a "wheel within a wheel," the principal process by which they derived their motive power. "This in essence is nothing more than a

gigantic static generator," he explained. "One of these wheels turns in a counter-clockwise direction, the other in a clockwise direction. And when these two counter-rotating forces reach a speed equivalent with the [rotation] of the Earth or of the planet involved, and in direct relation to its mass, then a static charge is forthcoming to the extent that it is repelled away from the Earth.

"We know that the Earth has a magnetic field surrounding it. Now, when you are inside a magnetic field, you are in truth inside of a gravity field, because gravity, electricity, and magnetism are one and the same....

When you go into a gravity field, every atom, every molecule, every electron of your body is acted upon simultaneously with every atom and molecule of everything around it. And therefore, these people, being inside a gravity field, have no sense of motion, no knowledge whatsoever of quick turns, of stops and starts, because they in reality, from an atomic standpoint, are a part of the craft itself....

"Now, another thing that our scientists do not understand is how these craft can travel with such fantastic speeds in our atmosphere without getting hot and disintegrating and burning up. **Well, here again, it's because the forces of nature have been harnessed in this vehicle, and an ionization effect as such takes place with this static charge surrounding the craft—the skin of the craft—whereby you have several inches of complete insulation on the outside of the craft by means of a vacuum, in which your craft travels through a vacuum and carries its vacuum along with it all the time.** *{I somewhat* *disagree. Whatever is inside the field remains there until the field is turned off. When the field is reactivated, it encloses what is touching the craft at that time and remains there until the next opening of the field. Moving on:}* It can never get hot, it can never get cold; it remains the same temperature always.

"So, you see, they have thought of everything. And they tell us that our science knows this, but it's being kept from us.... *{In my Chapter 8, "Secrecy about UFOs," there are numerous examples of cover-ups of UFOs that were identified as such. So, while the government may ask for proof that UFOs exist, it is they (the military) who are keeping all this confidential. The text continues:}* Now, when I took this information to Germany, I was received with open arms [by] various other physicists and scientists who were associated at that time, or who had previously been associated in one way or another—with the teaching staff or in an advisory capacity—with Heidelberg University, and who were in complete accord with what I had to tell them. And checking it out mathematically, Dr. Hermann Oberth said, 'My God, it will work! Why haven't we thought of this?'" [6] [Note 6: Lecture, Reno.]

Oberth was one of the few scientists to have the courage to speak out on the controversial UFO topic. In 1962, he wrote an illuminating article in which he refers to the Wiesbaden conference in general—and to Anderson in particular:

"As far as the so-called 'contact' persons are concerned (those persons who allege they have been passengers in UFOs or have spoken to space people)," he wrote, "I had expected to encounter swindlers, hysterics, or schizophrenics. But I must say that among these contact persons, Carl Anderson, particularly, made quite a congenial, reasonable, and clean-cut impression. Skeptics should understand that I studied medicine and began my professional career as a doctor in a military hospital for three years, where I also had the care of mentally ill persons [and] I would bet a hundred to one that some of the contact persons are normal and have seen and experienced *something....*" [7] [Note 7: "Dr. Hermann Oberth discusses UFOs," *Fate*, May 1962, pp. 36-43.]

Oberth was skeptical that aliens would look like us. But it is interesting to note that in the article his technical explanations for UFO propulsion parallel many of those revealed by Anderson in his 1960 presentation (which Oberth translated into German after the conference) and, more comprehensively, during private meetings with Oberth and other scientists. "We cannot take the credit for our record advancement in certain scientific fields alone; we have been helped," conceded Oberth some years later. When asked by whom, he replied, "The people of other worlds." [8] [Note 8: Robin Collyns, *Did Spacemen Colonise the Earth?* Pelham Books, London, 1974, p. 236.]

Regarding UFO sightings per se, Oberth calculated that 11% of reports resisted conventional explanations. "They cannot be hoaxes or lies because they involve responsible men such as senior air force officers, or radar readings, or photographs from responsible sources.... Furthermore, the reports check out against each other so well that a common origin is to be concluded from them...." He then proceeds to describe the "state of the art"—at least, as it was in 1962:

"The discs always fly in an attitude as if the driving power were effective vertically with the plane of the disc; when they hover steadily over a certain area they are in a

horizontal position; if they want to fly fast they tilt and fly with the broad side facing forward. *{An example of this can be found in Chapter 2.13, sketch #19.}*

"In sunlight, which is stronger than the luminosity of the discs, they appear to have a metallic sheen. At night they appear dark orange or cherry red if their maneuvers are such that little driving power is required— such as hovering. Under such conditions they emit very little light. If more driving power is required, the luminosity increases and they appear yellow, then yellowish green, then green like a copper flame, and at the highest speeds or accelerations glaringly white. They also may suddenly light up brightly or darken, even disappear.... *{See Chapter 2.2 for Paul Hill's explanation of ionization as it relates to UFOs, which is similar to this. Continuing on:}*

"If we establish the working hypothesis that the UFOs are machines, we also have to assume the following:

"...They are flying by means of artificial fields of gravity. This would explain the sudden changes of directions. If an apparatus built by humans were capable of changing its direction and speed as suddenly as do UFOs, the passengers would be pressed against the wall so violently they would be crushed to death. **Artificial fields of gravity, however, would mean that the occupant would be rushing forward along with the vehicle and that between him and the vehicle no tractive force [the pulling force exerted by a vehicle, or machine or body] would even come into being.**

*{I would like to provide an **example**: "To make one complete rotation in 24 hours, a point near the equator of the Earth must move at close to **1000 miles per hour (1600 km/hr)**. The speed gets less as you move north, but it's still a good clip throughout the United States. **Because gravity holds us tight to the surface of our planet, we move with the Earth and don't notice its rotation in everyday life.**"* [247] *Now returning to the text:}*

The hypothesis also would explain the piling up of the discs into a cylindrical or cigar-shaped mother ship upon leaving the earth because in this fashion only one field of gravity would be required for all discs. *{In Chapter 18, I have sketch #50, which is an example of the magnetic hookup in a different situation. Let us continue:}*

"They produce high-tension electric charges in order to push the air *{and any small rocks or debris}* out of their paths, so it does not start glowing, and strong magnetic fields to influence the ionized air at higher altitudes. First, this would explain their luminosity. Even the poles of our electric influence or induction machines glow in the dark. Secondly, it would explain the noiselessness of UFO flight. Our jet aircraft have a high noise level because they move through the calm air and create violent turbulences. The UFO, however, does not create turbulences near it because the air has the same speed it has, and the speed of the air decreases gradually with distance from the UFO.... Finally, this assumption also explains the strong electrical and magnetic effects sometimes, though not always, observed in the vicinity of UFOs...." [9] [Note 9: "Dr. Hermann Oberth discusses UFOs," *Fate*, May 1962, pp. 36-43.] [248]

In this chapter, we briefly touched on a couple of stories relating to abductees and the sensation of cold that they experienced within the UFOs. The next chapter focuses on the abductees' relationship with water.

Chapter 5

Abductees—Water and Bases

5.1   What We Can See through the Eyes of Others

Abductee is a word which had little immediate importance in ufology until the Betty and Barney Hill case of September 1961. Like many strange happenings in relation to UFOs, this was just another item to be kept at arm's length in order for us to remain respectable ufologists. However, with UFOs being a worldwide phenomenon, another case came up and was translated into English and brought forward for comparison to the Hill case. This event which occurred in Brazil in October 1957 involved a man by the name of Antonio Villas-Boas. These two cases from two different continents caused the abduction scenario to be viewed in a different light because of a similar aspect: In both cases the physical features of the aliens were generally humanoid in appearance.

I wish I could say "well, that's that," but variations on a theme keep occurring, and more is being learned each day. While an abduction case may have been published, all facets of it may not have been taken into account (remember Charles Fort's cigar boxes), and water is a feature that may not have been addressed. In this neglected category, there have been abductees inside a craft, observing not only the interior layout and instruments, but also looking out the window to the field holding back the water. Yet, as far as I know, this is another topic which is not mentioned in any journals or books prior to the early 1990s.

5.2   Abductees and Water

Whether one believes in the reality of abductions or not, the physical aspects of the UFOs involved in abductees' tales remain consistent with descriptions from those who have not been abducted but have also seen a UFO. The passages given below are included to reveal the physical changes observed by the abductees on the water environment. Given their correlation with the physical changes observed by non-abductees, I feel that these changes offer our only chance to view the water-UFO phenomenon from an insider's perspective, in a physical sense, and therefore should be considered a necessary part of the overall study of UFOs. These accounts not only reinforce what others have shown us, but also take us places we would otherwise never go. In the very least, these stories show that while abductees have become astronauts, they have also become aquanauts.

Besides the text that shows a commonality of the abductees' descriptions of the field, I would like to introduce other abductee cases that have a water connection other than the one between the craft and water. I will start with text from the Hill case mentioned above, written by Betty Hill's niece, Kathleen Marden, and Stanton Friedman in their book *Captured! The Betty and Barney Hill UFO Experience*. The first text is:

> The dream contained some elements of the UFO encounter and abduction, but in many respects were entirely different.
>
> For example, in the week following Betty's first hypnotic regression she had two nightmares. One was about water—perhaps a lake and a shoreline. But she couldn't recall anything else. [249]

In all of the previous texts and books written on this case, I cannot remember water ever coming up. However, if one wanted to abduct someone and remain close to the area in a place where no one would find them, what better hiding place could there be than the bottom of a nearby lake or other body of water. Yes, this is sheer speculation, but it might fit the nightmare.

Also of importance was Betty and Barney's evaluation of the craft's interior temperature, which I talked about in Chapter 4, section 4.11, case dated 09-19-1961.

Water makes its appearance again when Betty questions if they are capable of going into water:

> This huge, glowing orange ball was part of Betty's and Barney's conscious recall of the events of September 19, 1961, but they were uncertain of its origin. Betty's hypnotic description of the departing craft is more detailed than Barney's, but again, their accounts contain correlating data. Betty stated, "It's a big orange ball and it's glowing and glowing and it's rolling just like a ball. **It must be...I don't know—water? Do they go underwater?** It goes down and then there's a dip, and then, zoom." [250]

From dreams we go to an exciting abductee story called *Witnessed*, which reads like a mystery novel. It was written by Budd Hopkins, who was a major investigator of the abduction phenomenon and died August 21, 2011, a year after my first book was published. The story is important on its own merits, but like the Hill story, water again comes into play.

The abductee, Linda Cortile (a pseudonym) is taken out of her New York City apartment building through a window to a hovering UFO on November 30, 1989. Two government agents on the ground are witnesses to this event and become major players in the story. Obviously, my interest is water:

> After she was escorted up and in, the oval turned reddish orange again and whisked away, coming in our direction, above us. It must have flown over the FDR Drive while we were sitting underneath it. **It then plunged into the river behind us, not far from Pier 17, behind the Brooklyn Bridge.** Someone else had to see what happened

that morning. I know what we saw, and we'll never forget it. [251]

So far all we have is craft entering water, which has become quite familiar to me. But now we enter a hypnotic regression session:

> The session had been harrowing but extremely valuable to both of us. The episode in which she apparently spoke in an "alien tongue" was something I had never encountered before, as was one particular detail from her daytime childhood abduction. (In this year's earlier incident, **the UFO in which she was being held came to rest, apparently, underwater. Through a large window she could see the murky bottom of what she took to be the East River; garbage and even a soft-drink bottle were visible in the UFO's lights.**) These recollections would have special relevance fifteen months later, after I received the first letter from Richard and Dan and began to reinvestigate the astonishing events of that November night. [252]

In my estimation, the field of the UFO had to be turned off in this instance. She was seeing the garbage and a soft-drink bottle floating by, which would not be visible if the field were in operation because of the rapid rotation and bubbles it produces. But why would they want the field off? Well, when the field is on, it ionizes both the atmosphere and water, causing them to glow. This would make the UFO visible even in the murky East River water. So what do we learn here? First, the craft is hermetically sealed even with the field off (no leakage in the interior of the craft). Second, as in the previous case, water is a good hiding place, especially in the confines of New York City where thousands of eyes might spot this big glowing light in the water. In addition to this, there was one more physical effect of the UFO that was observed as it passed over the witness and departed: "When it flew over us, I could feel the hair on my head, arms, legs, etc., stand straight up, and it wasn't from fear. **The static cling was incredible. The electricity was tremendous**. I wasn't amused at all." [253] This is a further indication of the electromagnetic effect of a UFO.

5.3   "Beam Me Up, Scotty"
—William Shatner as Captain Kirk in the TV series Star Trek

The concept of teleportation was thought of, in a sense, many years ago. The word is based on the Greek prefix "tele" (meaning "distant") and the Latin verb "portare" (meaning "to carry").

Definition of **teleportation**:

The theoretical transportation of matter through space by converting it into energy and then reconverting it at the terminal point. [254]

In 2007 the History Channel presented a TV show called *Star Trek Tech* in which teleportation was explored amongst other technologies that were appropriate to *Star Trek*, the series. The program brought up the fact that the first person to use the term "teleportation" was our original anomalist and ufologist, Charles Fort, in the year 1931:

> The look to me is that, throughout what is loosely called Nature, teleportation exists, as a means of distribution of things and materials, and that sometimes human beings have command, mostly unconsciously, though perhaps sometimes as a development from research and experiment, of this force. It is said that in savage tribes there are "rain makers," and it may be that among savages there are teleportationists. [255]

Yes, this was more of a natural view of the word but almost implies transportation via molecular or atomic levels, which science fiction writers have used for a long time and science seems to be getting comfortable with as well. Our next case includes this teleportation from the surface of the water to a hovering craft.

In *The Allagash Abductions* by Raymond E. Fowler, four men are abducted at one time. This occurred on the Allagash Wilderness Waterway in northern Maine on August 24, 1976.

The four men are in a canoe at night. They see a light near a mountain, and then one of the men starts flashing a light on and off at it. Next, they have company overhead with a beam from the craft coming down towards them that encompasses the entire party and the canoe. No water reaction is noted, but they are beamed up to and *through* the floor of the craft. (This also happened in a case of an abducted elk in the state of Washington, where witnesses saw the elk pass through what appeared to be the solid bottom of the craft.)

One interesting aspect of the Allagash case is the composition of the beam, which, unlike a flashlight making a full yellow circle on a wall, is *hollow*. Before continuing with the beam, there is a small but relevant piece of text that could have a relationship to the electromagnetic field of the UFO:

> Jack Weiner describes the shocking sight from his own perspective:
>
> ...It [the craft] was very large, as big as a house, at least 80 feet in diameter, spherical in shape and pulsing light. There were changing patterns in it that **reminded me of the science experiment that uses magnets and metal filings to illustrate the magnetic fields of magnets.**
>
> We all sat awestricken, wondering what it could be. [256]

Under regressive hypnosis the abductee describes the beam as a "tube," and through his eyes we gain knowledge of it:

JIM: I see the inside of the tube!

TONY: And where are you when you see the inside of the tube?

JIM: I think I'm still in the boat, but I don't see the trees anymore.

TONY: What is it like inside the tube?

JIM: It's sparkling—No, it's not a sparkle, it's... (lets out a long deep breath of air). It's a, it's kind of a, it's moving. It's, it's, it's moving.

TONY: Does it move left to right or up and down?

JIM: No, the walls, the walls move.

TONY: Towards you or away from you?

JIM: Nope, just in, in themselves. Like, when, an, not exactly like but when you blow smoke through a, through a light beam, you know? You can see it. Oh! I know how. Like when dust, when you see dust, when you see dust particles in a light beam. [257]

Star Trek never explained how the "beam me up, Scotty" transporter functioned, but the History Channel gave a fairly good explanation of how it might be done in reality by means of its show *Back Engineering a UFO* (a two-part TV series). Basically, a computer would have to separate the molecules of the body, and they would all be identified and held in a relative position next to each other. The same is done to the floor of the craft. On command, the molecules of the abductee are moved up, while the molecules of the floor structure are held in position but separated so that the abductee's identified molecules can pass through between them. The abductee's molecules are reassembled once inside the craft. Science *fiction* you say? Well, back to the regressive hypnosis, canoe to craft:

> JACK: Next? (Pauses, breathing heavily.) Something happens. Something changes. **I feel—the feeling changes. There's something different—it's—something's happening** (pauses, puffing out air). Ohhh—Whew-w-w—Charlie's screaming!

[Same session, later on:]

> JACK: I feel funny (pauses). I feel funny. I don't feel like I'm here (pants). I don't feel like I'm supposed to feel (pauses) like something's happening, and I'm really scared (pants heavily). Something is happening. [258]

Then, craft to canoe:

> JACK: The tube is the machine and then we're afraid. The tube is scary.
> RAY: Why?
> JACK: Because it starts to—it starts to move—and it's doing something. It's doing something. It's coming towards us and I don't know what it's going to do! And—It's—Oh-h-h! Oh-h-h! Oh-h-h! (Begins to hyperventilate and there is a long pause.)
> RAY: Does the tube bump into you?
> JACK: No. **It makes us come apart. It makes me feel like I'm flying apart. I don't like it! It makes me feel like I'm coming**

> **apart!** I'm going to be sick. And it's—and it's— happening so fast! And I can't stop it (puffs out air). And I feel funny. I feel real funny. I feel—I feel like—scared (pants).
>
> TONY: **Do you feel like you're coming apart? Do you get the sensation that you're coming together again?**
>
> JACK: **Yes.** [259]

The other witnesses told a similar tale of their transport to and from the craft, so this is not a one-witness show. The transporter beam does work, and thanks to our friends, the abductees, this knowledge is brought to us even though they were told that they would remember nothing. There are many other abductee accounts where the witnesses said they were beamed through ceilings, walls, and windows before going into the UFO, so, the Allagash abductees had an easier time of it because they just had to go through the floor of the craft.

Unfortunately, escaping was not an option for these four men as they were on a lake with no protection. However, there is another case I read about that was not water related other than the fact that the witness was near a pond. The physical sensations which she experienced were somewhat similar to those of the men in the Allagash case, but she was able to escape from the beam:

02-06~08-1991   *Chaville, France*

She then lived on the Rue Alexis Maneyrol, at the edge of the Meudon forest in Chaville (Hauts-de-Seine), a commune in the southwestern suburbs of Paris, France....She was taking a walk with friends...

After arriving at the northern edge of the Ursine Pond, they stopped to admire the stars....

At nightfall, the three strollers went back to their homes.... Sylvia went back alone to the pond....

As she walked along the forest edge, Sylvia was startled by a light that appeared on her left. She turned, thinking that it came from a street lamp. But she saw it was a luminous beam directed towards her, apparently coming from a circular object above a street lamp a little distance away. Suddenly, the beam enveloped her. **She felt an odd sensation in her body** and got scared. **She felt very light, as if the beam were lifting her up.**

"If I had to describe the sensation that I experienced in my body, I would say that **my body was seized with fizzing (sensations) (like a caress on the skin)**, and that **each spot of my nervous system was being sucked up**," she explained. "This gave (me) **a feeling of lightness and attraction**. I was scared, and I cried to myself: 'Sylvia, go home, something very serious will happen to you!'"

Panic-stricken, she made an effort to escape from the beam and ran in the direction of her home. For some moments, the beam seemed to be attempting to trap her again, reorienting itself by a small succession of jumps. Then the phenomenon stopped. [260]

5.4   Another Case History

Another abductee account involves two young staff members of a summer camp in Vermont named Buff Ledge, now known as Camp Norfleet, on August 7, 1968. The complete story is in Walter N. Webb's book entitled *Encounter at Buff Ledge: A UFO Case History*. I read it at first only because I am interested in all aspects of ufology and felt that since Mr. Webb is a researcher of repute, I would be in safe territory with it. It did not take many pages to find out that this case was in my ballpark—water.

The primary focus in Mr. Webb's book is the two witnesses— one male, the other female—and what was revealed while they were under regressive hypnosis. However, water, as

I said, is my focus, so here is a piece of the text from one witness before the actual abduction:

> Almost immediately the UFO reappeared and in another three seconds descended vertically to the same spot over the lake. After stopping momentarily, **it plunged into the water broadside**. Michael said he both saw and heard **the resulting splash**, which **seemed strangely small for such an impact**. (For reasons that will be discussed later, the alleged impact point would most likely have been around one to two miles from the observers.)
>
> Instantaneously, Michael reported, a steady gale buffeted the dock and three-foot, white-capped waves sprang up out of nowhere all over the calm lake surface. [This is caused by the vortex at the bottom of the UFO, which I described in Chapter 2.] At the same time, dogs and cats reportedly howled and screamed up and down the lake as if the animals were in pain. [Chapter 9 deals with animal and fish reactions to UFOs.] The wild scene was unreal and unearthly. The witness recalled hearing the trees behind him on the bluff creaking and branches breaking from the force of the wind, and in order to maintain his own footing on the dock, he said he had to really "lean into it." Michael surmised that all of the effects were too widespread to be caused by just the UFO's splash alone. But exactly what forces could account for the effects were as unanswerable as everything else that was purported to have occurred during this weird experience. (Taking the reported effects into account and applying the Beaufort wind scale, I estimated a wind velocity of around 35 to 40 miles per hour.)...
>
> After perhaps two or three minutes, **the disc emerged with another splash in the same level attitude** and hovered momentarily. Instantly,

all the disturbances ceased; the wind, waves, and animal noises all stopped at once. The lake was calm again. [261]

The intent here might be considered a scouting mission where the UFO is trying to find an adequate place to park in the lake in order to perform the often lengthy abduction process. This may be assumed because it is only after exiting the lake that the UFO moves towards the witnesses and the abduction begins. However, whether or not the craft returned to the water was never brought out in the text of the account, and it would be speculation on my part to assume so. However, it would be a handy way to continue the examination of the witnesses without leaving the area and without being seen.

Many observations were made by the witnesses in this case, and the investigation was done in a very professional manner.

5.5   Abductees That Know They Are Underwater

In my earlier chapter "Physical Influences of a UFO on Water," I used observers' testimony to demonstrate how a craft's protective field moves water at various stages: as it comes to the surface, as it is level with the water, and as it hovers above. However, there are other witnesses besides those on land and on ship—abductees in a craft.

I received a photograph, purportedly taken by a diver, from a Turkish researcher several years ago. I have no way of verifying whether it is a valid photo of an underwater UFO or not, and because of the black and white format of this book, it would be very difficult to distinguish the UFO if I tried to reproduce it here. However, it reminded me of an abductee's tale I had read. In the photograph, an almost black UFO is in dark blue water, but a vivid green light is visible. This made me think of the following story which comes from the book *Connections: Solving Our Alien*

Abduction Mystery, co-authored by Beth Collings and Anna Jamerson. The following text is by Ms. Jamerson:

> Since the therapist was not an investigator, she did not know the questions to ask me to find out what else had happened during this abduction, and I didn't volunteer any information. I was terrified of the gray beings, and wanted to leave their presence as soon as possible. I also saw a huge ocean liner coming at me bow first—I felt like I was below it, maybe in the water. It was all black on the bottom of the prow and white above. I kept seeing the red color and sometimes the blue.
> **At one point I saw myself in a long tunnel with a small green light at the end of it. The tunnel had soft sides.** I was thrilled, awed and amazed that I was not walking down this tunnel—I was just moving down it. Floating! I even wiggled my feet at that point. She [the hypnotherapist] never asked me to go to the end of the tunnel, so I don't know where, or to what it led. [262]

Jamerson comments that her therapist did not know the questions to ask her to find out what else had occurred during her abduction. I find this akin to a doctor operating on a patient without knowing the patient's physical problem. Also she adds that the therapist never asked her to go to the end of the tunnel, so she did not know where or to what it led. And the trail stops here. If only someone experienced in ufology could have attended that session and asked the right questions, the craft's destination, possibly a base, might have been revealed, not to mention a possible description of the interior of the craft.

At a conference held by the Fund for UFO Research in Maryland, I spoke with Ms. Jamerson regarding additional information on the water's behavior. She told me that the book contained everything she remembered about the incident and that she was not certain she was underwater.

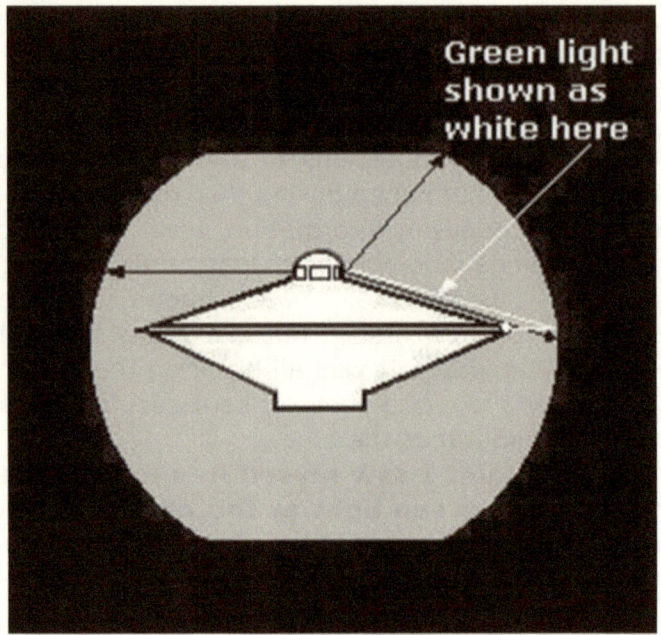

Sketch #34

However, after rereading her text and comparing it to the text in *UFO Contact from Undersea,* which I will address later, her description leads me to believe she was.

What initially got my attention was her comment about seeing herself in a long tunnel with a small green light at the end of it. The tunnel mentioned will come up again in subsequent cases, but the "small green light" that the witness saw conforms to sketch #34 where the light (white in this sketch) may be reflecting off a wall of water. As can be seen in this same sketch, when the abductee looks out the window, she is looking at a curved wall created by the craft's protective field. The small light reflecting off the wall from the UFO's exterior represents the light she saw at the end of the tunnel. In this account she goes on to say that she was not walking down the tunnel but instead was floating down it, even wiggling her feet. This floating scenario in ufology is common in abductee literature and recalls a chapter in Paul Hill's book *Unconventional Flying Objects* titled "High-Acceleration Loading on Occupants":

> At this point it must be made perfectly clear that, while UFO fields have been called force fields because of their effect on mass, UFO fields are actually **acceleration fields**. This is directly analogous to gravity fields which cause weight because they are **acceleration** fields. [263]

An example of what Hill is describing is similar to the NASA aircraft known affectionately as the "Vomit Comet." The aircraft climbs in a steep vertical arc, moving up and then curving downward. It is at the point where the aircraft is in a dive that an unsecured person in the windless environment of an aircraft is weightless or in free fall (matching gravity's acceleration). This is very similar to what skydivers experience, except in their case they are affected by the air resistance caused by their bodies falling through the atmosphere and winds moving horizontally at various altitudes. Mr. Hill describes the field in this manner:

The Superposition of Fields

> Acceleration field strengths, like all other field strengths, can be added algebraically (added or subtracted), or, better stated, they can be added vectorially. This is called the principle of superposition. This principle together with the equivalence (interchangeability) of acceleration and gravitation is one of the foundations of general relativity. What we do is perfectly analogous; we note the equivalence of vehicle acceleration and force field acceleration and that they can be added vectorially. What it means, practically, is that if the occupants of an accelerating vehicle can be properly located relative to a UFO-acceleration field (force

field), the effect of UFO-maneuver acceleration on the occupants can be neutralized.

From a practical standpoint it may be desirable to adjust the neutralizing field to neutralize most but not all of the vehicle acceleration. Then the operators would know by the feel approximately what was going on when not looking at the accelerometers or speedometer. If the vehicle accelerates in a straight line, the vehicle acceleration is space-wise constant. Since the neutralizing fields have an acceleration which is a space variable, there may be some feel to the acceleration in any case. [264]

In Paul Hill's portrayal of the "partial" field, the abductees would feel like they are floating, as in the case of Anna Jamerson, but have enough weight (acceleration) to maintain contact with the floor in order to actually move about the craft.

In addition to the underwater experience quoted previously, Ms. Jamerson also had an encounter with another form of water, fog:

> The next memory was that of our being on the beach together, after school one evening. I must have been in junior high school at the time. It was not unusual for the whole family to go to the beach after school, have a quick swim, and then have a picnic supper on the beach. It was unusual that just my mother and I had gone. Everyone else must have been busy doing other things. A storm came up, wind and rain forcing all the other swimmers from the ocean. My mother and I continued walking along the beach and noticed a fog bank coming in off the ocean.
>
> The next memory I had is of being enveloped by the fog and then being on a craft. Two gray beings took my mother away, telling me that she had other duties to perform. [265]

The fog, I believe, was steam, created when the craft's field made contact with the water. I have collected many eyewitness accounts that suggest the field is hot. There have been several cases where the weather was clear, yet a UFO came out of the only batch of fog on the whole visible ocean.

Returning to our discussion of the tunnel, we have a description of it put another way by abductee K. Wilson in Part Seven of her 10-part monograph, which is titled *The Alien Jigsaw*:

> **07-??-1995:**
> Now, suddenly, we have banked sharply to the right and we're making a hairpin turn—going back in the same direction from which we just came. Now we're heading downward at a very steep angle. We're moving very fast—very fast—swoosh! Now there is water around us—I see a tube [another word for tunnel]. We must be in a tube that is open to the top. Like one of those rides at a fair. The water all around us is splashing upward and it has a yellow tint to it. This is weird. [266]

So once again the field reminds the witness of a tunnel. The entrance to this "tunnel" is not visible because surface water is being pushed away and around the craft by its rotating force field. The yellow tint of the water brings to mind the Shag Harbour case with its yellow foam (Chapter 18). She continues:

> This is interesting because I didn't get wet, so the open top of the tube craft must have been glass or something see-through. Even more interesting is that I was taken to a naval ship out in the ocean. I'm positive about what I saw and I'm positive it was a naval vessel. (I've been able to visit a couple of naval ships docked in Norfolk, Virginia.) [267]

The witness is looking through the cupola above her as the water is coming up the sides of the UFO and turning back into it, as depicted in sketch #35. After this sketch, Wilson goes on to say:

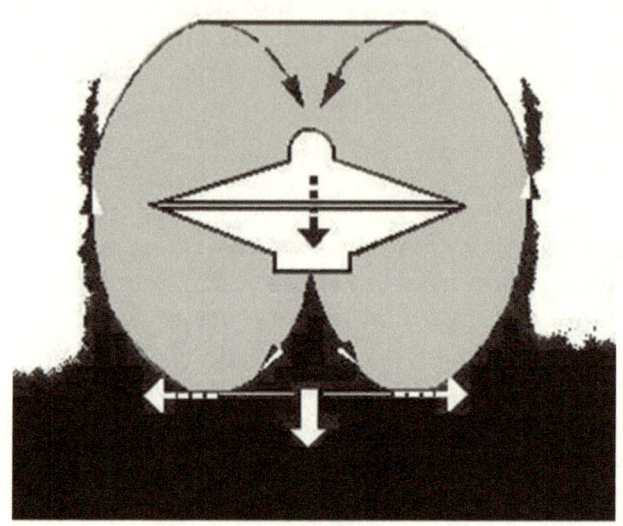

Sketch #35

08-??-2006:
I looked upward and could see out to the outside of the craft and saw that it was either raining very, very hard or we were underwater. There was actually so much water I could not see out of the window of the craft. **There were a lot of bubbles and the "swooshing" of water over the windows [as we] moved downward**. [See sketch #35] We were being returned to the places they had taken each of us from. That was my last memory of what I believe happened to me during the (at least) one hour of missing time that day. [268]

The field's rotation is made clear by the movement of the water and, here, also by bubbles, which may be caused by the heat of the field or cavitation.

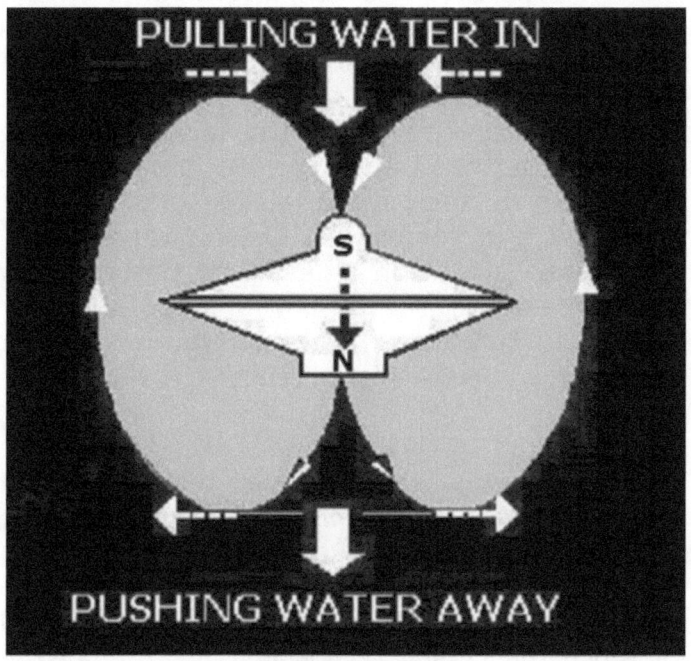

Sketch #36
Pushing away as it enters—Pulling in as it goes under

In *UFO Contact from Undersea*, Dr. Sanchez-Ocejo conducts a lengthy hypnotic session with witness Filiberto Cardenas, who was abducted in Miami, Florida, on January 3, 1979. From the doctor's summation:

> In an attempt to further establish the underwater "tunnel" mentioned by the witness, it was learned that this "tunnel" was not of solid rock as we think of a tunnel, and it did not have any projecting stalactites or stalagmites as already mentioned by the witness...**He could not describe its entrance**

opening in any frame of reference known to him.... [269]

Here again the subject is not aware he is entering this "tunnel" because, although the craft has yet to submerge, its rotating field has reached the surface and, through friction, is pulling the water up past the craft. To visualize this, it would be best to go back to sketch #35. Dr. Sanchez-Ocejo added: "...he [the abductee] had difficulty describing the walls. They were not of solid finished material like brick or stone, but were translucent and flowing like firmed water." [270]

"Firmed water," like "soft sides," is an attempt by these abductees to describe something only someone in their experience could have seen: ocean water held back and rotated around by a UFO's protective field. The doctor continues:

> The ships flew in an "airspace" that opened in front of the craft as they proceeded and may have closed again behind it, and no water or anything else actually touched the surfaces of the vehicle itself. The swiftly flowing "water" was close but did not touch the crystal (windshield) in front of the witness's face. It appears to us that the witness is describing a ship-generated force field of some kind which separated the mass of the material ahead of it and produced a capsule of "space" around it in which the craft flew. [271]

Again we have the abductee's view of the immersion into water in *The Andreasson Affair, Phase Two* by Raymond E. Fowler. The abductee's full name is Betty Andreasson Luca:

FF-??-1950:
Finally, Fred proceeded to ask her what she could see through the canopy after the craft dived into a body of water. Through this traumatic segment, Betty had to be calmed several times.

BETTY It's going so fast, oh! It feels like I'm staying still. I'm tilted a little. Oh, there's water coming up! Ooooooh! Ooooh!

FRED What's the matter?

BETTY Ooooh! I'm crashing into that.

FRED No, just take it easy. Take it easy. You'll be fine. [After Betty calmed down] Okay, go on.

BETTY It's like the water is rolling around and around, and it's all white up there. Round and round, and it's stopping now and it's just water. But now I'm going the opposite way. Ooooh! [272]

What has been happening to the witnesses here is essentially what was explained in Chapters 3 and 4 except that now it is being seen from the *inside* of the UFO by the abductees. At first the field just touches the surface of the water, moving it away from the craft. The field is invisible so the abductees are surprised because they think that they have a few seconds before hitting the water, but it is suddenly around them forming a tunnel shape. In K. Wilson's case, the top of the UFO is not yet submerged so that it is interpreted as an opening in the tunnel (Sketch #35), but soon she assumes that it must be raining heavily and that she is protected by an unseen clear glass or plastic. Then, due to the amount of water she observes, she senses that she must be underwater. Sketch #36 shows the field both pushing the water away as it does in its descent, and then, as the craft becomes totally submerged, the scenario changes as the water is being pulled to the rear of the craft, back into the center of it at the top. As the field displaces the water initially, it then replaces what has been displaced by pulling it back in. Of course, this is a continuous procedure.

Another incident involving submergence of a craft occurred in Gulf Breeze, Florida, where a wave of UFO sightings began in late 1987. Living there at the time was Ed Walters, a prominent building contractor and devoted family man who had a difficult time with his many UFO experiences. Because of his job, he always had a camera handy to record his work's progress, and as a result of this, he was able to take pictures of multiple UFOs, this when some people might be lucky to get even a single photo of one. These photographs became the focus of much of the attention in this case. Needless to say, his credibility was questioned as was the authenticity of his photographs, so MUFON (Mutual UFO Network) was called upon to investigate his photos and the reports of his alien encounters. Besides the local Florida case investigators, two eminent people in ufology were also pressed into service: Dr. Bruce Maccabee to analyze the photos and Budd Hopkins to examine Walters' claim that the UFOs were pursuing him.

I was initially interested in one of the photos taken by Walters of a UFO sucking up water on the Santa Rosa Sound near Pensacola Bay which appeared in the book *UFOs Are Real: Here's the Proof*, co-authored by Walters and Dr. Bruce Maccabee (See Sketch #13 in Chapter 2.). However, after finishing my first edition of *UFOs and Water*, I read another book by Walters, this one co-authored with his wife, Frances, and was startled to find that he was probably an abductee himself. The following excerpt from this text, *The Gulf Breeze Sightings*, was the trigger that set me on a course to study this publication more thoroughly:

> The investigators asked if there was anything I could remember that seemed strange. Any strange recurring dream? In other words, was it possible that I had been abducted in the past? The idea they suggested was that the UFO could block your memory of an abduction and the event could only be remembered under regressive hypnotism.
>
> I told them of a dream I'd had several times, but I made light of it, as if I thought it were nothing. Actually this dream was very vivid and occurred at least once a week. The dream would begin with me rising high in the sky and looking over a coastline. I

could see the sandy beach with waves breaking on the shore. Sometimes I would recognize the beach but most often not. **Then I would quickly descend and pass beneath the water into the ocean. I would gasp for air, in fear of drowning, but as I went deeper and deeper, I realized I was inside a container with a large diamond-shaped window. Through the window I could see the water and fish. Shortly thereafter I saw a lot of bubbles passing in front of the window, followed by rising sand, which soon completely covered the glass.** That's all I remember of it.

I didn't tell them all the details or the frequency of the dream. Nor did I tell them of the other times in my life when an unexplained loss of time had occurred. I didn't want to encourage the MUFON investigators to think anything other than what I wanted to be true—that last night had ended the phenomena. [273]

Once again, we have an account of immersion showing the rotation of the UFO's field upon entering the water, and even more interesting...rising sand or silt! Yes, the UFO went right to the bottom of whatever body of water he was taken to and was drawing up the sand/silt from the bottom—past the window. Further text about the UFO's operation in the water is given later in the book, and the missing time he mentions definitely puts this case into the abduction category:

When I was thirty-three[11] years old, we had only recently returned from living in Costa Rica and set up housekeeping in Corpus Christi. I went out on a solo fishing/canoeing trip, expecting to spend the day relaxing. If I managed to catch a fish or two, fine.

---

11 In an e-mail from Dr. Bruce Maccabee dated August 14, 2011, he told me that Walters was born in 1947, so 1947 + 33 = 1980 (approximate date).

The wildlife along the coastal barrier was spectacular. I should have taken a camera instead of a fishing pole. The water cut cleanly beneath the bow of the canoe as I slowly passed hundreds of brown pelicans nesting in and around the trees of the protected islands. The waterway I was in was deep, about fifty feet, and quite often I could see porpoises breaking the surface of the water.

At about noon I was drifting with the current and decided to see what Frances had fixed me for lunch. As I reached into the bag, the bottom of the canoe hit something, then was suddenly still, as if I had run aground. But the sound was like metal on metal. Before I could lift my paddle, the canoe drifted free and continued on.

I looked back, but saw nothing in the clear water that could have snagged me. Turning around, **I saw a stream of bubbles stretched out in front of the canoe, racing ahead of me**. My attention was no longer on the banana nut bread in my hand, but was glued to the strange bubbles. This was no porpoise.

As the bubbles burst, I could smell the scent of chlorine. The current continued to push me forward, but the bubbles had stopped in my path. I paddled to the left to avoid them. The bubbles moved left. I veered to the right. The bubbles moved right.

Only thirty feet away I could make out a green glow beneath the water. Very big and getting bigger. [Here again we see the ionization of the water by the UFOs' field. Green coloring of the field underwater is common in many cases. And we continue:] I tried to back-paddle. The current carried me forward, closer to whatever was rising ahead of me.

What seemed like moments later I found myself sprawled in the bottom of the motionless canoe. I looked up to find I was beached near the mouth of the channel, miles from where I had last been. A barge passed by into the Gulf. My food had flies on it, and **my watch read five o'clock.** [274]

Walters mentions that he could not account for the long period of time from "about noon" to when his "watch read five o'clock," which, in this case, is a good indication of a probable abduction.

So here, through the last five abductee accounts, we have seen the operation of the field, not in a scientific manner, but through the abductee's eyes, where we get a momentary description of the abductee's own inspection of the unknown through the windows of the craft. The scientist trying to examine this new phenomenon for the first time has only limited access via the testimony of the abductee. Additionally, abductees may be able observe the appearance or construction of the craft and its interior, including the strange lighting in the rooms which they have reported. They may have access to the instrument panels, which in several cases they are instructed to use; the table(s) where medical examinations or procedures are performed; and even the aliens themselves. On the other hand, though, scientists cannot observe any of these things firsthand. They know only what they can learn from the witnesses.

5.6   Abductees and Underwater Bases

The idea of the existence of alien underwater bases has been broached several times in ufological books, journals and magazines. The premise of bases was justified only because UFOs were seen going into or coming out of the water in particular locations. Not many authors ventured more than a supposition about what was happening outside the witnesses' field of vision, and examples of such assumptions (in bold print) follow:

01-10-1958 *Atlantic Ocean off the coast of Curitiba, Brazil*

The quantity of UFO and USO sightings off the Argentine coast **suggests** that **the idea of a submarine UFO base** in the South Atlantic may not be too crazy after all. There has also been some activity off the Brazilian coast further north. [275]

03-25/26-1973 *Caribbean Sea off the coast of Venezuela*

Discussing the widely held **belief about underwater bases**, the press interviewed a Sr. Julian Hidalgo Montezuma, a night watchman, who said he recently saw UFOs over Macuto and Naiguatá. He said he had seen several "orange lights," tubular in shape, "like our rockets," rise up out of the water far out at sea and vanish into the sky. On other occasions he had also watched them emerge and then fly off eastwards along the Venezuelan coast. Sometimes he has seen them emerge several times in one night, and, like many others in Venezuela, he **advanced the view** that **there might be a vast submerged "mother-ship"** out there. [276]

03-02-1975   *San Matías Gulf off Valdés Pen., Argentina*

Many UFO investigators and students of the phenomenon feel that much evidence points to the **possibility of a submarine base** for UFO operations somewhere beneath the waters of one of these deep gulfs. The water here is over 500 feet deep in places. The area is sparsely populated and inhospitable to humans and is well protected naturally from disturbances of civilization. [277]

03-28-1976   *Adriatic Sea off Porto San Giorgio, Italy*

Numerous reports, both from witnesses on land and fishermen at sea, have told of a very bright, oval-shaped object which emerged from the sea. Then the object gained altitude and headed quickly inland. **According to the**

press, the sighting was the confirmation of the presence of a mysterious **alien base submerged in the waters** facing San Benedetto del Tronto. [278]

03-05-1979   *Atlantic Ocean near the Canary Islands*

During the following weeks, the Spanish Air Force conducted extensive investigations, which was confirmed by the Commander of the Second Division of the Air Force Staff. Llopis was also interrogated and he said that he was "quite **convinced that it had been a UFO, which had its base under the ocean** near the Canary Islands." But this time the Air Force kept quiet about the results of the investigation. [279]

12-05-1984   *Off the island of Elba, Italy*

Only a brief report was made by two pilots of a civilian jet airliner (flight IG-635) coming from Pisa and going to Cagliari regarding the observation of a large, burning object which was seen going down toward the sea. This sighting was also thought to be related to the tragedy of the in-flight explosion of an Itavia [Airlines] DC-9 going to Ustica on June 27, 1980, in a sort of ufological **hypothesis** about the Ustica massacre with so much about an **alien base under the Tyrrhenian Sea**. [280]

02-09-1999   *Pacific Ocean 180 miles south of Lima, Peru*

On Thursday, February 11, people in the cafes around Pisco's central plaza were talking nonstop about the second UFO display the previous night. **Rumors** spread about the **presence of an undersea base** a dozen kilometers offshore. Some said it was an alien base; others claimed that it was "an OTAN (Spanish acronym for NATO) secret base."

According to *UFO Roundup* correspondent Pilar Valencia, Lima's "**tabloid press went wild** on days 1 and 2 of the flap," printing the Pisco **undersea base story** and

speculating that the UFOs were in Peru to observe "the end of the world." [281]

Of course, the above cases represent the opinions of witnesses, reporters, and researchers. However, the more interesting text comes from those who have *been there*. This is not the normal experience of people who sight UFOs, but of those who have been invited into these craft or, much more frequently, abducted by the occupants of these craft. Unfortunately, I am not an abduction expert or researcher, but I have read a great deal of abduction literature written by those in the field who concentrate their efforts in that direction. From the few water stories available to me, the idea of an alien underwater base has become a possibility.

In this day and age our Moon is figuratively very close to us due to technological advances, yet we would not consider it a place for long-term habitation due to its lack of air and water. However, it is a point that we could use to continue on to other spheres in stages, like connecting dots, until a habitable planet is reached. Obviously as soon as we have developed the *Star Trek* technology to determine the suitability of each planet as we pass it, the sooner we can dispense with the continuous stopovers as we expand our reach within the universe. Nonetheless, space travel is still a long and time-consuming trek. So how about stopping at a planet along the way to rest and perhaps become a temporary tourist, observing the life forms there while taking a break from our routine? This place must be where the native intelligent life forms have little or no access to us. On planet Earth, such a place could be the oceans, where humans would have to deal with aggressive life forms and conditions that would make such areas uninhabitable. Would we construct a major city there? I think not, as other stopover points might be more suitable. In that case, all that would be necessary is a cheap motel or military-style base in which to rest, and return to, from our exploration of their n^{th} rock from their sun.

So here we have the possibility that our planet may be the wayfarers' stopover point instead of a potential place of cohabitation. There is no mystic city on Earth, such as in the movie *The Abyss* (1989). Merely a curiosity en route, our planet is just a trip to the zoo on the way through town.

To illustrate this point, I offer the following excerpts from some of the abductees' tales. The problem with some of these accounts is that many of them are not fully explored because the researcher or psychiatrist using hypnotic regression is not acquainted with the overall field of possibilities, as was illustrated in the Jamerson case in *Connections*, and therefore allows pieces of information to slip by as meaningless. My collection is small, so I offer it only as a path upon which future researchers can expand.

> **01-03-1979:**
> The curve in the tunnel mentioned may have been simply a redirecting of the force field for a change of vector that was controlled by the extraterrestrial pilot sitting to the right of the witness in this small vehicle. This rush of **water not touching the ship** has been described by another UFO abductee who was **taken in a spacecraft to an underwater base.** [282]

In this last sentence, we have what Ivan T. Sanderson is alluding to in his book *Invisible Residents* as an underwater alien civilization. I do not feel this is a correct portrayal of the aliens' presence in underground or underwater bases. Unlike the situation posed by the movie *The Abyss,* bases described by abductees do not usually have the impression of a city, while the interior of some alien crafts can be an entirely different setting and has been described as enormous and similar to a small city or mall. At the bases, however, there are accounts of a more militaristic Spartanism, even the presence of military personnel, whether it is ours, theirs, or even theirs posing as ours. I would suggest that these installations are used as jumping-off points or repair stations en route to various parts of the universe. The advantage of a base on a planet where meteors burn up in its atmosphere is obvious. There has been a lot of discussion about building a base on the Moon as an intermediate point on the way to other planets, but the unprotected moon, as close as it is, would be extremely dangerous, even if the bases were underground. So, the debate continues.

The fact that stations can be built both underground and underwater is presented in Richard Sauder's books: *Underground Bases and Tunnels* [283] and *Underwater and Underground Bases*. [284]

The photocopy which follows is a U.S. Navy document from 1966 that talks about the construction of military bases beneath the sea floor in the middle of the ocean. These would be actual bases built by humans. We have the technology to do this. Therefore, it would seem an easily accomplished construction project for the talents of an advanced extraterrestrial civilization.

NOTS TP 4162

Vol 1

47966

MANNED UNDERSEA STRUCTURES – THE ROCK-SITE CONCEPT

by

C. F. Austin

3 1788 00040 4999

Research Department

ABSTRACT. Large undersea installations with a shirt-sleeve environment have existed under the continental shelves for many decades. The technology now exists, using off-the-shelf petroleum, mining, submarine, and nuclear equipment, to establish permanent manned installations within the sea floor that do not have any air umbilical or other connection with the land or water surface, yet maintain a normal one-atmosphere environment within. This presentation briefly reviews the past and present in-the-sea-floor mineral industry. The methods presently practical for direct access to and from permanent in-the-sea-floor installations are outlined, and the specific operations and types of tools indicated. Initial power requirements and cost estimates are included.

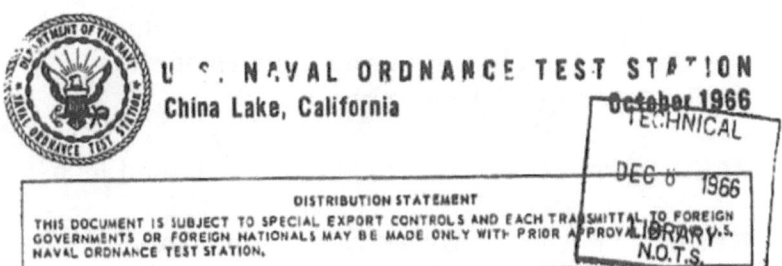

U. S. NAVAL ORDNANCE TEST STATION
China Lake, California

October 1966

TECHNICAL

DEC 6 1966

LIBRARY U.S. N.O.T.S.

Photocopy from *Underwater and Underground Bases* 285

Bases are one of the things that only abductees have had the privilege of seeing, so, drawing from my minuscule number of abductee reports related to water, let us continue with a few more cases. The first one, from the *Andreasson Affair, Phase Two*, is a direct continuation of what Betty was expressing in a preceding excerpt (pp. 194-195) pertaining to the water movement. However, as she continues, we see a different tunnel from the one which is formed by the craft in the water. Here she exits and discovers new topography:

> **FF-??-1950:**
> If Betty's latest hypnotically relived experience had any basis in physical reality, where had she been taken and why? The only known planet in our solar system with large bodies of liquid water is our own planet earth. Because of the apparently short time that elapsed between Betty's kidnapping and the alien craft entering the water, it's hard to conceive how it could have traveled anywhere except to one of our own oceans.
>
> **FRED**   Okay, just relax, relax, relax. You're doing fine. Don't be scared. Just go on. [Betty again calms down]
> **BETTY** **That whole window is water, like in water.** I don't know how to explain it.
>
> At this juncture, Fred had Betty, without affecting her trance, open her eyes and draw what she was seeing. Then he had her continue.
>
> **FRED**   Okay, close your eyes. Just relax and let's go on.
> **BETTY** And it looks like we're coming above some water. **We are out of that water**, and we're into some place that looks like ice all around.

LARRY Where are you looking now, Betty?

BETTY Out of the window.

FRED You see through this window. Can you see the outside from this window?

BETTY That's what I think I'm seeing.

FRED Do you see any clouds?

BETTY No, it's just a big, big—**looks like a big cave or tunnel of ice** with icicles all over, but there's a light around it. [286]

I will jump ahead chronologically here in order to bring in a case from *The World's Greatest UFO Mysteries* by Nigel Blundell and Roger Boar that might have some relevance to the one above:

11-??-1980:
The rat figures communicated without speaking, but she said one of them did talk to her. **"He said they came from Antarctica," she recalled. "There is a tunnel that goes under the South Pole, that's why they come out of the water.** Others are extraterrestrials." [287]

An unknown ice tunnel in the Andreasson case and again a tunnel at a specific geographical point—the South Pole. And, from *UFO Contact from Undersea*, we continue:

LL-??-1972:
And then...this is the way they went...and they came this far...in...to a rock formation under water...got this close...and when it was close to the wall here, he used exactly the same beam and **the water was displaced between the rock and the wall and an opening appeared and we went in and it closed back again** and we found ourselves in a very deep cavern...and this was very...there was no water there...just a cavern...and they propelled the ship onto an area...and I could see no less than 25

or more of the same (ships)...Other similar ships... smaller ones and a real big one...but many, many small ones and many similar ones. And we stepped and walked on solid ground...and when that took place, it was just like being home...and we then started talking. [288]

And again from *UFO Contact from Undersea*:

01-03-1979:
"He is told that they currently are in the pyramid in the Pacific Ocean; the others are in the Atlantic Ocean and one 'deep in the earth.'" [289]

Here, in *Connections*, abductee Beth Collings is joined by researcher Budd Hopkins:

UNDATED:

Budd: Well, you told me a lot about this dream. There's something about—there is something about....soldiers? What's this about soldiers?

Beth: They're not soldiers.

Budd: What are they?

Beth: Guards.

Budd: How do you meet these guards?

Beth: At the door.

Budd: What's the door for?

Beth: It's in the tunnel.

Budd: **Where's this tunnel?**

Beth: [Sharply:] **Underground! We're underground.** I don't like this. This is terrible! It's cold as shit down here! Oh!

Budd: How'd you get there, underground?

Beth: I don't know. We're down underground. We're down here and Anna's not happy. She is definitely not a happy camper!

Budd: What's happening underground?

Beth: We have to go into this big room. It's like a hangar or something. Big, huge hangar. [Sighs, voice shakes:] Oh-h-h, boy! Wow! I don't think Anna thinks this is neat at all. Oh, well....A guy puts something on me. [290]

In *The Alien Jigsaw*, K. Wilson shares the following:

04-29-1992:
This base was next to a large body of water. There was an outside portion of the base, but most of it was underground. The outside or above ground portion consisted of two huge warehouse-hangar type buildings. They were beige in color and had no windows. The area around these buildings was concrete. Not much to look at from the outside with the exception of there being seven to eight large cigar-shaped craft hovering a foot or two above the concrete ground...I suspect these cigar-shaped craft can travel underwater...

There are a lot of military personnel and others around me. I remember being here before. This is some type of military base, but I'm not sure which one...Probably Navy, but I'm not sure if this is our military...We are now walking inside a structure. I think we are in a huge, underground structure. There is a place under here where water can come in. It is like a man-made river system about twenty feet wide. It is to my right. There are some military people here and they are telling us about a newly designed craft. [291]

K. Wilson goes on to relate:

06-17-1992:

I had another encounter involving a cigar-shaped craft...These craft have amazing aerodynamic abilities. They can withstand super high speeds and do rolls, sharp banks, loops, you name it. They also have cloaking abilities....

I noticed that we were hovering and getting ready to land on a huge object in the ocean. At the time I was living in Oregon and believed it to be the Pacific, but who knows how far we actually traveled. I believe this object we landed on was a known (to them) landing platform. It may have been an enormous craft that was in the water and all I could see was the top of it or it may have been just a flat surface. [292]

This said cloaking ability was well-presented at the July 2005 MUFON International Symposium in Denver, Colorado, by Elaine Douglass in her "Invisible UFOs: Case Reports and Videos Suggest the Reality."

5.7 Amazing the Information We Have at Hand. If Only We Would Listen...

In conclusion, I submit that an abductee's presence onboard an extraterrestrial craft is as much a benefit to us as a spy in a foreign government is to one of our covert agencies. Here we have one of ours on their craft reporting all that can be seen, both in the craft and outside the windows, not to mention their bases on this planet. Each bit of information is what Leonard G. Cramp (*UFOs and Anti-Gravity*) calls a "piece for a jig-saw" [293] puzzle. However, a word of caution: When an abductee is telling his or her story in a regressive hypnotic session, it would be of benefit if ufologists who are familiar with the area being revealed could be consulted to extract the greatest amount of detail in their realm of knowledge.

The excerpts presented from these abduction cases are only a minute part of the total story. I heartily recommend reading the books mentioned, as all are quite interesting and pertinent to the overall abduction story.

Chapter 6

Radar and Sonar

RAdio **D**etecting **A**nd **R**anging
and
SOund **N**avigation **A**nd **R**anging

6.1   Show Us Your Instruments' Documentation

Science wants instrument readings to prove there are such things as UFOs, but, since we are not informed by the UFO pilots of their intended destinations, it is virtually impossible to have instruments in place to record them. Even when one has the ability to use instruments, such as radar or sonar, conceived by scientists and designed by engineers, and a UFO is tracked by one or both of them, the cry goes up that the instruments must be faulty. An example is the 1952 Washington, D.C. sightings where three separate radars picked up the same targets. After the radars were checked and found to be operating normally, in came a scientist who was on the government's payroll who claimed that "temperature inversion" was the culprit. However, after the media furor was over, this was contradicted by yet another government agency, the weather bureau, which said that a temperature inversion could not have been the reason for the sightings. Naturally, this made the last pages of the newspapers but has since been proven true.

We cannot analyze the skin of a UFO or its internal structure, nor can we comprehend its engine. However, we can measure a UFO's speed, which in many cases is amazing, as well as its depth in the ocean, which shows the craft's remarkable strength and speed in such a dense medium. Unfortunately, the military does not want to share its information with us, so we, as civilian researchers, need access to sophisticated radar and sonar equipment like the military possesses in order to gather this information.

This chapter is for the accumulation of cases that deal with radar or sonar so that we can see from them what to expect in instrument readings. Admittedly, the greatest number of cases will come from radar inasmuch as it encompasses aerial UFOs as well as those on the water's surface.

6.2   Radar

Definition of **radar**:

> Any electronic system used to detect and locate objects at distances and under conditions of lighting or obscuration that would render the unaided eye useless. It also provides a means for measuring precisely the distance, or range, to an object and the speed at which the object is moving toward or away from the observing unit. [294]

6.2 a.   Radar Only (operating and detecting)

Radar and sonar present a problem in that they do and do not pick up a UFO that is otherwise seen by crew members on ships. If we knew why this occurred, it might be yet another piece of the puzzle in understanding the operation or composition of the field surrounding the craft. Furthermore, there are several instances where radar is rendered useless due to the close proximity of a UFO. So this brings up the consideration that the power generated by the field is not only harmful to biological entities but also interferes with instruments. Our first and most important radar cases are those that "see" the object:

02-10-1951   *Atlantic Ocean off Newfoundland, Canada*

"I found out a few months later that Gander **radar did track the object** in excess of 1800 mph. I did not see the

213

reports made by other members aboard the aircraft. I did talk to the Air Force at Wright-Patterson AFB in 1957 but did not look at the report. **They said they had it and many similar reports**." [295]

11-??-1954 *Strait of Florida between Cuba and Florida, USA*

Ten days after the sighting, the film was shown to the Cuban military High Commands at the Cuban Navy building. Our witness was present and confirmed to us that also the **radar** of the frigates **detected the 3 objects** when the UFOs appeared on the horizon until they disappeared in the west. [296]

11-05-1957 *Gulf of Mexico, south of New Orleans, LA, USA*

At 5:10 that morning (November 5), veteran **radar men** on the *Sebago* had **picked up a strange flying object**. According to Commander Waring, the UFO had raced around the cutter for ten minutes. Tracking it constantly, the radar men had seen it stop in midair, then accelerate swiftly. At one point, its speed was almost 1,000 mph. [297]

04-04-1959   *Indian Ocean*

A similar luminous spot appeared abaft [toward the stern; back; behind] the beam and a third one of lesser intensity was observed abeam [to or at the side of a boat]. **Radar showed four circular targets** at 20 miles and 3 larger ones at 40 miles. The observation lasted from 1810 to 1930 G.M.T., and only one lightning bolt was observed above the 3 luminous sources. [298]

11-20-1963   *Off the eastern coast of Scotland, UK*

"Capt. Murray alerted Stonehaven radio, put his vessel about and made for the spot where the light had vanished. The collier had **two radar contacts** on her screens, but when she reached within a quarter of a mile of them, they disappeared. The *Thrift* searched for three hours, circling the area several times and was joined by lifeboats from Aberdeen and Gourdon, a Shackleton from RAF Kinloss, which dropped flares on to the surface, and a B.P. transporter. They discovered no traces of wreckage, however." [299]

??-??-1964   *625 miles east of Bermuda*

On this night in 1964, the lookout, whose duty it was to scan the sea for anything and everything, called down to the wheelhouse through the nearby voice-tube that he had spotted a "light on the horizon." The **radar** officer below reported a clear screen, no bogies.

Sometime later, the lookout once again called down that it was [a] ship. Moments passed and he repeated his statement, adding that it was getting closer.

Still, **radar** wasn't tracking anything....

"Later, **the CIC (Combat Information Center) guys said they tracked it** at 1,800 mph as it went out. There was nothing in 1964 that could go that fast," he says, adding he heard later that the ship's captain had radioed the sighting to a base on Bermuda, and three jets had been scrambled. [300]

01-12-1965   *Near Lynden, Washington, USA*

"The Air Force contacted me next day and after a thorough interview admitted that they had **located a UFO on radar** that night." [301]

08-04-1965   *St. Louis Bay, Duluth, MN, USA*

Nautical UFO. An object splashed down into St. Louis Bay. At about [the] same time, **7-10 UFOs were tracked by USAF + RCAF radar**, moving at 9000 mph, between 5,200' and 17,000'. USCG investigation. [302]

12-14-1966   *Porsangerfjorden, Norway*

The Norwegian cargo ship *Moder* **picked up a motionless airborne object on its radar screen** while in the area of the Porsan Fjord [sic—Porsangerfjord] in northern Norway on the 14th of December. Shortly after the **radar contact**, the ship was lit up by an unknown object which fell softly into the water at a distance of one hundred meters (about 328 feet). [303]

Note that the text reads "fell softly into the water," which I envision as a slow, controlled descent and not the normal fast entry. This is important to realize as we discuss the field's rotation and visibility to radar later in this chapter.

??-??-1970   *Atlantic Ocean near the Bermuda Triangle*

"...so at this time I was switched from the bridge watch to ah...I was put into the bridge and was handling the helm at the time. And the **radar man kept on coming in and giving reports** that 20 some odd miles ahead of us, **there was an object that was spotted**, and ah...what was odd about the object...it was 20 miles away, but was like about 20 feet off the water." [304]

04-23-1976   *Atlantic Ocean 700 miles SW of Bermuda*

Suddenly the destroyer emerged from the fog and the **radar** shack erupted with excitement. **A sudden large "blip" appeared on the scope**, and now half of the ship's complement had been awakened and was crawling the decks watching. [305]

??-??-1978 *Adriatic Sea off San Benedetto del Tronto, Italy*

The medical examiners in charge of the case estimated that the time of death was two nights earlier, and this coincided with the sighting of an object on the sea moving at very fast speed. It seems that some seamen witnessed the sighting. In addition, **there was a radar record** from an unknown military base or ship. [306]

6.2 b.   Radar Only (operating and NOT detecting)

We expect the instruments we build to function properly and consistently. However, next we encounter something that is not just a problem at sea but also in land cases: fully functional equipment that does not detect a known sighted UFO.

04-10-1977   *Mediterranean Sea off Israel*

The bright silver object was on bearing 082 degrees, sextant altitude 25 degrees. It remained steady for three to four minutes when it unexpectedly zoomed downward with a tremendous speed, continuously emitting the bluish-white light, heading directly toward the sea. The object reached the sea's surface at 9:45 and was on bearing 087 degrees. It remained on the surface for about two minutes still giving off its light. **The ship's radar was unable to pick up a return from the object**. [307]

11-09-1978   *Adriatic Sea off Silvi Marina, Italy*

The crew...saw a red sphere that rose from the water at high speed at an angle of about 45 degrees; the observation lasted only four or five seconds. The light remained stationary in the sky at an altitude of approximately 300-400 meters and then it went out, leaving a trail of bluish color in the sky. **The ship's onboard radar did not signal anything**, and in that section of the sea, the depth was only 23 meters. Only about two hours after the sighting, there were magnetic anomalies in the ship's onboard radar, but not during the observation of the light. [308]

??-??-1978/9   *Gulf of Oman near Iran*

Photography was attempted and failed. **The object was not picked up on ship's radar.** Several members watched the object for several hours through binoculars. [309]

07-26-1980   *Atlantic Ocean off the coast of Brazil*

Atilio Sacarpate is a machinist from Argentina residing in Natal who was a member of the crew of the *Toche* (sp?)-*Seahorse* that was leading the tugboat because it did not have radar. Being in command, Atilio saw the approaching of a light which was moving at high speed and was **not being captured by radar** and was not recorded in the naval charts as a lighthouse. The two ships were near Pititinga, from where the coast could be seen. [310]

02-18-1996   *Sea off Iceland*

I asked if the balls of light were connected to the triangles and he said, "No, they are totally separate and independent." The caller was asked if the triangles were showing on the ship's radar. He replied, "No, the **radar is not picking them up**; the whole crew are standing on the deck watching them." [311]

MM-??-1998   *Southeast of Coronado Island, CA, USA*

After visual confirmation, I notified the Operations Specialist that I had a visual contact on the water bearing at 090 degrees. I asked him if he could spot a contact on his **radar** screen, to which he replied, "No, **I have no contact at that bearing**." (Having spotted objects on numerous occasions during our many sojourns at sea, I had a pretty good sense of directional association.) I then repeated the bearing and asked him to check again to which he replied, "No, **I have nothing on my screen**." I then asked the aft lookout if he could see the contact I was referring to. He reported that he could see the object as clear as day and asked me for more information on it. [312]

6.2 c.   Radar Only (rendered inoperable)

In these cases we have not only the failure of the radar but also, in some other cases, failure of the compass and other electronic instruments as well. Usually this outage only lasts while the UFO is in the area near the instruments.

05-23-1968   *Atlantic Ocean near the Azores*

The USO matched several course and speed changes and rendered **compass, radar, and radio equipment inoperable**. [313]

EE-??-1968   *Unknown waters off Lüda/Dalian, China*

In early 1968, four coast artillerymen of the naval garrison at Luda, Liaoning Province in north China, discovered a golden luminous object of oval form that flew along leaving a thin trail in the air. It climbed steeply at an incredible speed. A short time later the object disappeared in the sky to the right of the fleet. The moment **the egg-shaped object began its ascent, all communications failed and so did the radar, almost causing an accident in the fleet**. [314]

10-24-1969   *Pacific 350 miles south of Valparaíso, Chile*

The incident involved a Chilean Naval destroyer and was witnessed by crew members and the commander of the vessel. Up to six UFOs, including one large object, were observed. The objects were **verified on radar** and observed visually. As the main object moved over the ship, **the vessel's power went out**. [315]

12-??-1977   *South Georgia Island near the Antarctic*

It hovered at the altitude of four to five kilometers. The trawler's **radar station was immediately rendered inoperative**. The object hovered over the area for three hours and then disappeared instantly. [316]

11-07-1978   *Adriatic Sea off Silvi Marina, Italy*

Other sources told of a red flash seen coming out of the sea. In any case, **the onboard radar stopped working** and the witness notified via radio a colleague aboard the fishing boat *Aquila*, who confirmed the **radar malfunction**. [317]

11-13-1978   *Adriatic Sea off Giulianova, Italy*

The radar had stopped functioning. At that point an object emerged from the sea. This was described by other fishermen as an opaque, pear-shaped ball. The object, which was green, disintegrated in the air with consequent luminous effects. [318]

04-10-1979   *Atlantic Ocean off New Bedford, MA, USA*

According to Marreiros, owner and skipper of the 78-foot vessel, **the encounter also disabled the vessel's Loran system** for about 15 minutes after the experience, and the trawler was **also unable to pick up the object on the vessel's radar** system throughout the time. [319]

07-05-1979   *Gulf of Alaska, south of Seward, AK, USA*

"...a super bright object came down through the clouds and became stationary close to the water. At the same time, **a target appeared on our radar** at about two miles distant, a really strong target. All of a sudden **the radar heading was knocked off of its setting** and one of the crew had to reset it. That just never happens. It was a bright glowing object with a bowl-like shape. It faded out and simultaneously disappeared from the radar screen, then **reappeared visually and on radar. The radar heading was knocked off when the object approached the boat**. The object was only a couple of feet off the water. The incident lasted for about five minutes and the radar was picking up a strong signal." [320]

SP-??-1967   *Golfo de Penas, Chile*

Admiral Jorge Martinez, Chilean Chief of Naval Operations

Chilean Chief of Naval Operations Says, "UFOs Are Real"

> SANTIAGO DE CHILE—Dr. Virgilio Sanchez-Ocejo reports, "The former Chilean Chief of Naval Operations, Admiral Jorge Martinez, declared, 'UFOs are real.'" The ex-CNO made the shocking statement during an interview with a Chilean television network. The interview was conducted by a journalist from Teletrece. The retired admiral admitted that he personally witnessed two UFOs at sea when he was a young lieutenant and saw a very luminous white object in front of his patrol boat. When he took command of his first destroyer, then Captain Martinez saw another similar **object emerging from the water, creating a strong interference with the navigation system and the ship's radar**. Guillermo Jimenez, another former Chilean officer, claimed that the sonar system onboard his ship detected two submarine-like objects that caused similar malfunctions with shipboard electronics. "They displayed the same size and metallic resonant characteristics that ordinary subs typically show; however, these objects were too fast to be submarines," said Jimenez. Both Jimenez and Martinez described how **the gyrocompass systems were out of control. "The radar displays went completely blank**

and the gyro was spinning very rapidly, as if a strong magnetic force was present." Some UFO investigators call this phenomenon "UUO" or unidentified underwater objects. Interviewer Rodrigo Ugarte told us that there were many other officers that described similar events but did not want to go on the air. [321]

6.2 d. Radar Can Illuminate Plankton

The following two stories should probably be part of "Water Wheels," found in Chapter 10, "Negative Aspects of Water Ufology." However, I am including them here due to the fact that radar creates *some* of these glowing mysteries and shows us how one thing affects another, which in turn makes us wonder if some part of a UFO can affect our radar in the manner previously discussed in *Radar Only (rendered inoperable)*.

11-30-1951   *Unknown position Persian Gulf*

But of far greater significance was **the appearance of the strange lighted sea waves in certain waters when a ship's radar was activated.**

The M.V. *British Premier* was in the southern part of the Persian Gulf on Nov. 30, 1951. The ship's master, Capt. F. G. Baker, made this report:

> ...The ship's **radar apparatus had been switched on** with a view to checking her position when in the same instant...the most brilliant, boomerang-shaped arcs of phosphorescent light appeared in the sea, gyrating in a clockwise direction to starboard and clockwise to port, but all sweeping inwards toward the ship from points situated from five to six points on either bow and some two miles distant, and conveying the impression that they ricocheted from each other on meeting at the ship's bows and then turned and travelled away astern to

similar points which were equidistant on either side and about four points on each quarter. Duration: 15 minutes. [322]

02-09-1953   *Malabar Coast off Mt. Delby, India*

Another case of radar triggering occurred 15 months later, 10 miles west of Mt. Delby, off the Malabar Coast. The **ship's radar was operating** that night of Feb. 9, 1953, when Capt. M. J. Paice of the S.S. *Strathmore* recorded the same ricocheting action of the light sweeps:

> Between 0130 and 0200 white patches of light were observed on the sea surface. Milky white patches were first noticed on the starboard beam about two cables away and appeared to "flash" about once every second. Later they moved closer to the ship, being as bright as a phosphorus patch, although there was no indication of phorphorescence [sic] in the water even when the ship's wake broke into patches. The patches had made different movements, each one continuing for a minute or so—rotary, clockwise and anti-clockwise—toward the ship in waves and away from it in waves parallel to the ship's course.
>
> During the entire time of observation, the period of reaching maximum brilliance and fading was about one second, giving a regular flashing appearance. At 0152 the waves reached their maximum brilliance, appearing to travel from the starboard quarter to the port bow. **On switching off the radar,** the phenomenon ceased abruptly close to the ship, but it was still faintly discernible on the port beam about two cables away. At 0157 the **radar was switched on again**, the phenomenon reappeared close to the ship but only faintly, and then disappeared altogether. **Nothing**

was observed on the radar screen during this time that was out of the ordinary.

Evidently the light-wheels respond to a ship's radar transmitter which can turn them on or off. But is it the nature or the *specific frequency* of the radar beam that activates the light beams? And, to take the theory one step further, is the radar energy activating the *Notciluca* [sic— *Noctiluca*] organisms in a time-phasing manner so as to form the visible beams? Because ships carry navigation radar, we can no longer assume that they encounter the light-wheels accidentally, not when their passage, with radar operating, can trigger the appearance of the phenomenon in this area. [323]

6.3   Sonar

Definition of **sonar**:

> Technique for detecting and determining the distance and direction to underwater objects by acoustic means. Sound waves emitted by or reflected from the object are detected by sonar apparatus and analyzed for the information they contain. The term sonar derives from the phrase "sound navigation and ranging." [324]

A newspaper story that appeared during World War Two is interesting. The men in the U.S. submarines who listened to underwater sounds while in the depths of the ocean were surprised to hear noises such as honks, beeps, purrs, grunts and a whole assortment of other sounds. These strange noises, creating strong underwater vibrations even when inaudible on the surface, were often confused with the hum of enemy propellers. Recognizing that strange underwater noises can be caused by fish (biologicals as used in the movie *The Hunt for Red October*, 1990), the Fish and Wildlife Service teamed up with the Navy and produced recordings of fish noises so that those who listened in submarines could be trained to know the difference between an ichthyologic burp and a Japanese propeller. It is nice to know we can detect small fishes and, obviously, large undersea submarines. Now how about those UFOs?

6.3 a.   *Sonar Only (operating and detecting)*

01~08-??-1947   *Unknown position in the Pacific Ocean*

...they mapped what they thought was an underwater mountain top or a solid object of some sort. It was very large and **they made several passes over it with their sonar**. My father was not in the sonar department. He was used sometimes as a helmsman and the captain's driver in port. I am not sure, but he may have been on watch or on the bridge—I don't know for sure. He says after several passes the object accelerated to a fast speed and disappeared into the depths and they lost it. [325]

04?-??-1952 *Atlantic Ocean, Cape Chidley, Labrador, Canada*

At this time **we both heard the sound of some type of very intense, high-pitched whirring,** like a runaway electric motor. Savino exclaimed, "Hey!" As he pointed into the ocean right in front of our position, what became apparent to both of us were two white lights in the water.... Suddenly to our left, having come up the ladder to our deck, appeared a Navy type. He yelled and asked us if we had seen anything in the water. When he got next to us, we pointed over the side and asked, "Is that what you're looking for?" He exclaimed, "Damned! Yes, that's it!" **We lost track of it awhile back, but still could pick up the sound of its engine**, so we knew it was very close...The bosun's mate looked at me and said, "Laddie, that's no submarine. It's been following us for the last two hours about 2 miles off our stern (the rear of the vessel)." [326]

The only way that the crew could have tracked the UFO underwater is by the use of sonar, but the ability to track said object was lost as it came closer to the ship because the UFO's engine noise became intermingled with that of the ship's.

06-01-1958   *Altafjord, Norway*

...a silent "unknown aircraft" with no identifying markers crashed into the Alta Fjord [Altafjord]....The frigate KNM *Arendal* and the submarine KNM *Sarpen*, along with divers, searched for the aircraft fruitlessly for over a week. The *Arendal*, however, **did get a sonar reading of a mobile object**. [327]

02-14~21-1960   *Golfo Nuevo, Argentina*

What was it all about? Was there really a submarine there? The tight-lipped navy obviously thought so. **Ships had picked up the "object" with sonar gear** three weeks ago, had tracked it into the Golfo Nuevo. Now they were determined to bring it to the surface and get a good look at it. [328]

??-??-1962   *Bass Strait, Australia*

Later, **radar** personnel advised they had **picked up a metallic object travelling underwater** which was quite large and the ship had given chase. It eventually disappeared off **radar** at over 100 knots. [329]

In the above case the use of the word "radar" is not correct. It should be "sonar" as the object is underwater.

??-??-1963   *Atlantic Ocean off the coast of Puerto Rico*

A **sonar** operator on one of the small vessels, otherwise listed as a destroyer, reported to his bridge that one of the submarines had broken formation and gone off in what appeared to be pursuit of some unknown object. This operator did not, of course, know if this was a "plant," since the maneuvers they were engaged in were exercises designed to train personnel in detection of enemy craft, and in such exercises, decoys must of course always be employed. However, this operator's report was not at all within the limits of any such simulation. Trouble was that said **unidentified subaqueous object** was traveling at "over 150 knots"! [330]

SM-??-1963 *Atlantic btwn. Puerto Rico & Key West, FL, USA*

Lorimer A. Grant in a letter describes an experience he had as a U.S. Naval officer in late summer 1963. He was on the USS *Sarsfield* sailing from Puerto Rico to Key West. Sailing at 24 knots, the **sonar picked up a large object at 300 to 400 feet underwater**. It seemed to be about 200 feet long and 40 feet wide. The ship circled the object. Checks were made by radio and they were informed that no known submarines were in the area. The object was radioed to surface and identify itself. When it did not, the ship circled and dropped concussion grenades on it. The object did not change its speed or course. [331]

03-??-1966   *Gulf of Tonkin, China-Vietnam*

The captain called in a helicopter from the carrier we were operating with. It hovered over the object and **lowered a sonar boom**. The pilot was on the loudspeaker of the bridge so we were hearing his description of the object. Silver, metallic sphere, no seams, marking, or rivets, just **an independent ball floating just below the surface**. At this point the pilot decided to have his door gunner open fire on the object. I was trying to film all of this with my 8mm movie camera. The distance was too much for my camera. [332]

11-21-1972 *Hermansverk, Norway*

At the same time other odd events were occurring. Aircraft experienced unexplained electronic problems. Yellow and green objects were seen flying along a mountainside. **Navy vessels registered sonar contact with something in deep water.** [333]

11-08-1978 *Adriatic Sea off Porto d'Ascoli, Italy*

...saw a red and yellow spherical light that came out of the sea. It seemed to go back in after completing a parabola in the sky. The light was moving continuously and slowly and did not reflect its own brightness on the sea. About one hour later, the **radar** of another fishing boat, the *Andrea Padre*, **reported something submerged** that was moving and that seemed to follow the boat. [334]

Note that the fishing boat reports a submerged object. Again, this had to be with the use of sonar, not radar, as the object is below the surface.

6.3 b. Sonar Only (operating and NOT detecting)

10-??-1978 *Zone of the Channels, Chile*

A sailor guarding his post onboard a Chilean Naval vessel saw a metallic object "sliding" just below the surface of the water next to the vessel, at a speed much faster than that of a submarine. According to the witness, the **sonar onboard the ship did not detect the object.** Another strange fact was that the craft did not appear to disturb the water in its path, moving smoothly without any sort of visible turbulence. [335]

6.4 Radar and Sonar

The account that follows is a well-written report by a well-respected ufologist. The text here is a small representation of the total case, and I would heartily recommend reading it in its entirety on my website.

02-28?-1963   *30-50 miles off Spitsbergen, Norway*

He told her he was present when an unidentified target was **picked up on his ship's radar and then tracked by sonar after it entered the water!**...Tom happened to be in charge of the shift at the time of the UFO incident. Besides himself, other personnel in the room comprised **three radarscope operators** and **two sonar operators.**....

Each of the three radarscopes in the room displayed a different height level in the atmosphere. At approximately 0315 hours, Preston recalled, a stationary "bleep" appeared abruptly on the highest-level scope. The target's vertical height was approximately 35,000 feet, and it was located somewhat west of the zenith (overhead point) at perhaps 70° elevation. The bleep indicated a seemingly hard, solid object giving off a strong reflection; the size of the target on the screen, according to the witness's best recollection, implied an actual diameter or length for the object of between that of a jet fighter and a 707—in other words, said Preston, roughly 100 to 120 feet across....

After a few minutes, Tom notified his senior officer who came into the radar room, looked at the target on the scope, and then withdrew. **The officer proceeded to radio the nearest ship to learn if it also "painted" the same target. It did. Thus, a radar set malfunction was ruled out.**....

As the target descended, the two sonar operators aimed their pulses in the general direction of the dropping object. Almost immediately (in a matter of seconds) following loss of radar contact, **both sonar operators received audible "pings,"** indicating a strong echo from a fast-moving submerged target at a range of probably 20,000 yards (roughly 10 miles)....

The underwater target appeared to be traveling in the same general azimuth and at the same descent angle (at least initially) as the airborne object, implying that the two unknowns were one and the same! The target's speed was considerably reduced, "down to hundreds of miles per hour" but "still moving damn fast," remarked Preston, and **it was now moving along a zigzag path away from the ship**....

The **image on radar** gave all the outward appearances of reacting to the jets' approach and then successfully **eluding further detection by submerging in the ocean** and eventually retreating from view. Another example of apparent intelligent behavior: The target appeared to follow the fleet's evasive "Z" maneuver. [336]

07-MM-1966   *Pacific between Seattle, WA and HI, USA*

He called **sonar**, who during the excitement had **reported contact underwater** at the same bearing. The captain announced into the 1MC, "This is the Captain. I have the conn." The reply came back instantly from the helm, "Aye, Aye sir." I knew that the helmsman was passing the word in the control room that the captain had personally taken control of the boat. I also knew that rumors were probably flying through the vessel.

The captain called down and ordered someone to **closely monitor the radar**. His command was instantly acknowledged. As the five of us stood gazing out over the sea, the same ship or one exactly like it rose slowly, turned in the air, tilted at an angle and then vanished. I saw the chief snapping pictures out of the corner of my eye.

This time I had three images from which to draw conclusions. It was a metal machine, of that there was no doubt whatsoever. It was intelligently controlled, of that I was equally sure. It was a dull color, kind of like pewter. There were no lights. There was no glow. I thought I had seen a row of what looked like portholes, but could not be certain. **Radar reported contact** at the same bearing and gave us a range of 3 nautical miles. The range was right on, as the craft had moved toward the general direction that we were headed. [337]

04-23-1976   *Atlantic Ocean 700 miles SW of Bermuda*

Hedison immediately checked with the ship's radar and sonar men. **Radar reported no "blip" on the scope and sonar reported no engine, crew nor mechanical underwater sound denoting a vessel in the water other than their own sound**. The lieutenant then viewed the radar screen for himself and concurred that there was indeed no visible "blip" to indicate anything at all in the vicinity. [338]

6.5   Now You See Them (on Radar or Sonar), Now You Don't

The next sketch (#38) came about because of the reversed field, which was discussed in Chapter 3. This brought about the need to try to discover a reason for the discrepancy of the instruments seeing and then not seeing the target. The inconsistency of identification through our electronic devices led me to consider one thing: The fault is not with our equipment, but with some quality of the field itself. Since this field is reversible, I tried to view the radar or sonar beam hitting the field and any resultant effects thereto.

In this composite sketch, I have vertically divided the craft into two sections, with the left side rotating as a craft coming out of the water seems to operate. In this drawing it is easy to see that the beam would be reflected back, as the rotating field is coming at the radar/sonar beam.

On the right-hand side, with the field in reverse, it is quite a different story. The radar or sonar beam is given a boost in the general direction that it was heading in the first place and therefore does *not* return to the receiving station. This gives the craft a mysterious cloak of invisibility, or what the military would call "stealth," making it invisible to radar and sonar, although it might still be seen visually. A couple of cases illustrating this concept follow:

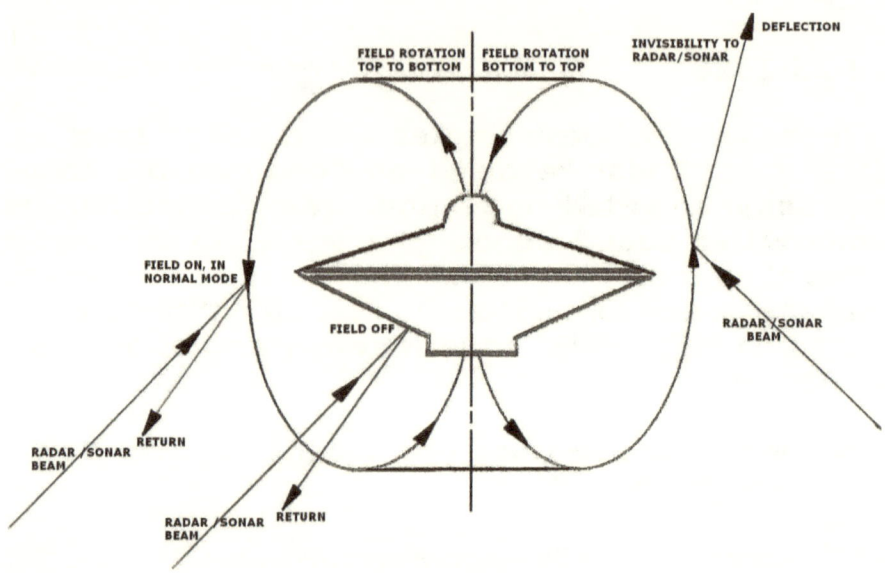

Sketch #38

03-15/16-1968   *East China Sea, southwest of Okinawa*

Suddenly I saw a huge light beneath the water moving rapidly from the northeast and closing the ship. I reported this to the officer of the deck through my headset. By this time the huge light had passed in front of the ship, and both the starboard and port lookouts confirmed my report. It was definitely round and appeared to be revolving. **The ship was not equipped with sonar detection, and radar saw nothing on the scope** beings [sic-being that] the object was deep in the water. The lookouts and the OOD [officer of the deck] continued watching the light as it moved with incredible speed toward the southwest. When it was nearly out of sight on the horizon, a bright light suddenly appeared above the ocean, and **radar** immediately picked up a blip from out of nowhere on the scope. [339]

A "huge light beneath the water" means that the field is on, ionizing the water and causing the illumination. On exiting the water, the field is in what I refer to as its normal mode (my concept of the field as it **exits** the water), as in Chapter 2, and reflecting the radar beam back to the receiver.

The next case is a repeat of one that we had in section *6.3 b. Sonar Only (operating and NOT detecting).* Note that the field is in reverse, geared for submergence, and the UFO is not disturbing the surface of the water, as was discussed in Chapter 3. Also, the sonar's beam is being deflected away from the ship's receiving equipment by the reversed field.

10-??-1978   *Zone of the Channels, Chile*

A sailor guarding his post onboard a Chilean Naval vessel saw a metallic object "sliding" just below the surface of the water next to the vessel, at a speed much faster than that of a submarine. According to the witness, the **sonar onboard the ship did not detect the object**. Another strange fact was that **the craft did not appear to disturb the water in its path, moving smoothly without any sort of visible turbulence**. [340]

Another consideration in these cases is that water disturbance due to a UFO may not be visible from a distance. Therefore, a craft with the field off could still gradually sink into the water without being noticed by a ship in the distance.

In future water-UFO cases, it would help to equate water reaction with the radar or sonar reports to verify the relationship of the field's rotational direction and whether radar or sonar is detecting the object.

The list of radar and sonar cases will continue to grow as we move into the next chapter which deals with aircraft carriers and other large ships.

Chapter 7

Aircraft Carriers and Other Large Ships

Among the water cases contained on my website, it is not unusual to find ships of one kind or another. However, the one type of ship that has a commonality with UFOs is the aircraft carrier. There are very large UFOs that carry their own fleet of scout craft, just as the carrier carries aircraft. Although there are very few accounts where witnesses onboard carriers have actually seen a UFO going into or coming out of water, I feel this category is important because, in several cases, physical influences, as seen on radar and sonar, electronic shutdowns, heat, and so forth are still evident. In this chapter, I would like to add what is either seen in relation to the ships or what is personally experienced by the crew members. The number of sightings on these major ships does show that the military, the Navy in particular, is aware of these objects. Radar and sonar cases concerning these aircraft carriers and large ships were excluded from Chapter 6 so that they could be included in this chapter as a group. In addition, some of the cases here mention military secrecy, which is actually the topic of the next chapter. Those comments will be enclosed in the following symbols: ► ◄.

I will start out with two World War Two sightings in which we can see the beginning of the military's recognition of UFOs. A recent book by my friend Keith Chester, called *Strange Company,* [341] is an in-depth study of the war years and what were called "Foo Fighters" in their day, many of which are now recognized as UFOs.

The term Foo Fighters was used to refer to unexplained phenomena seen in the skies over Europe and the Pacific Theater during World War Two. At first, Allied pilots thought they might be German secret weapons, but it was later discovered that the Germans also reported such strange phenomena. The Robertson Panel commissioned by the U.S. government in the early 1950s noted that many Foo Fighters would have been called flying saucers if that expression had been considered during the war.

Most of the cases referred to in Chester's book concern aircraft, but I will take you down to the surface of the water and show how the military's estimation has changed since those confusing days of combat on the seas in WW2. In those years, the recognition of UFOs was a learning process, as crew members of ships and planes explained to their disbelieving superiors the unbelievable things that they had seen either on their instruments or with their own eyes. With the end of the war and in view of the accumulated reports, I am sure that the upper levels of the military had become aware of UFOs. Also as we journey through the war years, we will see a change between the officers and crew regarding secrecy, the focus of the next chapter.

7.1   Fear in Wartime of an Unknown Enemy

08-05-1942   *Pacific near Guadalcanal*

Two days before the WWII Pacific Guadalcanal invasion by American troops to retake this Japanese occupied island, an unforgettable and astounding event involved the nearby U.S. Navy Fleet. One Navy witness, a chief at the time aboard the destroyer USS *Helm* #388, had an excellent observation of an incredible encounter with an unknown, unidentified intruder. At 10:00 a.m. **the fleet received a radar report from one of the cruisers, and a little later a visual sighting of the object was made from their destroyer.**

The object was approaching the fleet on the wrong radio beam, which changed daily. Since the object was coming in on the wrong beam, it was considered to be enemy or hostile. All ships went to battle stations. **When the unknown approached to within 3,000 yards, the crews opened fire on it**. The unknown then made a sharp right turn and headed south from an approach heading of 320°. The UFO increased its speed and then circled the entire fleet once, now at about 3,500 yards away. The object was traveling at such a tremendous speed that the gun crews could not coordinate a lead point fast enough on the target to hit it. All of the gun crews were just firing wildly, trying to get a hit. The target then circled the fleet one more time, then headed south again roughly at the approach point.

Afterward, the gun control director estimated **the vehicle had reached speeds of up to 10,000 mph**. The whole incident lasted only about 5 minutes or less. The witness, who still does not want his name revealed, was on watch duty at the time. He had a pair of 7x50 binoculars and got a pretty good look at it. It appeared to be a fairly flat, silvery disc with a dome right in the middle of the top side. There was no trail or exhaust and no sound. Its altitude remained relatively between 3,000 to 4,000 feet. Distance remained close to 3,500 yards.

After the incident was over, all the crews and personnel were shaken by the incredible experience of encountering something so unbelievable in its speed and maneuvering capabilities. The captain of the destroyer stated that to at least pacify his crew, he would make a determined effort to find out what it was. Since it was only two days prior to invasion time of Guadalcanal, radio silence between ships was imperative and it was maintained. Any messages were relayed only through flying aircraft (PBY's).

Four days after the invasion took place, the captain called most of the crew together and relayed a message to them from Command Headquarters that the object they had encountered was neither Japanese nor German [duplicate words deleted], not an enemy, nor was it one of ours. For months afterwards, the witness and other crew members continued to inquire of the captain for any further information as to what the object was. He told them he simply could not find out any more details relative to the incident. Fleet ships involved at the time were three cruisers and seven destroyers; **all fired at the unknown intruder**. The crews were all apprehensive for days that more of the strange visitors might return in force to possibly attack them.

The speeds which this "aircraft" displayed and somehow eluded the thousands of shells fired at it gave the crews the uneasy feeling they were no match for it. This was very unnerving. [342]

Many of the elements of later UFO cases are already seen in this report. First, radar from a ship detects the intruder, and then a visual sighting is reported from yet another ship. Although I have never heard of a case where a UFO rides a beam to a vessel, this account might have been a mistaken impression of the crew member(s), brought on by fear of an oncoming attack by the enemy. The reason the crew opened fire on the incoming "unknown" was, most likely, just a matter of self-preservation (considering that period in history) and not a desire to intentionally shoot down an *alien* craft. The estimated speed of the craft was of concern but only applied to the consideration of combat: the possibility of an advanced enemy technology. An estimate of 10,000 mph in those years is inconceivable, but who knows what advanced technologies are being used by your opponent? In an important statement from the case above, the captain said that he would try to find out what the unknown object was and later communicated to his crew that, according to HQ, it remained unidentified. In short, the captain was probably interested in solving this enigma, also. We will note the changes in this attitude as time rolls by.

7.2   Closer to the End of the War

SM-??-1945   *Unknown position in the Pacific Ocean*

I can't recall any dates, but it was during the summer of 1945, at sea between Guam and Saipan. **Our air search radar on several occasions picked up a large group of blips approximately one hundred miles away on a course that would bring them directly overhead**. We challenged them but got no response. **Their speed on our plot indicated over one thousand miles per hour.** Visual lookouts were alerted but saw nothing. These groups passed overhead at approximately two thousand feet and showed up on the radar screen as definite solid (not fuzzy) objects. When our visual spotters reported there was nothing in view, I personally went out and searched the sky with no luck. I believe these were reported to the Naval Bureau but can't be sure. Our gear was carefully checked and calibrated and was in good order. These were definitely not ionized clouds, inversions, sea gulls, mechanical defects, or any of the other common explanations. After many months of operating radar gear, one knows when they have a definite solid object on the screen, and, for my money, these were solid although invisible objects. [343]

The king of the sky in those days was the P-51 Mustang, moving at around 450 mph. Yet radar has the audacity to tell us that there is a craft that can do over 1,000 mph? Well, at least that was slower than the 10,000 mph reported in the other case. I guess if they can do 10,000 mph, it is not much of an effort to slow down a bit.

World War Two ends, and there is a new assessment and a new way of doing business. Communication between officers and crew, especially in times of warfare, is about to change. My intent in this chapter is to show the correlation of water-UFO reports with those of the common aerial variety. So we begin:

7.3   Radar Only (operating and detecting)

Note that the two previous cases from World War Two also belong in this category.

FF-??-1951   *Unknown waters off the Korean Peninsula*

It was a night in '51. Under lowering clouds, a United States task force of fourteen ships was cruising near Korea. Down in the Combat Information Center [CIC] of a CVE-class carrier, intercept officers and radar men were keeping a routine watch.

Suddenly a strange blip appeared on the CIC radarscopes. Some unknown machine, larger than carrier aircraft, was circling the fleet.

In minutes, Navy interceptors were boring up into the clouds that hid the intruder. At first, the CIC men had thought it was some new Red aircraft spotting the fleet by radar. But an hour passed with neither an attack nor message to bring enemy bombers. Even before this the CIC men knew, from the object's speed and maneuvers, it had to be a UFO.

As the hours went on, fresh pilots replaced the first group. Again and again, flying by instruments in the misty dark, they risked collision for a look at the unknown craft. But the UFO stayed deep in the clouds.

Down in the CIC, puzzled intercept men watched the mysterious "target." What could explain the hours of circling up in the overcast? Could "they" see through those tight-packed clouds by some unknown device—or a different kind of vision? What was behind this long surveillance—curiosity, or something more ominous?

Near the end of the seventh hour, another squadron was launched. Abruptly, the UFO stopped circling. As the tense CIC men watched, it swung in behind the nearest Navy plane. "Target joining up on wingman!" the lead pilot reported.

"Close in for visual on target!" ordered the CIC, fearing an attack on the wingman. Though the clouds made it almost hopeless, the leading pilot turned. Swiftly, the UFO speeded up, leaving the plane behind. **In less than ten minutes, the radarscopes showed it was two hundred miles away** [which would indicate a speed of 1,200 mph].

The signed report later certified by Board members was given to NICAP by one of the pilots involved, now a lieutenant commander on duty in this country. **The unknown machine, officially logged as a UFO, was tracked by radar operators on all fourteen ships**. Its long surveillance of the task force remained a mystery. [344]

??-??-1953   *Near Guantánamo Bay Naval Base, Cuba*

Letter sent to Chester Grusinski [See Chester Grusinski's own UFO sighting in case dated 10-??-1958, coming up on page 241.]:

An undated letter was received by Mr. Grusinski from William H. "Boots" Pierce, Cdr., USN (Ret.), which read in part, "Unfortunately though, I had debarked from *Roosevelt* by 1958 and was instructing at Pensacola. I did not make the cruise mentioned in your letter, but I did hear about the sighting at Gitmo [Guantánamo]. **In fact, *FDR*'s log will show a UFO sighting during 1953 when we took her to the Med** [Mediterranean], **both a visual sighting from the bridge and on CIC's radars.** It was an interesting thing of which I was a part." [345]

07-26-1956   *Port at Rio de Janeiro, Brazil – USS* FDR

"These objects...finally vanished with tremendous speed...When first sighted the objects were suspended in midair over one another with several hundred feet between them. **A radar fix placed them several miles from the *FDR*'s position at a height of 2000 ft.** Two rows of bright, counter-rotating lights could be seen through the middle of each object that made them appear to be round in shape. For several minutes they never moved. Then the upper one released a fireball object that dropped into the top of the lower one. Within seconds they both vanished with tremendous speed. Radar contact was almost impossible. It was estimated the objects could have been between 75 and 100 feet in length." [346]

10-??-1958　　　　*Caribbean Sea near Cuba - USS* FDR

"I saw a bright white ball of light. [Note: Chester Grusinski had handwritten 'cigar shape' in the margin.] It headed straight for us, getting bigger and brighter. It was spherical, approximately 75 to 100 feet long. It turned red orange and **I could feel the heat on my face.**… [Note that once again the heat of the UFO's field is felt by a witness.] "Another one for the same time period, the 25th or 27th of July, 1956, **a cigar-shaped object was seen and tracked on radar. And the CIC radars said it went off the scope in two or three sweeps**. That was witnessed by John C. Hau, Chief Warrant Officer." [347]

UNDATED—12　　*Gulf of Mexico - USS* Lexington *(CVS-16)*

Glowing Orange Object Buzzes USS *Lexington*
Date of Sighting: (1962 to 1966—exact date unknown)
Time of Sighting: around midnight
Date Reported: February 12, 2005
Number of Witnesses: several crew members (exact number unknown)

Description (in witness's own words with some editing): While serving aboard the USS *Lexington* CVS-16 (aircraft carrier), I was an elevator operator for aircraft. One night around 12 midnight **I saw a large orange glow in the sky. I called flight deck control. They said they had it on radar** and tried to get it to identify itself. When they got no reply, they sent 2 F4 jets up. You could see their afterburners and the F4s did not even come close to catching it. The object (light) took off out of sight. As soon as the jets returned, it came back. This pattern repeated itself 3 times. After the 3rd time, guess they thought it was no threat. It followed us for about half hr [a half hour] I guess then left. ▶We were told the next day to say nothing about the sightings.◀ I wish I had at least written down time/day/year, but did not. Even years after I did not know who to report it to. Then as time elapsed more, I figured no one would believe me. I just thought that I would report it as I remembered it. We were operating in the Gulf of Mexico out of Pensacola, FL. I worked the hanger [sic] deck elevator. I was in the V3 division/aircraft handlers. [348]

The following account, which is rather lengthy, is much more than just a radar case:

LL-??-1963   *Off the coast of Sardinia, Italy - USS* FDR

A Moment in (Recent) History
by Jon Baughman

Letter from Jordan

On August 3 Grusinski [See case dated 10-??-1958 on page 241] received a letter from Harry A. Jordan, who lives in Omaha, NE. Ironically, Jordan has also been looking for crew members from the *FDR* who might have witnessed the strange occurrences. He found Grusinski's name on a posting on a Canadian website.

In his letter, Jordan said, "**I detected a very large UFO during my mid-watch on radar off the coast of Sardinia** in late '63. I provided the Senate Intelligence Committee with an affidavit detailing my experience. During this particular cruise there were several photographers on board to document the NATO fleet operations. They were present on board the night of my contact on radar. ▶I was told in no uncertain terms to keep my mouth shut for 20 years. ◀"

The complete text of Harry Jordan's experience can be downloaded by going to: http://www.majesticdocuments. com/witnesses/jordan.html

Jordan on the *FDR*

From 1961 through December 1965, Jordan served [in the] U.S. Navy aboard the destroyer escort USS *Laffey*, destroyer USS *Loeser*, and [the] aircraft carrier USS *Franklin D. Roosevelt* CVA-42: **He served in the operations intelligence division on all three ships to display, analyze, report, and record radar, radio and electronic emission data**. For this he received extensive training at the Naval Station, Newport, RI.

While on the *FDR*, he served two cruises in the Mediterranean between 1962 and 1964. After discharge from active duty, he worked as an instructor at [the] Jones Point Naval Reserve, Alexandria, VA, on the use of navigational equipment particular to radar intelligence, thus was an expert in his field. He served as a recognition expert on detecting surface or air objects as far away as 300 miles. He trained both enlisted men as well as naval officers.

His account, firsthand

Here's Jordan's firsthand account of the UFO encounter, the text taken directly from his testimony before a U.S. Senate UFO hearing in the summer of 1998. It is quoted word for word:

> Late one night while operating off the coast of Sardinia at approximately 0200 hrs. Zulu time, I was standing midwatch on a SPA-8 repeater and height-finding equipment. My IFF box was enabled and in the process of challenging an aerial contact approaching the ship from a bearing of 012 degrees relative to our course. The aerial contact was detected at 600 miles according to the calibration rings on my scope. **I made several adjustments to the calibration settings because I didn't believe I detected such a contact at that distance. The aerial contact was at a height of 80,000 then dropped to 65,000 feet in about 20 seconds where the contact then hovered for about ten minutes.**

During this period the watch officer woke the division commander and informed him of our situation. U.S. Naval Task Forces operating in the Mediterranean in those days were being subjected to frequent flyovers by Russian Bear aircraft with electronic warfare capability. It was my job to stay alert for such a scenario. The division commander came into CIC, observed the radar returns, expedited his own ECM scan of the contact, and then informed the orderly standing watch at the captain's stateroom. The captain was appraised [sic] of the situation, ordered a course change into the wind, and increased ship's speed to launch aircraft. During this time I was continuing to observe the contact and watched the Phantom F-4B's once they attained sufficient altitude. The Phantoms went to afterburners, vectoring to a course straight for the bogey on my scope. After a period of about 15 to 20 minutes, the F-4B's were within 200 miles of the bogey due north of our course. They turned on their conical scan radar. The bogey disappeared from my scope. The pilots did not detect any contact on their scope, were turned by the Air Ops C.O., and headed back to the ship. The captain had kept the ship on a heading into the wind during this entire operation, so aircraft could be launched or recovered at will.

The Phantoms returned to the ship and were recovered. About five minutes after the recovery of the last aircraft and during the process of turning the ship back to our original course, the contact winked back on again! This time it closed from a distance of 600 miles from the ship to right above us in about 1 minute. **The bogey was traveling around 3600 miles an hour at an altitude of 30,000 feet. I never saw any heat signature.** Someone in CIC yelled out, "Damn it, what the hell is that!" I couldn't determine who made the comment. Lookouts on the port and starboard lookout wings of the conning tower could not see anything above the ship with their binoculars. The situation was bizarre and watch personnel looked scared but calm.

Watch logs are always kept particularly when a critical situation is underway. General Quarters practice drills are another matter, but this wasn't a practice drill. The captain did not call General Quarters. I recall vividly when it was first determined how unusual the situation had been. ▶My division C.O. (Commander Gibson—USN) asked me what I had put in my log if anything. He said to me, "Jordan, this never happened." (I was reminded of that comment when I first saw the movie [*The Hunt for*] *Red October*.) I replied, "Yes sir." I never mentioned anything to fellow sailors or anyone else until years after my Honorable Discharge from the Navy.◀ I was Honorably Discharged from the Navy in December of 1967 at [the] Jones Point Naval Training Center in Alexandria, Virginia.

On watch that night were a plotter on the DRT, one on the VG, myself on the radarscope, and the watch officer. ▶It was my distinct impression from hearing the airmen aboard during chow time that they were confused about why the heck they would have such an unscheduled launch and not be told what had been going on that night. It was their duty to arm aircraft, etc. To me they had been obviously kept in the dark on this one operation. I never said anything to them about what happened that night.◀ During active duty I stood lookout watch night and day. I logged well over 1000 hours operating radar equipment and standing watch during my active duty with the sixth fleet. What I experienced that night will never be forgotten. What I have experienced since discharge from active duty has became [sic] more fascinating as the years progressed. (End of quote)

More testimony

But that isn't the end. Jordan had some other interesting comments for the senators at the hearing, again, quoted here:

> In 1985 I met Dr. John Kasher, a physics professor at the University of Nebraska in Omaha. He was a consultant at Lawrence Livermore Labs in California and presently a summer consultant on plasma physics at Huntsville for NASA. He is currently the State Director for the Mutual Unidentified Flying Object Network of Nebraska. Dr. Kasher and I have personally investigated over 50 cases related to UFO's. These cases involve people from age 5 to [the] 70's.
>
> We have videotapes of private interviews, physical trace cases, and lab reports. Some of these cases involved events which occurred on the ground simultaneous with STS-NASA missions. There are uncut, unenhanced videotapes produced by live-real-time NASA select 6 camera shots taken aboard the space shuttle. At present, cameras aboard the shuttle are monitored. The video signals are delayed several seconds and sometimes minutes before being fed to public video feed networks. Immediately after the STS-48 mission UFO flap in which Dr. Kasher and I were involved, NASA began to delay the live feed. Their premise for altering the video coverage of each mission was that they didn't want the world privy to the medical condition of the astronauts. This is pure hogwash as never has the medical condition of the astronauts been public knowledge through video monitoring to public channels. We have a copy of the original NASA directive which went out from NASA to all ground facilities the day after [the] STS-48 UFO flap.

The UFO's which currently fly on station when the shuttle is up are being fired upon by high performance aircraft using state of the art particle beam weapons. The very character of a particle beam [is that it] does not need an atmosphere to be seen. It imparts its own coherence upon the vacuum of space. We have videotapes of STS-missions as recent as STS-80 during the month of March 1997 wherein one can clearly see these beams coming up from the earth through the cloud cover below the space shuttle. What concerns us most here from our personal field experience is that out of all our private interviews with eyewitnesses, not one instance of a negative or sordid nature has been told to us. Virtually all close encounters on the ground have been very passive, bizarre, and extraordinary to ordinary people. Whatever "ordinary" means these days is another matter. It is with great concern for our fellow Americans and citizens of the world at large we question the reasons for why UFO's are being fired upon. What the federal government of the United States of America knows about UFO's is not being shared. I question the current status of this posture and reasoning behind it. These matters might well be beyond knowledge of the Senate, Congress, and perhaps even the President of the United States. If public hearings were held regarding these scenarios, there will be some red faces at the Pentagon and NASA regarding the issues of UFO's and the reality of their occupants. Many decent, law-abiding Americans have had real experiences and are prepared to come forward to share proofs. The truth needs to be brought forward without further hesitation. (End of quote)

In a sense, this testimony by Jordan speaks for itself. It needs no interpretation by this writer.

In recent months, because I occasionally write about the UFO mystery, a few *Broad Top* readers have privately inquired of my colleagues, "Is he all right?"

Don't be concerned about me.

I am concerned about many of you.

I worry about people who are too absorbed in the trivial, mundane, and uneventful events in their lives that they don't have time or make time to ponder on grander ideas and mysteries. [349]

SM~FF-??-1965   *Atlantic between the Azores and Spain*

...air traffic controllers were watching **3 bogies on radar**, but had no visual contact. Bogies did not respond to any communication. They were following fleet at distance of 40-50 miles for over an hour at approximately the speed of the fleet....

The task force mission was to protect the aircraft carrier. They were protecting from sub attacks.

(initials deleted—witness) does not remember the aircraft carrier's name.

The aircraft carrier also reported three bogies on their independent onboard radar.

Albany's radars were set on sector scan to focus on bogies.

The jets reported over the radio that they had radar contact with bogies.

Jets never got closer than 20 miles (thus no visual?).

Jets move about ¼ inch per radar sweep at 1350 knots. The bogies began to move away in approximately ¾ inch jumps per sweep. Bogies maintained triangular formation. Bogies disappeared from radar within 7 sweeps. Jets were still visible on radar. [350]

??-??-1989 *Indian Ocean steaming toward the Gulf of Oman*

U.S. Naval Officer's UFO Experience

In 1989, I was an officer on the aircraft carrier USS *Midway*. After having been off the ship for a couple of weeks to help with the birth of my first child, I joined the ship in the Indian Ocean steaming toward the Gulf of Oman. During this time period, oil tankers had been attacked while going through the Strait of Hormuz.

One day, I was standing watch in the CDC (combat direction center). It was my job to overwatch [sic—watch over] the activity of a crew whose job it was to track the position and movement of all surface contacts (ships) in our vicinity. We never like anybody getting within three miles of us, and we are especially wary when there is a possibility that we may be attacked. I was watching my own radarscope, which happens to be two feet in diameter, very closely when **the sweep of the radar showed three bright returns equally distant from each other**.

On the next sweep, there were three blips again. However, they had moved quite a distance. Normally, if a single blip moved that far in a single sweep, I would have taken it as a random high wave crest. But, these three blips were quite strong returns, and they maintained an exact triangular formation. If an aircraft flew low enough, it could sometimes be detected on my "surface" radar which is aimed low to detect ships.

My first thought was that these were three aircraft. But, I had seen aircraft on my screen before, and **these blips were moving far faster than anything I had seen before, far faster than even our fighter jets.** My second thought was that the blips were missiles. This concerned me a little to say the least. I immediately contacted the other half of CDC which keeps track of the air traffic. I was very certain that they would not have allowed anybody to slip within our perimeter, which happens to be fifty miles in the air. I asked if our guys were firing any missiles for practice.

I thought this might have been a possibility, especially since the blips weren't moving directly at us. They were coming closer but on a tangent. The air guys didn't have a clue about what I was asking them but said that we were conducting no tests. When the three blips got to what would be their closest point on the tangent, two of the blips made a ninety-degree turn away from the ship. The third, however, turned directly toward the ship. I was rather excited at the time, but if I had to hazard a guess, I would say that about a minute had passed between the time when I first noticed the three blips and when they turned. I was extremely agitated, and since the guys on the air side apparently weren't detecting my bogies, I called to the lookouts on the 1MC (intercom).

We have actual people on the superstructure with binoculars. I told them that we had something coming in at us at a high rate of speed, and I told them what direction it was coming from. I yelled at the commander in charge of the complete CDC telling him what was going on, but the bogie was coming in so fast that there was no time to react. When the bogie got to within one mile of our ship, **it suddenly disappeared from my radarscope**. I called to the lookouts, but they had seen nothing.

I reported everything to the commander, but his reaction was strange. There was no reaction at all. While contemplating all the things that had just happened, I realized that I couldn't have been tracking missiles. There are none on earth that I know of that can travel at the velocity I saw on my radar screen. Sure, maybe a shuttle in the vacuum of space, but **these bogies had to be flying within thirty feet of the surface of the water for my radar to detect them**. A few minutes had passed when the three blips appeared again at nearly the same starting point as the first time. I hollered to the commander letting him know they were back. I called to the air side again. Basically, everything repeated almost identically.

After the second run, I was extremely disturbed. Something unidentified could fly to within a mile of us without full detection, without our lookouts seeing it, and with our total inability to do anything about it even with all the might of an aircraft carrier at hand. The commander seemed strangely unaffected.

So, I left my station, walked over to the commander and expressed as calmly as possible the facts and why they were so distressing to me. I said, "Something just got to within one mile of us, and it got there faster than anything I've ever seen. It made course changes that should have been impossible. **And, the only way our lookouts couldn't have seen it, is if it dove into the water**. How is this possible?" I told him that I was a skeptic but that the only explanation I could think of was that these things might have been UFOs, like flying saucers from another world.

He spoke rather quietly and said that while I had been off the ship, our CDC had detected an unidentified aircraft and that we had launched, what we call, our alert fighter to intercept and identify it. He said that the alert fighter had gotten to within visual range; **the pilot saw a metallic glint when the object accelerated away from him and dove into the water.** The pilot flew over the area, but there was no hint of a crash. One of the carrier's escort ships was sent to the area, but not a trace of anything was found. He said he believed that also had been a flying saucer.

I am still an officer in the U.S. Navy, so I am reluctant to reveal very much personal information. I will give corroborating details if necessary. I'm not saying that what I experienced was proof of extraterrestrial visitation, but I am certainly open to the possibility. [351]

SP~SM-??-1994   *Pacific Ocean - USS* Constellation

UFO Sighting from USS *Constellation* Aircraft Carrier, Spring-Summer 1994
by Anonymous to Mr. Wes Penre

If you're into homework, you'll find my story can be substantiated by confirming the whereabouts of the USS *Constellation* (CV-64) (aircraft carrier) during the spring-summer of 1994. I worked in C.D.C. (combat direction center). **We tracked** [I assume on radar.] **an unidentified contact traveling at impossible speed towards our ship.** It was first reported as an inbound missile, but was too fast. When we expected impact, about one second of silence went over comm. channels, then screaming from the island watches stationed outside. [The island is the main vertical superstructure on the right (starboard) of the flight deck.]

This confusion lasted approx. 10-15 seconds, then they were all quiet. They advised a large light hovering above the island [as above] and as soon as it showed, it went straight up vertical and went out of sight. The watches were immediately relieved and debriefed by an IS (intelligence specialist) and I (personally) made a log entry into a blue log book pulled from a safe filled about 5-7 high with other log books....

And to you nay-sayers, if you don't believe we are being observed...I wished I lived in your dream world.

Combat Operations Specialist (OS3)
USS *Constellation*. Go Connie! [352]

7.4   Radar Only (operating and NOT detecting)

02-??-1952 *Atlantic Ocean 400 miles off New Jersey, USA*

Hooray! I am not blind as my CO called me in circa Feb. 1952. I was on air watch aboard USS *Fort Mandan* LSD-21.

We were located approx. 400 miles east of New Jersey on maneuvers. I spotted exactly what Chester saw. [Referring to Chester Grusinski's UFO sighting in case dated 10-??-1958 on page 241 of this chapter] I reported it as the gondola of a blimp. The CO said there were no blimps within 500 miles. Then I heard him whisper, "Blind SOB!!"

The drawing Chester made of the UFO with windows is just what I saw.

As far as I can remember, and I do remember it well, it was a partly cloudy day, cold and windy. The surface was usual for the N. Atlantic, a little rough but not bad. I remember the clouds as I thought maybe I saw some weird kind of reflection of a blimp from over the horizon, especially since **they told me there was nothing on radar** and not a blimp around.

I still distinctly remember that it looked like a long blimp gondola and it had windows. I did not see the gas bag; I figured a cloud covered it. I would estimate it at ½ to ¾ miles away. I followed it a bit and then reported aircraft off the port beam. When I reported, I took my binoculars off my eyes. After the report, I could no longer find the object. I figured it went into the clouds. I was on the bridge about 15 feet from the captain about 65 feet above the water.

Ten years ago I was stationed aboard USS *Shadwell* LSD-15, the same class and age ship, and stood on the same spot and remembered that day.

I never heard of a UFO, so at the time, I merely believed that I had seen a blimp. I don't know of anyone else that saw it, however, this month I am going to a reunion of that ship's crew, so I will ask around. [353]

??-??-1966   Pacific Ocean - USS Yorktown

This is a story about a curious event that took place about 1966. We were days out of Pearl Harbor en route to Japan for a port call, then on to Vietnam. I was one of four lookouts on duty on the 07 level, two forward, two aft. I was on the forward station.

We were steaming west with our four destroyer escorts on station approx. 2-3 miles out, each on our quarter beams. It was about 2200 hours and the weather was balmy with clear skies. There was no activity, just a straight run to Japan.

The other lookout and I looked up and saw a light in the sky coming towards us from the port side! We were surprised because CIC had not said anything to us about an approaching aircraft. It looked like a bright landing light coming at us. We called down to CIC on the sound power phones and reported the aircraft. They responded with

"negative, no plane on radar." Meanwhile the light kept coming closer and getting bigger.

The light is now almost overhead and rather large. Then it stopped. It just stayed over us for several minutes with no sounds coming from it. CIC kept insisting there was nothing showing up on radar. The bridge watch was also reporting the light as was the two aft lookouts. We were all on the same phone system.

After several minutes of hovering over us, the bright light went much brighter and lit up the entire task group almost like daylight. We looked at each other and said, "We don't have anything that can do this and they [the Russians] don't have anything like this. What the hell is it?" Now the whole ocean and destroyers were lit up for a number of square miles. The really bright light went off leaving the original stationary light as it was. Then it started to move from port to starboard and then shot away and faded from sight within several seconds. All this without a sound.

I estimate that between the guys on duty aboard the *Yorktown* and four destroyers, there had to be a least a hundred guys that witnessed that incident. Now I have to tell you I'm the kind of guy that has to kick the tires before I can tell you what I saw. All I know is what I just related. I'm curious [to] know if any readers of this site saw that or if Brian S. knows if we or the Russians had anything flying those days without sound and [able to] evade radar. And no, I've never seen anything like that before or since. [354]

10-02-2007   *Atlantic Ocean off Virginia, USA*

During a routine weather observation aboard a naval vessel during flight ops, I witnessed an object to my east moving at an extreme speed and changing directions rapidly before disappearing. We were about 100 miles off the coast of southern Virginia. It was a white light about 4 times the size of a star. Imagine an airplane at night with its lights not blinking, but flying [at] about 5,000 mph. Moving at a 45-degree angle from east to west, it abruptly changed course (it made a right angle) and disappeared. I am a meteorologist for the Navy and observe the sky on a nightly basis. I have never seen anything like this. **It was not caught on our radar**. It appeared to be about 5-7 miles away, but I suspect it was much farther away than that. [355]

7.5   Electromagnetic (EM) Effects

In this section, we see another manifestation of the UFO upon the vessel. EM effects are quite common in UFO cases, both on aircraft and land vehicles. With aircraft in close proximity to a UFO, such effects are revealed by the sputtering of the aircraft engines. The same sputtering has been noted with concern to land vehicles. However, in many other cases the engine is completely shut down only to amazingly restart itself after the UFO departs. Here, though, we do not have a small car. Instead, we have large ships like aircraft carriers which are similar to floating cities. Three cases follow as examples:

??-??-1954   *Unnamed location -* USS FDR

Although I would like to divulge detailed information to you, I must first think of my family. As an officer on the *FDR*, I too witnessed a daytime event involving a cigar-shaped craft that covered a span of at least 100 feet, a larger 300+ foot cigar, and two 50-foot (roughly) round discs. Without stating my position, **I will state that we attempted to engage the craft with weaponry and, in the ten minutes of viewing and the 4 minutes of this attempt, the weapons would not allow themselves to be prepared for engagement.**
We also attempted to prepare launch of aircraft at the conclusion of the event. I was informed of the object when it arrived by [a] fellow officer who engaged me in conversation concerning the proper procedure in dealing with this type of event. At the 14 minute mark, the craft rapidly traveled upward as **two small, round objects exited the water**.

At the 21 minute mark, a much larger cigar arrived from directly above the *FDR* and the two smaller disks entered the cigar, and then the cigar once again departed upward at 23 minutes roughly. **At [the] 26 minute mark, the cigar returned and dove into the water.** [356]

07-02-1971   *Atlantic Ocean near the Bahamas*

U.S. Aircraft Carrier Stopped by UFO
By Jim Kopf

This encounter occurred in 1971 while aboard the aircraft carrier USS *John F. Kennedy* CVA-67 (now CV-67) in the Bermuda Triangle [southeast of Florida—from NUFORC's version of the case]. I was assigned to the communications department of the *Kennedy* and had been in this section about a year. The ship was returning to Norfolk, VA, after completing a two-week operational readiness exercise (ORE) in the Caribbean. We were to stand down for 30 days after arriving in Norfolk, Virginia, to allow the crew to take leave and visit family before deploying to the Mediterranean for six months.

I was on duty in the communications center. My task was to monitor eight teletypes printing the "Fleet Broadcasts." On the top row were four teletypes each printing messages from four different channels. On the bottom row were four more doing the exact same thing except the signal was carried on different frequencies. If one of the primary receivers started taking "hits," I would be able to retrieve the message from the bottom one. I also notified Facilities Control of any hits so they could tune the receivers. On the other side of the compartment (room) was the NAVCOMMOPNET (Naval Communications Operations Network). This was the ship-to-shore circuit with the top teletype being the receive and the bottom as the send (known as a duplex circuit). Next to this was the Task Group Circuit for ship-to-ship communications (task group operations or TGO).

It was in the evening, about 20:30 (8:30 p.m.), and the ship had just completed an eighteen-hour "Flight Ops." I had just taken a message off one of the broadcasts and turned around to file it on a clipboard. When I turned back to the teletypes, the primaries were typing garbage. I looked down to the alternates which were doing the same. I walked a few feet to the intercom between us and the Facilities Control. I called them and informed them of the broadcasts being out. **A voice replied that all communications were out. I then turned and looked in the direction of the NAVCOMMOPNET and saw that the operator was having a problem. I then heard the Task Group operator tell the watch officer that his circuit was out also.** In the far corner of the compartment was [sic—were] the pneumatic tubes going to the signal bridge (where the flashing light and signal flag messages are sent/receive[d]). There is an intercom there to communicate with the signal bridge and over this intercom we heard someone yelling, "There is something hovering over the ship!" A moment later we heard another voice yelling, "IT IS GOD! IT'S THE END OF THE WORLD!"

We all looked at each other. There were six of us in the Comm. Center, and someone said, "Let's go have a look!" The Comm. Center is amidships, just under the flight deck, almost in the center of the ship. We went out the door, through Facilities Control and out that door, down the passageway (corridor) about 55 feet to the hatch that goes out to the catwalk on the edge of the flight deck (opposite from the "island" or that part of the ship where the bridge is). If you have ever been to sea, there is a time called the time of no horizon. This happens in the morning and evening just as the sun comes up or goes down over the horizon.

During this time you cannot tell where the sea and sky meet. This is the time of evening it was. As we looked up, we saw a large, glowing sphere. Well, it seemed large; however, there was no point of reference. That is to say, if the sphere were low, say 100 feet above the ship, then it would have been about two to three hundred feet in diameter. If it were, say, 500 feet about [sic-above] the ship then it would have been larger. It made no sound that I could hear. The light coming from it wasn't too bright, about half of what the sun would be. It sort of pulsated a little and was yellow to orange.

We didn't get to look at it for more than about 20 seconds because General Quarters (battle stations) was sounding and the communication officer was in the passageway telling us to get back into the Comm. Center. We returned and stayed there (that was out [sic-our] battle station). **We didn't have much to do because all the communication was still out**. After about 20 minutes, the teletypes started printing correctly again. We stayed at General Quarters for about another hour, then secured. I didn't see or hear of any messages going out about the incident.

Over the next few hours, I talked to a good friend that was in CIC (combat information center) who was a radar operator. **He told me that all the radar screens were just glowing during the time of the incident. I also talked to a guy I knew that worked on the Navigational Bridge. He told me that none of the compasses were working** and that the medics had to sedate a boatswain's mate that was a lookout on the signal bridge. I figured this was the one yelling it was God. It was ironic that of the 5,000 men on a carrier that only a handful actually saw this phenomenon. This was due to the fact that flight ops had just be [sic – been] completed a short time before this all started and all the flight deck personnel were below resting. It should be noted that there are very few places where you can go to be out in the open air aboard a carrier. **From what I could learn, virtually all electronic components stopped functioning during the 20 minutes or so that whatever it was hovered over the ship. The two ready CAPs (Combat Air Patrol), which were two F-4 Phantoms that are always ready to be launched, would not start.**

I heard from the scuttlebutt (slang—rumor mill) that three or four "men in trench coats" had landed and were interviewing the personnel that had seen this phenomena [sic]. I was never interviewed, maybe because no one knew that I had seen it.

A few days later, as we were approaching Norfolk, the commanding and executive officers came on the closed-circuit TV system that we had. They did this regularly to address the crew and pass on information. During this particular session the captain told us how well we did on the ORE and about our upcoming deployment to the Mediterranean. ▶ At the very end of his spiel, he said, "I would like to remind the crew that certain events that take place aboard a naval combatant ship are classified and are not to be discussed with anyone without a need to know." This was all the official word I ever received or heard of the incident. ◀

Being young and excited about my visit home and going to the Med, I completely forgot about it until years later when my wife and I went to see *Close Encounters of the Third Kind* at the movies when it first came out. In fact, the friend that had been the radar operator was with his wife and went with us. As we walked across the parking lot to my car, I ask[ed] him if he remembered what we had experienced years earlier on the ship. He looked at me and said he never wanted to talk about it again. As he said it he turned a little pale. I never talked about the incident again. When I discovered "Aliens and Strange Phenomenon" on MSN and started reading the posts, I started thinking about it again. Now I seem obsessed in finding out all I can about this phenomena [sic].

Jim Kopf
Mt. Airy, Maryland [357]

WW-??-1979   *South China Sea*

They were approximately 100 feet long and 40 feet across with about 20 yards between them. They glowed the typical bright, bioluminescent green and at first we thought it was some kind of marine life, but as the first one passed beneath the stem of the ship, **we lost power, steerage, the gyro and compass went wacky AND the chronometers and wristwatches stopped!** [358]

7.6   Can They Travel Fast Underwater?

Tales about a UFO's speed in the air are too numerous to mention, and only now are we on the threshold of such speeds in linear flight with recent tests of our high-end propulsion systems in experimental aircraft. However, when it comes to water, the tales are sparse and with few comparisons to speed. In Chapter 2, I laid out how the field operates and that what it does to air it can also do to water. The following case is one of the best I have come across in showing that a UFO can easily outdistance a jet fighter using its afterburner while the UFO is still submerged. Keep in mind that 50 knots (57.6 mph) is the guesstimate of the top speed of nuclear subs, while the F-14's cruising speed before afterburners is 633 mph and their top speed with afterburners is 1,544 mph.

09~10-??-1974   *Pacific Ocean 15 miles off Guam*

Please note that all names have been removed for confidentiality.

Dear Sir:

While aboard the USS *Reclaimer* ARS-42 while off Guam, I was [a] witness to an underwater event that I have remembered for years. Oddly, it wasn't till this last Xmas season that I saw, heard, or paid much attention to underwater UFOs. Here is the account of the event as best as I can recall:

It was the midwatch. The OOD was a warrant [officer], Boatswain Mr. Xxxxx. I was a GMG/DV SN standing the petty officer of the watch aboard the oceangoing tug/salvage ship USS *Reclaimer*. We were steaming off Guam, waiting till first light to enter and dock. The starboard lookout, SA Xxxxx Xxxxxxx, called me to report a contact. What we saw was a single red flashing light. We remarked we were being pulled over by Barney Fife (Andy of the Mayberry show) (The

Reclaimer had a top speed of maybe 18 knots [20.7 mph].) for speeding. It was just like the red light on a 60's model police car, constant flashing. The water was clear. There was something...a shape, craft...supporting the light. Nothing really we could make out. I called the OOD, reported the contact. It was late, maybe 2 a.m. We watched it change course a little. It continued to parallel us. Boatswain Xxxxx contacted Guam control since there were Russian trawlers operating in the area. **Guam control acknowledged that they too were tracking something and were sending a wild weasel F-14 to investigate. I thought this odd to send a fast mover to inspect a slow underwater vessel. The jet came out of the sky like a greasy snowball [and] dropped to the deck. This thing [F-14] keyed on that [USO] and accelerated. The F-14 went to full afterburner right off the deck, pulling a rooster-tail[12] behind it. It rattled the ship severely as it went by. This thing [USO] pulled easily ahead of the jet, crossed the horizon in seconds while underwater... no visible wake. It just...I mean this thing was FAST.** Well, we just stood there. Wow, you know...what do ya say. We speculated about what it was. [359]

If you have served on a carrier or other large ship and have had a UFO experience, I would like to hear about it and post it to my website with the rest of those that have come to light. Being a field investigator for MUFON, I will guarantee confidentiality of your personal information. My

---

12 rooster-tail: A rooster-tail is usually created on the water by a fast-moving boat. It is a spray that kicks up into the air behind the boat due to the converging trails coming off both sides of the boat forcing them upward into a tail. It could be construed as being 90 degrees or less to the direction of the boat in an upward direction. If a jet gets close enough to the water, the wing vortices kick up a spray of sorts behind the aircraft which could be called a rooster-tail. [360]

mailing address is on the bottom of my website's Home Page.

In the next chapter, it is not only what we see on the ship per se, but also what we hear...and what we are told not to discuss.

Chapter 8

Secrecy about UFOs

Loose Lips Sink Ships

Secrecy is extremely necessary during our normal tribal wars since giving the enemy even a small bit of information could help them to eventually pinpoint a ship's location or approximate location, its firepower capability, speed, and so forth. However, releasing data about an unknown craft that is possibly extraterrestrial and has a history of worldwide sightings to its credit is hardly a necessary component of confined-area warfare secrecy, yet the stories that follow demonstrate that this is the case. It may be understandable if a captain of a ship feels that his upward movement in the ranks is threatened by releasing a report to unbelieving superior officers, but if and when he does, the onus of secrecy falls on the higher authority, and the question then arises, why keep all this covered up? This is especially remarkable where it has been aired on radio and television and written by our government that:

8.1 "UFOs Do Not Constitute a Threat to National Security."

But how do they know this? Are we in a conspiracy *with* them? Why then do we shoot at these UFOs and scramble our fighters to intercept them, and, in several cases, participate in dogfights with them? One such dogfight has become very famous. It is not water related but did have both ground and airborne radar involved. In Iran during the Shah's reign, Iranian F-4 Phantoms were sent up to intercept these intruders on September 19, 1976, at 1:30 a.m., near Tehran:

As the F-4 closed in on the object, the jet lost all instrumentation and communications (both UHF and intercom). The pilot broke off the intercept and headed back to base. "When the F-4 turned away from the object and apparently was no longer a threat to it, the aircraft regained all instrumentation and communications."

A second F-4 was scrambled at 1:40 A.M., and **the electronics officer acquired a radar lock-on at 27 nautical miles**, 12 o'clock high position with the rate of closure at 150 knots. As the range decreased to 25 nautical miles, "the object moved away at a speed that was visible on the radar scope and stayed at 25 nautical miles...." The visual size of the object was difficult to discern because of its intense brilliance. The light it gave off was that of flashing strobes arranged in a rectangular pattern and alternating blue, green, red, and orange in color. The sequence of the lights was so fast that all the colors could be seen at once.

As the pursuit continued on a course to the south, another brightly lighted object emerged from the first object and headed straight toward the F-4 like a missile, at a high rate of speed. **The pilot attempted to fire an AIM-9 missile at the UFO, "but at that instant his weapons control panel went off and he lost all communications** (UHF and interphone). At this point the pilot initiated a turn and negative G dive to get away. As he turned the object fell in trail at what appeared to be about 3-4 nautical miles. As he continued in his turn away from the primary object, the second object went to the inside of his turn then returned to the primary object for a perfect rejoin." [361]

So, if UFOs are not considered a threat to national security, why the secrecy? One reason why the leaders of the world's nations might be concerned is that we could suddenly decide to give up our culture in preference for that of the aliens', thus leaving the leaders without the power structure they have built over the years. Another concern might involve the technology gained from crashed alien craft, which would give us a military advantage if we could reproduce the materials and methods of production used by the aliens—their technology. Of course, this might take many years as it is obvious that they are more advanced than we are. Whatever the reason, the knowledge of UFOs is still held in secrecy, with only mediocre cases being released as the *real* "Top Secret" files. I do not feel it necessary to go on a witch hunt here, as it gains us virtually nothing but the hunt. However, exposing cases that show the secrecy in action does add to the pressure on those who hold the power. To that end I have collected cases in the water realm that have this emphasized in the text and include some excerpts from them here.

8.2   Early On—Before UFO Secrecy

The previous chapter mentions the idea of the military mind being awakened to the reality of UFOs. Examples of this continued awakening process follow. In the beginning, what is seen is unknown and frightening, and then it goes on to be considered a secret weapon of a dastardly foe. And time marches on:

08-05-1942   *Pacific near Guadalcanal - USS* Helm

The best estimates on the craft's diameter, as speculated by the various ships' commands, was about 90 feet.
The witness still feels the exact details and collected naval information on this particular incident is no doubt still under high security wraps. The date of the experience he places at either October 9th or 10th, 1942. [Date is incorrect.] Due to the length of time involved, he said he has lost track of his other shipmates, but felt the information was definitely on file somewhere in Navy Intelligence files in Washington.

The interview with this witness is on a tape cassette in our possession. [362]

Note that within the total text on file for this case, there is no mention of threats to the crew or secrecy concerning UFOs. This, of course, was the period of the unknown "Foo Fighters," explained in Chapter 7. In short, you might think of the upper levels of the military as being in school, learning that we are not alone.

8.3   But Then...

The year 1945 brought about the end of World War Two and with it, the promise of peace and prosperity. More data was collected and then forwarded upward for review. However, two years hence, what is considered to be the beginning of the era of the flying saucer begins. Two major news stories set off to replace the horror of the atom bomb and substitute the new "enemy" from outer space. First was the Kenneth Arnold sighting in the state of Washington that is mentioned in just about every UFO book and needs no further elaboration here. Secondly, thereafter came the name that to this day is known worldwide: Roswell.

So from a military point of view, and I do mean military at the top of the ladder, the Foo Fighters of the war years continued into this new era, seemingly renamed in 1947 as flying saucers, and therefore cannot be dismissed as combat fatigue. If the crash at Roswell is true, and it is beginning to look that way, then the idea that the military would be interested in back-engineering an alien craft is valid as it could now be considered a major weapons platform in preparing for any future conflict. Starting with post-war cases, we have:

02-10-1951   *Atlantic Ocean off Newfoundland, Canada*

[Aircraft: Navy R5D = C-54 = DC-4 Skymaster]

"When we landed at Argentia (Newfoundland), we were met by intelligence officers. The types of questions they asked us were like Henry Ford asking about the Model T. You got the feeling that they were putting words in your mouth. **It was obvious that there had been many sightings in the same area, and most of the observers did not let the cat out of the bag openly. When we arrived in the United States, we had to make a full report to Navy Intelligence.**

"I found out a few months later that Gander radar did track the object in excess of 1800 mph. I did not see the reports made by other members aboard the aircraft. I did talk to the Air Force at Wright-Patterson AFB in 1957 but did not look at the report. They said they had it and many similar reports." [363]

8.4   The JANAP-146 Documents

Evolving from directives in World War Two were the JANAP instructions. JANAP stands for **J**oint **A**rmy-**N**avy-**A**ir Force **P**ublication, and JANAP-146 included instructions for civilian and military personnel on how to report sightings of enemy aircraft, surface vessels, submarines, missiles, and unidentified flying objects. JANAP-146 was primarily concerned with potential enemy craft and not with flying saucers but included UFOs as a possible category because they might perhaps be identified as enemy craft.

These instructions went through changes over the years, and I believe that the current JANAP regulation is 146 (E). However, the regulation in effect during Chester Grusinski's tour of duty was JANAP 146 (C) which went into effect on March 10, 1954. An excerpt from Grusinski's case, dated 10-??-1958, appeared in Chapter 7 on p. 241. It can also be found in its entirety on my website or additionally on *Flying Saucer Review*'s website. Still, Grusinski's reference (below) to 146 (E) is well taken as that one became effective on March 31, 1966 (revised again on May 17, 1977). It added that photographs should be sent to the Director of Naval

Intelligence and also included special reporting instructions for unidentifiable objects. Nonetheless, both versions of the JANAP documents make reference to the reporting of unidentified flying objects.

Grusinski was stationed aboard the USS *Franklin D. Roosevelt* from 1958 through 1960, during which time the ship had UFO sightings. He is also aware of UFO sightings by some *FDR* crew members which took place before and after his time on the ship. Several years later, through the Freedom of Information Act, he obtained copies of the ship's logs, but information about UFOs was not available. On August 25, 1994, Grusinski said:

> But we can be sure that the Regulation JANAP 146 (E) will have prevented details of any of these cases getting onto the warship's log-books. However, I possess letters of testimony from a number of crew members. I also enclose herewith, along with a photo of the carrier USS *Franklin D. Roosevelt*, prints of two of the photos of the flying disc taken on September 20, 1952, by Wallace Litwin during "Operation Mainbrace" and a sketch of the scene on our flight deck on the night in 1958 when, off the coast of Cuba, I saw the big "cigar." This sketch was made by my companion and fellow-witness William Scott. Along with the photo of our carrier, I also enclose a copy of the official U.S. Navy's MERINT radio-telegraph procedure chart for reporting UFOs and a copy of the letter of promulgation issued on March 31, 1966, by the Joint Chiefs of Staff in connection with JANAP 146 (E) instructions for the reporting of vital Intelligence sightings (incidentally, still in force today). [364]

What follows is the JANAP 146 (C) document [365] which would have been in force during Grusinski's tour of duty.

JANAP 146(C) COMMUNICATION INSTRUCTIONS FOR REPORTING VITAL INTELLIGENCE SIGHTINGS FROM AIRBORNE AND WATERBORNE SOURCES – 10 MARCH 1954

JANAP 146(C)

THE JOINT CHIEFS OF STAFF
JOINT COMMUNICATIONS-ELECTRONICS COMMITTEE
WASHINGTON, D. C.

10 MARCH 1954

LETTER OF PROMULGATION

1.   JANAP 146(C) COMMUNICATION INSTRUCTIONS FOR REPORTING
VITAL INTELLIGENCE SIGHTINGS FROM AIRBORNE AND WATERBORNE
SOURCES, is an unclassified publication.

2.   JANAP 146(C) COMMUNICATION INSTRUCTIONS FOR REPORTING
VITAL INTELLIGENCE SIGHTINGS FROM AIRBORNE AND WATERBORNE
SOURCES, is effective upon receipt and supersedes JANAP 146
(B), COMMUNICATION INSTRUCTIONS FOR REPORTING VITAL
INTELLIGENCE SIGHTINGS FROM AIRCRAFT (CIRVIS) and all other
conflicting instructions. JANAP 146(B) shall be destroyed by
burning. No report of destruction is required.

CHAPTER II

CIRVIS REPORTS

SECTION I - GENERAL

201. INFORMATION TO BE REPORTED AND WHEN TO REPORT

     a. Sightings within the scope of this chapter, as
outlined in Article 102b(1), (2), and (3), are to be reported
as follows:

          (1) While airborne (except over foreign territory -
See Article 212).

               (a) Single aircraft or formations of aircraft
which appear to be directed against the United
                    States, its territories or possessions.

               (b) Missiles.

               (c) Unidentified flying objects.

(d)   Submarines.

(e)   A group or groups of military surface
vessels.

(2)   Upon Landing.

(a)   Individual surface vessels, submarines, or
aircraft of unconventional design, or
                  engaged in suspicious activity or
observed in an unusual location or following an
                  unusual course.

(b)   Confirmation reports.

202.   SIGHTINGS NOT TO BE REPORTED

Reports are not desired concerning surface craft or
aircraft in normal passage, or known U.S. military or
government vessels and aircraft.

JANAP 146(C)

CHAPTER III

MERINT REPORTS

SECTION I - GENERAL

301.   INFORMATION TO BE REPORTED AND WHEN TO REPORT

a.   Sightings within the scope of this chapter (as
outlined in Article 102b., (4), (5), (6), (7) are to
            be reported as follows:

(1)   Immediately (except when within territorial
waters of other nations as prescribed by
            international law)

a)   Guided Missiles
b)   Unidentified flying objects
c)   Submarines
d)   Group or groups of military vessels
e)   Formation of aircraft (which appear to be
directed against the United States, its
            territories or possessions).
f)   Individual surface vessels, submarines, or
aircraft of unconventional design, or engaged
            in suspicious activity or observed in an
unusual location or following an unusual course.

(2)   When situation changes sufficiently to warrant

270

an amplifying report (see Art. 409).

SECTION III - SECURITY

308. MILITARY AND CIVILIAN

    a. All persons aware of the contents or existence of a MERINT Report are governed by the Communications Act of 1934 and amendments Thereto, and Espionage Laws. MERINT reports contain information affecting the National Defense of the United States within the meaning of the Espionage Laws, 18 U.S. Code, 793 and 794. The unauthorized transmission Or revelation of the contents of MERINT reports in any manner is prohibited.

    b. Military commands and activities in making local distribution Of MERINT reports and in subsequent communications regarding the contents of any MERINT report shall handle such communications in accordance with current security regulations.

402. EXAMPLES BY TYPE

    The following are examples of the types of "MERINT" reports. Specific application of Military or commercial procedure has been avoided since the means of transmission would determine the procedure to be used. Of primary importance is the expeditious handling and accuracy of the reports.

    a. "MERINT" REPORT

        (1) A radiotelegraph transmission:

RAPID US GOVT

TO

COMEASTSEAFRON 90 CHURCH ST. NEW YORK

MERINT 5126 N 14230W 3 UNIDENTIFIED FLYING OBJECTS HEADED NW AT 17000 FEET CIGAR SHAPE 50 FEET TO SW AT

2 MILES VERIFIED BY NAVIGATOR VISIBILITY UNLIMITED 211513Z JONES NKLN

This version required all marine vessels and aircraft to report information of vital intelligence to the U.S. Navy. Before this version of JANAP, all intelligence reports were expected only from military and civilian aircraft.

8.5   The MERINT Document

MERINT is an acronym which means **MER**chant Ship **INT**elligence. Another change that came out of the JANAP-146 (C) version was that sightings reports made by waterborne sources would be identified by the acronym MERINT. Sightings made from airborne sources would be identified by the acronym **CIRVIS** (**C**ommunications **I**nstructions for **R**eporting **V**ital **I**ntelligence **S**ightings). An agreement between the United States and Canada created the CIRVIS/MERINT reporting system to extend the early warning defense system to the entire North American continent.

The purpose of these reporting procedures is for the accumulation of intelligence from commercial ships and aircraft of foreign objects in an area where a military presence is absent. This enables us to know what is near our shores and in our airspace.

A Naval publication from the Office of the Chief of Naval Operations (OPNAV 94-P-3B), known as the "MERINT Radiotelegraph Procedure," follows. [366] It is to be used to report hostile, suspicious, or unidentified objects. At the top right side of the MERINT document, you will notice a drawing of a Buck Rogers-type rocket and next to it our more familiar circular UFO—not an unidentified airplane. So, clearly, UFOs are in a separate category from guided missiles, aircraft, submarines, and surface vessels.

FOR EARLY WARNING IN DEFENSE OF THE NORTH AMERICAN CONTINENT

MERINT
RADIOTELEGRAPH PROCEDURE

1. WHAT TO REPORT

Report immediately all airborne and waterborne objects which appear to be HOSTILE, SUSPICIOUS or are UNIDENTIFIED.

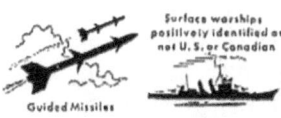

Surface warships positively identified as not U.S. or Canadian

Guided Missiles

Aircraft or contrails which appear to be directed against the United States, Canada, their territories or possessions

Submarines

Unidentified Flying Objects

2. SEND TO ANY

United States Naval Radio Station
Canadian Naval Radio Station
United States Coast Guard Radio Station
United States Commercial Radiotelegraph Station
Canadian Department of Transport Coastal Station

Receiving station will relay to military destination

3. HOW TO SEND

* MERINT MERINT MERINT (Coastal Station) DE (Own Signal Letters) K (Own Signal Letters) DE (Coastal Station) K
EMERGENCY (For U. S. or Canadian Naval or Coast Guard Radio Stations)   or
RAPID US GOVT COLLECT (For U. S. Commercial Coastal Stations)   or
RUSH COLLECT (For Canadian Dept of Transport Coastal Stations)

4. SEND TO ONE DESTINATION

ComWestSeaFron Navy SFran
NavyCharge Halifax
NavyCharge Esquimalt

Select destination nearest to your receiving station

5. SEND THIS KIND OF MESSAGE

Content—	Example—
a. Begin your message with the word "MERINT"	MERINT
b. Give the reporting ship's name and signal letters	SS TOLOA WHDR
c. Describe briefly the objects sighted	TWO UNIDENTIFIED SURFACED SUBMARINES
d. Give ship's position when objects are sighted, also TIME and DATE	5034N 4012W 07143C GMT
e. If objects are airborne, estimate altitude as "low", "medium", "high"	(not applicable)
f. Give direction of travel of sighted objects	HEADING 270 DEGREES
g. Estimate and give speed of sighted objects	15 KNOTS
h. Describe condition of sea and weather	SEA CALM
i. Give other significant information	ELONGATED CONNING TOWERS

6. SEND IMMEDIATELY

a. DO NOT DELAY YOUR REPORT DUE TO LACK OF INFORMATION
b. EVERY EFFORT SHOULD BE MADE TO OBTAIN ACKNOWLEDGMENT FROM RECEIVING STATION THAT MESSAGE HAS BEEN RECEIVED.

* The International urgency signal (XXX XXX XXX) may be used as an alternate to clear circuit.

Authorized by Secretary of the Navy

OPHAV 94-P-3B

273

8.6 The Veil of Secrecy Tightens

So now UFOs are mentioned on reporting documents and regulations. Welcome to the club, folks. In just a few short years, we learn that perhaps a glimmer of understanding has permeated the military mind to the point that it is not just a learning process but has become, for some unknown reason, a secretive process. Let us continue now with post-war cases which show that our military is still interested in those things that supposedly do not exist.

07-26-1956   *Port at Rio de Janeiro, Brazil*

[Ship's name: USS *Franklin D. Roosevelt* (CV-42)]
    These objects...finally vanished with tremendous speed...
    "All involved were told to keep a tight lip on what we saw. A report was filed, signed and sent to the Defense Department, along with a color sketch by me of what was watched that night by many." [367]

10-??-1958   *Caribbean Sea near Cuba*

[Ship's name: USS *Franklin D. Roosevelt* (CV-42)]
    "I saw a bright white ball of light. It headed straight for us, getting bigger and brighter...."
    As this was going on, the man on watch was yelling into the intercom for an officer to get up on deck.
    Grusinski continues, "I could see silhouettes of figures looking at us. They had no features and you could tell they weren't human. Then the bottom turned cherry red and it vanished in a flash."
    Soon after the incident crew members who talked about it were transferred. Then the CIA came onboard to investigate a so-called "gambling problem."
    "It was a massive cover-up," notes Grusinski. "There was no gambling...."

Strangely, these incidents have been removed from the ship's logs. [368]

UNDATED—12   *Gulf of Mexico - USS Lexington (CVS-16)*

Between 1962~66
The object (light) took off out of sight. As soon as the jets returned, it came back. This pattern repeated itself 3 times. After the 3rd time, guess they thought it was no threat. It followed us for about half hr [a half hour] I guess then left. **We were told the next day to say nothing about the sightings.** [369]

LL-??-1963   *Off the coast of Sardinia, Italy*

[Ship's name: USS *Franklin D. Roosevelt* (CV-42)]
"During this particular cruise there were several photographers on board to document the NATO fleet operations. They were present onboard the night of my contact on radar. **I was told in no uncertain terms to keep my mouth shut for 20 years....**"
Watch logs are always kept, particularly when a critical situation is underway. General Quarters practice drills are another matter, but this wasn't a practice drill. The captain did not call General Quarters. I recall vividly when it was first determined how unusual the situation had been. My division C.O....asked me what I had put in my log if anything. **He said to me, "Jordan, this never happened...."** I replied, "Yes, sir." I never mentioned anything to fellow sailors or anyone else until years after my Honorable Discharge from the Navy. [370]

01-12-1965   *Near Lynden, Washington, USA*

"**The Air Force** contacted me next day and after a thorough interview admitted that they had located a UFO on radar that night. **They told me not to talk to anyone— above all, not to newspapermen.** This was mostly for my own protection, they said." Because he believes the UFO secrecy is wrong, this Federal officer would have openly registered his feeling, but it was decided "not to have the ----- (agency) or me involved," the signed report ends. [371]

SM~FF-??-1965   *Atlantic between the Azores and Spain*

[Ship's name: USS *Albany* (CG-10)]
    The two air controllers (lieutenant two bars) smiled at each other and one turned to XX [witness's initials deleted] and said, **"This incident never happened."**
    (Initials deleted—witness) **suspects the air controllers were acting under an order not to discuss the incident but was never personally ordered not to discuss the incident.** [372]

07-MM-1966   *Pacific between Seattle, WA and HI, USA*

[Ship's name: USS *Tiru* (SS-416)]
    Before leaving the bridge, the captain took the camera from the chief and instructed each of us not to talk to anyone about what we had seen. He told us the incident was classified and we were not to discuss it, not even amongst ourselves.... He then asked me to read several pieces of paper that all said the same thing only with different words. I read that if I ever talked about what it was that I didn't see, **I could be fined up to $10,000 and imprisoned for up to 10 years or both. In addition, I could lose all pay and allowances due or ever to become due.** He asked me to sign a piece of paper stating that I understood the laws and regulations that I had just read governing the safeguard of classified information relating to the national security. **By signing, I agreed never to communicate in any manner any information regarding the incident with anyone.** [373]

02~03-??-1971   *Near the Azores*

The USS *Saratoga* [CVA-60] was on mission toward Scotland when a huge scraping was felt at the bottom of the ship, and it came to a shuddering, near complete stop. The bridge crew noticed a large blob of light moving rapidly away underwater. **The crew was told that they had hit a whale and not to talk about it.** The next night, a large blob of light was seen following the ship. The crewman/ witness says that the object was two times wider than the 35-40' wide ship itself. This latter pursuit lasted for about ten minutes. [374]

07-02-1971   *Atlantic Ocean near the Bahamas*

[Ship's name: USS *John F. Kennedy* (CV-67)]
A few days later, as we were approaching Norfolk, the commanding and executive officers came on the closed-circuit TV system that we had. They did this regularly to address the crew and pass on information. During this particular session the captain told us how well we did on the ORE and about our upcoming deployment to the Mediterranean. **At the very end of his spiel, he said, "I would like to remind the crew that certain events that take place aboard a naval combatant ship are classified and are not to be discussed with anyone without a need to know."** This was all the official word I ever received or heard of the incident. [375]

09~10-??-1974   *Pacific Ocean 15 miles off Guam*

[Ship's name: USS *Reclaimer* (ARS-42)]
Guam told us it was a classified matter not to be reported or discussed. The Capt., Lt. Cdr. Xxxxxxxx, came on deck...took the log book...end of story. [376]

LL-??-1974   *Near Vietnam - USS* Reeves *(DLG/CG-24)*

When we pulled into port a few weeks later, **all records in the Bridge, CIC, and Sonar logs about that time were torn out.** Whoever tore them out made one error; the pen imprint on the next page was still there. No one spoke of the incident again. [377]

04-23-1976   *Atlantic Ocean southwest of Bermuda*

The captain [of the U.S. Navy destroyer] **immediately advised all personnel on the bridge and on deck to forget the incident, and then, presently on General Communications, likewise advised the ship's company to forget the entire affair. Once again the following morning by means of General Communications, the captain reminded the crew to dismiss the previous night's incident**. Then according to Lieutenant Hedison, the captain entered the following remarks into the ship's log:

"At approximately zero-two-four hundred hours on twenty-three April 1976, Destroyer DD-000 on a course of two-nine-two, at 32° north longitude, 67° east latitude, did encounter an unidentified surface vessel." [378]

??-??-1978/9   *Gulf of Oman near Iran*

[Ship's name: USS *Nimitz* (CVN-68)]
Several members watched the object for several hours through binoculars. **Commanding officers informed those who witnessed the object that they were not to talk about it.** [379]

WW-??-1979   *South China Sea*

[Ship's name: USS *Tuscaloosa* (LST-1187)]
The skipper gathered us up in a huddle and asked that we not talk about what had happened.

The quartermaster asked the captain how he should make his log entry, and they went into the QM shack for a powwow which I was not privy to, **nor was I able to read the log later.** [380]

06-23-1983   *Sea of Japan - USS* Midway *(CV-41)*

At 1215 p.m., **the ship's captain, John D. White II, told the bridge watch to remain silent about this incident. He said this is a direct order, that this incident never happened....**
This is not just a usual UFO story; this is a case where the military actually tracked 5 UFOS, and **all of the bridge crews which were in formation at the time with the USS *Tuscaloosa* LST-1187 were ordered to keep quiet on this unusual phenomenon.** [381]

SP~SM-??-1994   *Pacific Ocean - USS* Constellation *(CV-64)*

I proclaimed my excitement and was met with instant rebuke. I was asked if I knew what the craft was. I advised an alien. I was advised it's unidentified, so log it correctly. I flipped through half the book for the next open page. I made my log entry and handed the book back. **I was then told with utmost seriousness and veiled threat, "Log it and forget it."**
From quietly talking to other shipmates, I found this is a normal occurrence in the military and that the punishment is extremely severe and non-public. So I'm sure you will understand my need for anonymity. [382]

It is so comforting and cozy to realize that not only does *our* military consider these supposedly nonexistent things sufficient enough to classify as secret but so do other countries:

11-??-1954   *South of Lundy Island, United Kingdom*

[British Naval sighting]
In the meantime, **Captain Chelwan has been ordered to treat his encounter with the U.F.O's as a military secret and to instruct his officers and men accordingly.** [383]

07-23-1958   *Cook Strait, New Zealand*

Officers of the *Pakura* [a freighter] who reported the object to authorities told the press that **they had been given instructions not to divulge a description of the craft.** However, Mr. D. Fife, who was the *Pakura*'s second officer and the officer of the watch when the "submarine" was sighted, described the vessel as "huge" and "traveling at speed" on the surface. It had features which made it unlike any he had seen before, he said. [384]

??-??-1962   *Bass Strait, Australia*

[Ship's name: HMAS *Voyager* (Australian)]
I asked my mate, who was on starboard watch, what he made of it. He looked at it through his night vision binoculars, simply said, "Yeah, I see it," and was silent until I goaded him to report it. He said, **"We don't report things like that; I have learnt that the hard way."** I could see he had had such experiences before, but despite my constant persistence, he would say no more about it. So I left none the wiser, wondering what the hell it was. [385]

[While searching for information about the ship in this case, I came across a second report which seems connected to this one.]

After half an hour **the crew were given orders not to discuss the matter with the press and to keep quiet about it.** [386]

02-28?-1963   *30-50 miles off Spitsbergen, Norway*

[Unknown British Naval ship]

The senior officer proceeded to go over the events of that morning, asking questions about the radar-sonar observations. **He told the six men that their conversations were being taped and explained that until more was known about the unknown target, they were to remain silent about what they had seen. "Gentlemen," the officer said, "we will remember that we have all signed the Official Secrets Act (or words to that effect)." Although there were no threats, the implication was clear that to divulge anything to anyone concerning the tracking of the UFO would be considered a breach of security.**...

When Preston came on duty once again at 2400 hours, he said **he was surprised to discover that a "spanking new book" had replaced the radar log used the previous morning.** [387]

01-19-1966   *Tully Lagoon, Queensland, Australia*

[Civilian sighting]

Flt. Lt. T. D. Wright, for Air Officer Commanding, Headquarters Operational Command, RAAF, Penrith, New South Wales, forwarded Moylan's report on Pedley's UFO sighting and Wallace's covering minute paper to the Department of Air, Russell Offices, Canberra. His communication, **classified "restricted"** and channeled to the Directorate of Air Force Intelligence (DAFI), also indicated, "This headquarters believes that the depressions of the swamp grass were caused by small isolated waterspouts." [388]

So am I to assume from HQ's "belief" in the above text that we are no longer allowed to talk about "waterspouts" because they are classified as "restricted"? Wait until the weather bureau hears about this.

SP-??-1967   *Atlantic Ocean south of Bermuda*

[Ship's name: HMCS *Annapolis* (Canadian)]

On cruise with the HMCS *Annapolis*, a lookout spotted a light and reported a "UFO." A second lookout reported the same. Officers confirmed this and even the captain was roused to see.

The object was an elongated cigar (c. 250 feet) with several windows in it, within which "people" could be seen walking around inside. **Seamen were told to keep their mouths shut about the incident.** [389]

10-07-1977 *Unnamed sea, Russian submarine repair ship*

"What was the attitude of the navy hierarchy?"

"Many officers were skeptical, but the reports were hard to deny. They were carefully recorded. On October 7, 1977, a submarine repair ship called the *Volga* was at sea when nine disk-shaped objects circled it. This lasted eighteen minutes. And all the time the radio, the onboard communications systems, all the electronic equipment went berserk. **The commander, Captain Tarantin, ordered his men: 'I want you to observe this carefully and to remember it! I want you to take pictures and to draw it so that when we return to the Soviet Union, no one will be able to say that your captain was drunk or crazy!'"**

"**Were such reports made public?**"

Azhazha shook his head no. "At the time, naturally, they were classified top secret. Now we have a more open attitude in this country. We are able to talk about such things. On a few rare occasions I did publish some UFO information, but I paid for it under Leonid Brezhnev. The official reaction was very harsh, very negative. My career suffered as a result, and the directorship of a scientific group was taken away from me." [390]

SP-??-1989   *Sea of Okhotsk, Russia*

"They tried to penetrate the device with an oxyacetylene flame, but without any result," Nicolai told me. "The naval officer informed Valeri that the object was later transferred to Vladivostok and then Moscow. I asked Valeri to try and obtain more information, but **he told me that the officer refused to speak anymore, as it was a top-secret matter.**" I have been informed that the Soviet Navy had specific instructions for reporting anomalous objects. [391]

07-27-1999   *Lake Backsjön, Sweden*

[Civilian sighting: Swedish military divers search lake]
After midday on the 16th of September, the search was discontinued. On October 1st, a report was finished for further distribution to the commanding officer and the military intelligence and security at headquarters in Stockholm. **Two pages were not classified, but transcripts from the security police's interviews and a detailed description of the techniques used during the search were classified.** UFO-Sweden has filed a request to declassify parts of the report. In all, the search cost 150.000 Swedish crowns. When I asked the colonel of the first grade, Yngve Johansson at FO 52, what people who see unidentified objects in the sky should do, he replied:
"I think that the public should report all the information and possibilities to us. We can then build up a picture of what happened. One incident does not give us the answer, but many could." [392]

8.7   The U.S. Navy and Flying Submarines

While one incident does not influence anyone, I would imagine that the U.S. Navy, having had many of these water-related UFO reports, considered a UFO to be a good weapons system and thought of imitating it. Okay, so UFOs go in and out of water. Why? We really don't know; all we have is speculation. Of course, is there any reason for us to want a craft like theirs? I mean aircraft do aircraft things and submarines do their thing, so what need is there for yet another expensive piece of machinery for warfare? Avoidance of radar and visual sightings? Well, it seems that not only was the emulation of these alien craft thought of, but the idea was funded as well. Think the Navy might be interested in those things that don't exist? Let us explore this next, keeping in mind that this was in the year 1964.

Navy Weighs Sub-Plane

This drawing, by the Convair Division of General Dynamics Corp. in San Diego, California, showed how a proposed combination seaplane-submarine would have appeared underwater. The company was studying the feasibility of such a craft under a $36,000 Navy contract. The craft would have been able to submerge by flooding the wing, tail, and hull compartments. Powered by batteries, it would have been able to travel 5 miles an hour underwater.

The following *Proceedings* magazine article comes from the U.S. Naval Institute, a private, professional agency which presents independent forums for discussions of national defense issues:

The Flying Submarine

During World War II, midget submarines were used extensively and often quite successfully by the British, Italian, German, and Japanese navies. These craft ranged from miniaturized versions of conventional submarines to modified torpedoes fitted with seats and controls for a two-man crew.

Because their primary mission was the destruction of ships in harbors, most of these craft had a relatively slow operating speed. Their other major limitation was a lack of range, and it was usually necessary to tow or carry them most of the distance to their objective and then retrieve them after their mission was accomplished.

Often, because of these limitations and the difficulty of recovery in enemy-controlled waters, the midget submarines were abandoned after accomplishment of their mission and the crews were either lost or captured. Thus, a report on the success (or failure) of the mission was not relayed back to the command in time to capitalize on the submarines' operation. This situation occurred on 19 December 1941, when three Italian "human torpedoes" seriously damaged the British battleships *Queen Elizabeth* and *Valiant* in Alexandria harbor. Both battleships were incapacitated for many months, but it was possible to keep them on even keels, and enemy intelligence, primarily from air reconnaissance, did not immediately learn of their complete incapacity.

The most needed improvements for the midget submarine appear to be increases in cruise speed and radius so that the submarine is able to return from missions without immediate assistance from other craft. The very low speeds required during attack were usually adequate. In fact, low speed and high maneuverability coupled with the ability to operate with the utmost of stealth are the prerequisites of a successful midget.

The usefulness of a weapon depends upon the type of warfare needed to combat the enemy. In the event of war, the Soviet Fleet is intended to destroy lines of communication and supply to our allies and to attack our territory with submarine-launched missiles. Our Navy's primary tasks are to protect our lines of communication and supply, to protect the country from enemy attack, to provide carrier-based striking forces for use against attacks on the enemy homeland, and to destroy enemy shipping. There is, however, a tremendous amount of shipping in the Soviet-dominated Baltic Sea, the essentially land-locked Black Sea, the Sea of Azov, and the truly inland Caspian Sea. These waters are safe from the depredations of conventional surface ships and submarines. It has been demonstrated repeatedly that a warship can only rarely penetrate the Dardanelles, Bosporus, or Kattegat if held by an enemy, and there is no reason to believe that the situation would be changed in a future conflict. If the Soviets believe the extremely dense shipping in the above bodies of water to be safe from underwater attack, then they will have no ASW surveillance or equipment in these areas.

Since it is probable that no conventional undersea craft would be able to enter inland waters such as those previously mentioned, it is necessary to develop a new concept of weapons delivery. It has been suggested that large seaplanes could carry midget submarines to their destination and later return to take them on board for return to friendly territory. However, the inherent disadvantages make this method impractical. A seaplane capable of carrying a small submarine would probably weigh a half million pounds or more, be prohibitively expensive, and be a conspicuous and desirable target for the enemy's air-defense systems. The possibility of success in retrieving the submarine would appear to be hopelessly small.

An air-towed midget submarine has many attractive features for such a role: detachable aerodynamic surfaces could be dropped after alighting, leaving a conventional small submarine to carry out an operation; there would be no need for in-flight power; and the tow plane could carry out its own diversion mission to protect the submarine. Such a weapon would have one serious disadvantage: being expendable, the undersea craft's crew would be faced with the demoralizing propositions of surrender or attempting to go through aroused enemy lines to reach friendly soil. If the wings were retained, the consideration could be given to a second flight by the tow plane in an effort to pick up the submarine; but again, the chance of success would appear to be slight.

Another alternative, a true *flying submarine,* offers more promise than either of the above methods. It could fly from a favorable location to its destination at minimum altitude to avoid detection by radar. At the completion of its underwater mission it could travel as a submersible to a location best suited for takeoff, become airborne, and return to base.

There are many alternate approaches to the design of a flying submarine, as has been made evident by the numerous proposals of recent years by reputable engineers. The basic mission and the requirement for compatibility of aerodynamic and hydrodynamic characteristics require that the performance of practical vehicles be rigorously limited to minimum aircraft and submarine capabilities. Size, speed in air and water, submergence depth, and payload must all be realistically established prior to serious consideration of any preliminary design. Each capability taken separately is extremely modest. It is the combination of these capabilities into a single craft which provides a remarkable vehicle.

The preceding discussion of proposed useful missions envisions an operating depth of about 25 to 75 feet, submerged speed of five to ten knots for four to ten hours, airspeed of 150 to 225 knots for two to three hours, and a payload of 500 to 1,500 pounds. It is believed these characteristics can be attained within a vehicle weighing only 12,000 to 15,000 pounds. The Bureau of Naval Weapons has recently awarded a contract to the Convair and Electric Boat Divisions of General Dynamics for analytical and design studies of the essential components and operational aspects of such a vehicle.

When an operating vehicle has been developed capable of achieving these moderate goals, it will then be time to consider a more versatile successor.

There are obviously basic design problems involved in any concept of a flying submarine no matter how modest its capabilities. The basic problem is suggested by the very term "flying submarine." The vehicle's density must be comparable with that of conventional aircraft of roughly equivalent performance, yet must be susceptible of increase to that required for operation in its alternate medium, water, in order to submerge, cruise, and hover beneath the surface with minimum power. The cockpit, engines, instrumentation, fuel, batteries or electrical power-generating fuel cells, electric motor, etc., must all be watertight. All other spaces would be floodable to minimize the inherent buoyancy, and consequently need not withstand the static pressures encountered during the vehicle's submerged operation.

The aircraft engine requires only moderate modification for effective waterproofing while retaining a self-starting capability after completion of the submerged phase of the mission. Both intakes and exhausts would be placed above the static water line. The intakes would also be located out of the spray pattern of the planing hull or hydro-skis, as is the case with any conventional seaplane. Electric torpedo powerplants would furnish extensive background datum for the alternate propulsion system. The battery-charging equipment would be far less elaborate than in conventional submarines, if batteries were chosen in preference to fuel cells. With the latter, no charging equipment would be necessary.

There remain numerous problems inherent in the various systems and components required in the flying submarine. Most, if not all, appear to be capable of solution through the application of existing techniques and engineering practices. These problems can generally be divided into six classifications: buoyancy, stability and control in both media, vehicle habitability and crew survival, structural considerations, availability of equipment, and the parameters of design and operation.

The first of these areas, buoyancy, with its directly related yet opposed requirement for high density, covers numerous items such as ballast tanks and their flooding and purging systems, and fore-and-aft fuel transfer to maintain longitudinal stability in flight and in the sea. The second, stability and control, requires investigation of a single system to operate in both media; the best aerodynamic / hydrodynamic configuration—i.e., a comparison of the merits of a conventional arrangement, delta wing with and without tail, canard, cruciform, etc.; take-off and alighting technique conventional versus vertical; and placement of surfaces for satisfactory maneuvering and diving from the surfaced condition. Habitability and survival problems require primarily the combination of systems already in existence in submarines, high altitude aircraft, and man-carrying satellites, with emphasis upon canopy design, provision of air supply and purification during submerged operations, and emergency escape for both submerged and in-flight conditions. Structural considerations require the determination of dynamic loads in flight and submerged conditions as well as the static loads imposed by deep submergence. Materials must be selected to avoid damage resulting from corrosion and galvanic action. The equipment associated with the aeronautical, electrical, naval and associated industries must be surveyed to establish the availability of items required in the prototype submarine as well as to list those components requiring development for this unique craft. The development of design and operational parameters requires the derivation of interrelationships between equipment weight; air cruising speed and range; water cruising speed, range and maximum depth, and vehicle weight.

The development of a practical flying submarine prototype will be both complex and laborious, but the potential returns are substantial and valuable. Consequently the concept of such a vehicle merits careful engineering examination rather than the overly optimistic accolade of a few imaginative enthusiasts and the simultaneous cold-shoulder denial of the hard-headed realist. [393]

There is a more recent attempt at building another type of hybrid vehicle, one which I have not included because, while it is different, it is not anything like the UFOs in this book and has no other purpose than delivering munitions to a target. I have yet to read a single case where a UFO "bombed" anything.

We will next consider reactions to UFOs by the natural occupants of the water world, fish, as well as reactions by some other animals which live on land.

Chapter 9

Fish and Animal Reactions to UFOs

Humans react to UFOs in a multitude of ways, ranging from fear bordering on terror to just plain curiosity. Many books have been written that contain examples of this subject matter. One such book is *CE-5: Close Encounters of the Fifth Kind* by Dr. Richard F. Haines. Its chapters—"UFO Responses to Overtly Friendly Human Behavior," "UFO Responses to Overtly Hostile Human Behavior," "UFO Responses to Human Thought," and "UFO Responses to Miscellaneous Kinds of Human Behavior"—all contain related cases. We therefore have links of the perceived threat by the mind in addition to whatever else exists in the physical experience. However, just like us, fish and animals will react to physical dangers, such as fire, flood, and predators. These reactions open a door to observe what causes this fear—how a UFO or its crew might influence life forms with mental or physical forces which come from the operation of the craft.

9.1 Reactions of Fish

The following comments were written to me by researcher Joan Woodward. More information about this topic can be found in her *Animal Reactions to UFOs: A Preliminary Investigation from the Animals' Perspective*, MUFON Special Publication, July 2005:

> Fish also have been reported to react to a UFO that is in or near the water. Fish have the same senses as we humans: vision, excellent senses of smell and taste that serve as a sensitive chemical sensing of their environment, touch, and hearing. Additionally, the lateral line of the fish is very sensitive to vibrations, even very low frequency

vibrations (beneath human detection). Fish are cold-blooded and so are vulnerable to any sudden changes in water temperature. Most reports suggest fish are trying to escape from what they might consider a predator.

UFOs that move into the water might be expected to affect fish by physical concussion and injury, by physical release of chemicals or gases by the UFO, by vibration (heard or felt), by temperature change, or by lighting effects. UFOs hovering over water bodies might be expected to affect fish more by vibrations or temperature changes and perhaps by light if it is projected into the water from the object. [394]

Visual sighting of fish below the water's surface depends on both the water's clarity and on how near the fish are to the surface of the water. I am certain that in most water-UFO cases, any fish reaction was either not visible to the human observer or was overlooked, as the human making the report was more occupied with his or her own observation and emotions. This chapter presents cases where fish were observed reacting to the presence of a UFO. There are certainly other reactions to a UFO's presence that are more devastating and can be found later in this chapter in the section entitled "Dead in the Water."

The following snippets of text taken from cases on my website are entered here for your consideration. Of course, the total case will give a better overall insight into what might have caused the reactions of the fish.

08-15-1663   *Ozero (Lake) Zarobozero, Sëmkino, Russia*

"As the fireball was coming over water, peasants who were in their boat on the lake followed it, and the fire burned them by the heat, not allowing them to get closer. **The waters of the lake were illuminated to their greatest**

depth of 30 ft., and the fish swam away to the shore, they all saw that." [395]

??-??-1947   Pacific, Humboldt Current off South America

Thor Heyerdahl (*Kon-Tiki*)
We saw the shine of phosphorescent eyes drifting on the surface on dark nights, and on one single occasion, we saw the sea boil and bubble while something like a big wheel came up and rotated in the air, while **some of our dolphins [cetaceans] tried to escape by hurling themselves desperately through space.** [396]

11-06-1973 Pascagoula River, Pascagoula, Mississippi, USA

...submerged craft. "It was large," he said, "maybe more than nine feet in diameter, and it was round and metallic.... He assured me...that the object made no audible sound, that his boat did not pitch or roll or vibrate, or have its engine stop or falter, and that he saw no wake or eddying on the water's surface during all the maneuvers.... [W]hen he resumed fishing after the object and the Coast Guard boat left the area, **he found that the mullet and trout had also disappeared. "I caught no fish that night or the next," he said. "The thing scared all the fish away!"** [397]

07-02-1974   Navegantes Beach, Santa Catarina, Brazil

On that afternoon, fishermen were working on the beach when they spotted a disc-shaped object with small lateral protrusions approaching very fast but silently. The object fell on the ocean, 100 meters distant from the Navegantes Beach coastline....

Two days after the crash, two fishermen...reported to their colleagues that they had seen a disc-shaped object half buried underwater. The ocean water around the disc was terribly hot, so they could not touch it....**For two weeks, all fish seemed to have vanished from the area.** [398]

02-26-1975   *San Carlos Reservoir, Gila, Arizona, USA*

During the presence of the object [which was moving slowly at an estimated altitude of 4 feet], Mr. G. noted that some fiberglass fishing poles were tapping against their fishing boat which was beached a short distance from the camper. **Also, fish were jumping out of the water in an area 100 feet north and 100 feet south of their location and within 50 feet of the shore.** The men checked the lines on the poles, but there were no fish on them and they could not account for the vibration which caused the tapping. Mr. G. later theorized that it may have been a combination of the sound made by the jumping fish and the vibrating poles which woke him initially. [399]

07-??-1989   *Aguadilla, Puerto Rico*

[This case is presented in full in Chapter 3.2.]

Cataquet's experience didn't end there. "I went up for some more air and went down again. This time, I swam under it [the UFO], some three feet from the sand at the bottom, and I felt it expelling water from underneath, as if expelling the water it was sucking in from the top. **I could see fish swimming around me and it** when suddenly the thing lit up for a minute. I was scared, but I stayed. Then, almost a minute later, it lit up and made a nasty squealing sound, as if you were stuck inside a bell and you could hear the sharp vibration. **The fish took off...I don't know why I thought they were telling me something, telling me to get out.**" [400]

08-23-2001   *Atlantic Ocean off New Jersey, USA*

"The cylinder was hovering about 1000 feet over the surface and the direct spot that it was hovering over was about 1500 feet away from my boat. During the encounter, which lasted only about 45 seconds, the object disturbed my compass and my GPS. The screen on my GPS said, 'Cannot locate signal.' **Small bait fish became excited and swam to the surface.** The craft emitted no light, but a pulsing, electrical sound came from the craft." [401]

9.2　More than Fish

While the primary focus of this chapter is on fish, Ms. Woodward's research mainly involved animals. She wrote:

> This study involved primarily mammals. In the presence of a UFO, the most common reaction of animals was fear. The second most common reaction was an alert or alarm reaction. Also reported were indifference or unawareness, calm interest, and unease. When animals did react, the UFO's altitude was usually estimated to be 500 feet or less, and, likewise, the UFO's horizontal distance was estimated as less than 500 feet, and commonly as near as 200 feet or less. In other words, relatively close encounters.
>
> The next most important feature involved in animal reactions was sound from the UFO. In situations where animals did react, the human observer(s) reported sound from the UFO 49% of the time. And it should be kept in mind that animals hear (and feel) vibrations that human witnesses may not detect. Conversely, in cases where animals did not react, all secondary features of the UFO were absent (sound, light beams, physiological effects, EM effects, wind generation, vapor/mist generation, and odor). [402]

Since I have also found some water-related cases involving birds and other animals, I thought it might be worth mentioning a few here.

9.2 a. Birds in Water Cases

Though not generally noted in UFO literature, birds do react to UFOs. In the following cases, examples of birds and water are highlighted in bold to show this reaction.

03-29-1938   *Morecambe Bay, England, UK*

A group of fishermen in Morecambe Bay, England, reported one such oddity in March 1938. **"I saw a sudden scurry of seabirds rise off the water,"** witness William Baxter told the Liverpool *Echo* (March 29, 1938)..."Out of the water there rose something large and black, like a big post. It was at least eight or nine feet high, and it rose and fell three times, then disappeared." [403]

04-09~10-1958   *Atlantic Ocean off Keta, Ghana*

Ducks and birds uproar

Mr. Neal then noticed a very bright light rather low in the sky and in the direction of the solitary coconut tree moving upward into the sky from the direction of the sea. **It was apparently this bright light that was causing the commotion among the birds and animals.**...

On the following night there was a repetition with the difference that the object appeared to be much higher. It came up from the sea moving in an easterly direction. After hovering for about an hour, it moved high up into the sky at about 4:30 a.m., and Mr. Neal went to bed. **Incidentally, the birds and animals again gave warning as on the previous night.** [404]

07-04-1969   *Duck Pond (a lake), Ossipee, NH, USA*

[Witness's name] was fishing alone at 7:00 a.m. (E.D.T.) on July 4, 1969, on the west end of Duck Pond (**fishing was good**) when he heard an abnormally large splash behind him. He turned and saw the splash are [sic—was] less than 100 ft. from him then looked up and saw [a] 20 to 30 ft. ea. [sic—in] "diameter," shiny, metallic-appearing UFO almost straight above him and going up "rather fast." The water in Duck Pond is clear, but mud was brought to the surface at splash [sic—splashdown] and [witness's name] said he heard [a] humming sound like [a] defective electric motor. **Birds on shore became quiet and fish stopped biting.** [405]

04-09-1970   *Aufhofen, Baden-Württemberg, Germany*

...a transparent ball some 40 cm wide...was floating, indeed gliding along...a mere hand's breadth above the ground.... "The thing...rolled along...beside me...halted, did a right-angle turn towards the little stream of water, and stopped there....From...the ball there emerged, downwards, something resembling a hosepipe...I am convinced that it reached into the water....Then [the ball]...passed across the road, quite close to me, to a point a pace or so into the field. And there it vanished, so absolutely silently, and at such lightning speed, straight upwards into the sky, that I was unable to follow it with my eyes.

"One thing, however, that did strike me was that, **shortly before the occurrence, all the birds had flown off in the opposite direction. Never before have I seen a crow move its wings so fast.**" [406]

05-??-1999   *Dillon Dam, Zanesville, Ohio, USA*

...it was around 10 p.m. All three of us were sitting by the water's edge with our fishing poles in the water....Anyway, suddenly, right in front of us, it sounded as if something "huge" was coming out of the water....No one said a word, but instinctively, we all rushed backwards about 15 yards or so....

The water rushed down off of the unseen object. You could tell the object was rising due to the sound of the water falling back to the lake. **We stood there, totally scared** to death for a few minutes, never heard a sound, no animals, no birds, nothing. [407]

9.2 b.   Dogs in Water Cases

As pets, dogs are usually protective of their masters. Nevertheless, during a UFO encounter, they can share the same responses of anxiety and fear as humans. Dogs may bark nervously, run away, and so forth. Specific examples of such behavior can be observed in the cases that follow:

07-19-1946   *Lake Kölmjärv, Sweden*

All of a sudden a humming sound was heard from the sky. "I looked up since I thought it was an airplane," says Knut today. "Instead, I spotted a rocket-like device diving towards the lake...." A tall column of water emerged and was soon followed by yet another cascade....The magnitude of the crash is underlined, however, by yet another witness (located in 1984) who...remembers everything clearly:
"The sound was horrible. I had never heard anything like it before—or since. My mother, who was washing clothes down at the shore, shouted at me to shut the windows because she thought it was a tornado coming in. **Our dog went crazy and ran away.**" [408]

09-??-1956   *Cabo Frio, Brazil*

"I was wearing only a bathing suit and had my three dogs with me. It was a dark night and when I arrived at the beach, I saw an object approximately 20 m wide and 3 m high moving from the sea towards the beach, similar to the operation of an amphibious-landing vehicle....I saw 2 persons get out....
"Then I pulled myself together in order to attempt another approach and **called my favorite dog 'Lion' to follow me. He ran, however, with drawn-in tail,** back to the hotel (as I later learned)....Both turned around then, went toward their vehicle, and entered again by a door. At the same time

I noted a fleeting shimmer of the interior illumination of the vehicle. This left the beach then with tremendous speed and flew over the sea...Only on the basis of the enormous speed and acceleration that I had perceived did the thought come to me that I had seen a UFO and not a navy vehicle.[409]

07-08-1962   *Unknown body of water, Nashville, NC, USA*

"It was a Sunday morning and the rest of the family had gone to church. I remained at home due to illness. I was about to put a pie in the oven when **I noticed one of our dogs running around to the back of the house**. Thinking the neighbor's dog was after it, I went to the front door to see.

"As I looked out the door, I noticed the oddest thing. The water in our front yard pond was splashing very high as if something had been dropped into it." [410]

01-19-1966   *Tully Lagoon, Queensland, Australia*

"I was driving the tractor through a neighbouring property on my way to my farm about 9 a.m. on Wednesday when I heard a loud hissing noise above the engine noise of the tractor. At first I ignored the sound, but suddenly I saw a spaceship rise at great speed out of Horseshoe Lagoon... It was blue-grey, about 25 feet across, and nine feet high....Travelling at a fantastic speed, it headed off in a southwesterly direction....But on my way home to Tully that night, I met Albert Pennisi, who owns the property where I saw the saucer. He said that about 5.30 a.m. on Wednesday, **his dog suddenly went mad and bounded off towards the lagoon**." [411]

01-09-1967   *Malta, Montana, USA*

The story all began a week ago Monday night Jan. 9. The Tremblays, who farm in the south country about 40 miles south of Malta, were preparing to go to bed shortly after nine o'clock when **they were warned of something strange going on by the barking of their dog**.

Wilfred, who was sitting on the edge of his bed, **looked out the east window of their home and immediately saw what seemed to be causing the dog to be alarmed**. It was a large object, rectangular in shape, with glowing red light all along the bottom and a larger amber light at the top, evidently toward the forward portion of the object in the direction it was moving. [412]

02-21-1967   *Sheboygan, Wisconsin, USA*

Object was seen at 8:40 p.m., coming from the southeast at a height of 80 to 100 ft....300 yards distant from witness... He also noticed something like heat waves behind [the] object. **His dog's barking caused him to look up the drive, thereby seeing the object**. He called the Sheriff's Department....

At 10:30 p.m. the night of the sighting, they were asleep or in a half sleep. Both awoke by a sudden strange noise of a high-pitched whine with an undertone hum being steady for 30 seconds or longer. **The dog outside was yelping and barking during this time.** [413]

02-23-1967   *Severna Park, Maryland, USA*

[Investigator's notes taken during a phone interview with the witness]

Greenish-gray object size of small plane, circled houses of witnesses, threw shadows on ground. When first seen by Mrs. Weston was about 30' off ground. Had short, stocky wings, fin at back, front rounded, mushroom-shaped, another fin in center of mushroom. As object came down very slowly, making circles, **family dog became highly agitated**. [414]

03-13-1975   *Near Mellen, Wisconsin, USA*

The next morning Jane went outside to see if she could find any traces from the previous night. She looked over toward the swamp near her home and saw the same or a similar object again—only this time there were no flashing lights or glow. The shape and color were the same, however, and it was hovering over some evergreen trees. She went back into the house to put on heavier clothing and when she came back outside, she brought the family dog with her. She was starting toward the evergreens when **the dog gave a big yelp and started to whine and paw at her ears, and then became completely still. Jane said that she could hear nothing and carried the dog back into the house because it refused to move.** When she came back out again, the object was gone. [415]

08-12-1976   *Ligurian Sea off Calignaia, Italy*

Around 10:30 p.m., three fishing enthusiasts **heard a dog barking** and later they noticed a strange glow in the distance. Shortly afterwards, this glow became more distinguishable, assuming the form of two lights coming from under the sea and no more than five or six meters from each other. [416]

06-22-1979   *Ligurian Sea Off Gorgona Island, Italy*

About 3 miles from the prow of the boat, they saw a huge object of cylindrical shape and black color. The object rose from the surface of the sea like a tower to a height of about 30 meters. At that point, the yacht approached carefully, while still being able to observe its shape and enormous dimensions. When they were about one and a half miles away from the object, it sank with a loud roar and a boiling of the waters, disappearing in less than three seconds....**Another strange detail was that during the occurrence of the phenomenon, the dog onboard the yacht barked and showed clear signs of fear and terror.** [417]

9.2 c. A Cat in a Water Case

This is my one and only water case involving a cat, but it is more powerful in a physical sense than most other close encounter cases dealing with this topic.

12-30-1972 *Small stream near Tres Arroyos, Argentina*

...he heard a loud humming noise...coming from overhead. Looking up, he saw a powerful light rapidly increasing in intensity and flooding the whole area. Within the area of light, he could clearly distinguish an enormous object....[A] powerful flash of light came from the under-part, blinding the witness temporarily. The flash completely enveloped the cat and then vanished...As soon as the flash of light ended, **the cat disappeared, even though she was suckling her kittens**. She did not reappear until February 16, *forty-eight days later*. Her back still showed scorch marks and burns. **She refused to go near the place where she had had her unpleasant experience**. [418]

Obviously, the emotional trauma left the cat with a scarred memory as well as physical burns on its body.

9.2 d. Animals in General in Water Cases

There are many stories of animal reactions in all areas of ufology. The use of the term "animal" applies not only to those in rural areas but also to those on farms where multiple animals of the same or different species are found.

08-15-1971 *Unknown location along New River, Guyana*

I was 15 and on my August holidays from school and travelling with my father and 2 other men on a surveying trip to the interior of Guyana. We were in the New River / Courentyne River area across from Suriname. **At dawn we were awake[ned] by the noise of all sorts of animals causing a commotion by the river.** We rushed from the

303

bushes to the river and saw a large, silver saucer hovering over the middle of the river. [419]

02-15-1977   *Swamp in Sundown, Manitoba, Canada*

A witness woke to the sound of agitated animals in her farmyard. She went outside and saw two sets of orange "windows" resting in a swamp about ¾ mile away. "Searchlight beams" were playing about the sky and seemed to originate from the swamp. [420]

There is scientific research on animal reactions (AR) to their environments, including sounds, lights, colors, magnetic lines of force, and so on. A gap occurs, however, when trying to relate this information to UFO sightings, which is what Ms. Woodward was trying to bridge in this AR report as best she could, given the limited tools available to her:

> In a small way the AR report addressed this sound issue. Where dogs reacted by becoming alert/ alarmed, the sound was described only one time as high-pitched (out of sixteen). When dogs reacted fearfully, sounds were described as high-pitched seven times (of twelve). Finally, four sightings describe dogs in pain relative to high-pitched sounds.
>
> It all cries out for more detailed data. So much data left out of the sighting reports. So many times quality/pitch of the sound is not even mentioned, not to mention any questions regarding the temperament or history of the animal so its reaction can be better evaluated. [421]

9.3   Lightning Strikes

Science has little on lightning hitting water and its ensuing effects. Aside from throwing a generated electric charge into some standing water, it would be virtually impossible to conduct a scientific investigation on the high seas for at least two reasons. First, it would be easy enough for meteorologists to find a possible thunderstorm on the ocean, but the severity of such a storm would prevent the distribution of instruments to measure from the strike point evenly over fixed distances. Secondly and most importantly is how do you "predict" where the strike point will be? This last question relates back to what I mentioned earlier regarding the fact that ufologists have no prior information about where a UFO might be seen or where it might land. *Here science and ufology have a similarity of inabilities.*

Upon hitting water, lightning spreads out both horizontally and down underwater in all directions from the strike point and weakens as it travels. Salt water is more conductive than fresh water and therefore lightning will travel further away from the strike point than in fresh water. Our concern here is "death" and the following can illustrate two reasons for death by a lightning hit (or a UFO's high energy electric field?).

> Ron Holle, research meteorologist, NOAA National Severe Storms Laboratory, Norman, Oklahoma, has this to say about lightning and water:

> Lightning can affect all parts of the body, but the usual cause of deaths is *heart stoppage* [italics added]. The electrical charge disrupts the heart's rhythm, stopping it. Usually, however, the heart will quickly resume beating. The electricity *is more likely to paralyze the brain's respiratory center* [italics added]. The victim will die from lack of oxygen unless someone nearby can quickly perform

artificial respiration to get the victim's breathing going again. This may have to continue for hours before the victim begins to breathe normally. [422]

When lightning strikes the wet trunk of a tree, it will turn the water into steam. This is very similar to what was discussed in Chapter 4 regarding the UFO's field possibly being a high energy force, similar to what lightning is. This could cause the conversion of water to steam as the field touches the water, and this is sometimes perceived as fog.

So again, electricity from the UFO's field could be a reason for this occurrence. As an example of electrical shock, I place here a bit of text from an account that can be found in its entirety in Chapter 14, "Sea Serpents or Water-UFOs?":

07-02-1893   *Puget Sound, Washington, USA*

One of the men from the surveyor's camp incautiously took a few steps in the direction of the water, and as he did so, the monster darted towards the shore and threw a stream of water that reached the man, and **he instantly fell to the ground and lay as though dead.**

"Mr. McDonald attempted to reach the man's body to pull it back into a place of safety, but **he was struck with some of the water** that the monster was throwing **and fell senseless** to the earth. By this time every man in both parties was panic-stricken, and we rushed to the woods for a place of safety, leaving the fallen men lying on the beach." [423]

The men did regain consciousness later.

9.4　Dead in the Water

Fish kills happen for a variety of natural reasons. Unless someone sees something odd, they happen without much human commentary. If a witness is at some distance from a UFO's entry into water or departure from it, his or her vision might not encompass any fish on the surface at that moment. If an odd light is seen entering or leaving the water and dead fish are observed days later, a witness might rightly or wrongly connect the events since the length of time for fish to die and rise to the surface could exceed the viewed UFO event itself, therefore accounting for the length of time between the UFO sighting and the appearance of dead fish.

Regardless of the sequence of events, ufologists are concerned with the cause of fish deaths because of possible links to UFOs. On the other hand, our interest in fish deaths is also connected to our awareness of health problems that could affect us directly such as such red tide and the like. In one case (02-19-1982 coming up on p. 311), where fish were taken to a laboratory for examination, the time between the UFO sighting and the discovery of the fish was two days, but an analysis was still accomplished. There was another case (12-30-1972 coming up on pp. 310-311) where an analysis was accomplished, but results of it were not forthcoming. Both health officials and ufologists are interested in the cause of death and are aware of bacteriological sources. However, when regarded within the context of a UFO event, compression, electrocution, and radiation poisoning fall more into our realm of interest.

When it comes to research in the areas of fish deaths and animal reactions, there is still a gap in that the necessary instruments, machinery, qualified scientists, and technicians cost far more than we individual ufologists can afford. We, unlike other researchers, do not usually get grants. In addition, I feel that ufologists should be granted access to information obtained from these laboratory examinations since any physical evidence collected can be valuable in our UFO research. A governmental office should not be allowed to restrict access to information for whatever reason it sees fit.

Though only a small percentage of reported cases mention fatalities, one should exercise caution when in the vicinity of a UFO. The causes of death in the cases that follow, like the nature and composition of the craft's protective field, are unknown, although one points to the possible conclusion that a concussion of the fish's air bladder by compressed water (shock wave) due to the rapid immersion of the UFO may be to blame. (See upcoming case dated 02-19-1982.) Death by other means could include electrocution, as this field has been called electromagnetic in the past, or radiation, which affected three people in the famous Cash-Landrum incident, as well as several other radiological cases. All of these possibilities are merely conjecture since we have but one laboratory result from all of these accounts.

9.4 a.   Dead Fish

If we think about a UFO and its powerful field, we must consider that if it is electrical, the electricity is traveling with it underwater and not just at the surface of the water. So, could the UFOs or strange lights sighted by the witnesses in the following cases be responsible for the deaths of the fish?

12-07-1954   Isla del Francés, Uruguay

...four male friends on a fishing expedition...saw a luminous object moving up and down in the sky. When it descended and appeared to land in the distance, they took a boat and rowed...to investigate....When the boat was within about 60 meters of the object...the...craft...took off and sped at an upward angle directly over the boat and flew out of sight.

The men investigated the spot where the object had been resting on the surface and found that **the water was hot. Next day they found many dead fish in the same area.** [424]

06-01-1958   Altafjord, Norway

...a silent "unknown aircraft" with no identifying markers crashed into the Alta Fjord [Altafjord]. At the impact site, 70 meters deep, a column of water rose up. The aircraft resembled a twin-engine, delta-winged jet. The witnesses were Bjørn Taraldsen, Nils M. Turi, Kate Julsen, and Rasmus Hykkerud. When others arrived half an hour later, **all they found were a number of dead fish**. [425]

07-08-1962   *Unknown body of water, Nashville, NC, USA*

"The water in our front yard pond was splashing very high, as if something had been dropped into it. I then walked over to the window and as I looked up, there it was....The object was circular, about 90 feet in diameter, and about 25 feet in depth. It was hovering about 75 feet above the ground and made no noise at all...." Mrs. McCain said water in the pond had splashed over the bank about eight feet, and the water is normally some three feet lower than the bank. **During the following week, dead fish kept floating to the surface of the water.** [426]

03-26-1966   *Atlantic Ocean off Florida Keys, USA*

...strange cigar-shaped object maneuvering in the sky quite low over the water...Then, what looked like a greenish volume of light, water or vapor, extended from the underside of the object down to **the surface of the water**, which they discovered **was strewn with dead fish.** [427]

10-04-1967   *Shag Harbour, Nova Scotia, Canada*

A row of lights was seen to submerge in Shag Harbour. [428]

Much has been written about the incident at Shag Harbour and more details about it can be found in Chapter 18, which is devoted entirely to that event.

While on a tour of Shag Harbour with researcher Chris Styles, I was told that a week after the UFO incident there (case dated 10-04-1967 above) that dead fish washed up on the shore of Shag Harbour. This is not in any of the UFO reports of the account, but Mr. Styles located a news clip (dated 10-12-1967) containing a picture that had this information. (I included it with the Shag Harbour case on my website.) Could this be yet another case of uninvestigated fish death with a "possible" UFO connection?

05-20-1968   *Moore Lake, Littleton, NH, USA*

...three young people whom Night Officer Victor Miller later described as "badly frightened" burst into the station house... shouting about a "red glow on the water" and a "thing" that had scared them while they were fishing....

When police went back to investigate further in daylight, Chief McIntyre says **they found horned pout strewn along the shore near the wharf. Only the heads, tails, and spines of the fish remained.** And since then other persons, such as John Smith, a shop teacher in Littleton's public school, who live near the lake have reported seeing red lights on the lake that night. [429]

This is a very unusual case in that no other account mentions skin and organ removal.

07-05-1969   *Valdez Channel, Alaska, USA*

Suddenly the water forty feet from the boat exploded into the air, and the water really turned orange for three hundred feet around the splash. A disk-shaped object with windows and lights shot up out of the water to about one thousand feet above. It hovered there. **The next thing I remember is the water around the boat was teaming [sic] with sharks**. [430]

In several cases where a UFO has been in the water, fish are killed. Sharks, being scavengers, would have a field day in a situation like that above.

12-30-1972   *Small stream near Tres Arroyos, Argentina*

...an enormous object...[A] shower of sparks shot from the under-part...It...descended still lower until it was no more than four to six meters above the ground....The site where the object appeared is surrounded by eucalyptus trees about ten to twelve meters in height. Most of the tops of these trees were scorched or completely burned....

A large quantity of dead catfish had been found in a small stream near the sighting. The remarkable feature is that **some of the catfish were gathered up and put into a refrigerator and on the following day were found to have turned dark red.** Five of the fish were sent to the Institute of Bromatology (science of foods) in La Plata. A report showing the Institute's analysis is unavailable. [431]

02-19-1982   *Lake Lacar, Argentina*

According to the witnesses, the UFO had plunged into the lake approximately 50 m to 80 m away from the coast (where the water is around 200 m deep).

For a minute, the witnesses silently observed the area where the object had submerged. Very slowly, the surface of the lake regained its calm. Without warning, the object shot out diagonally from the water, emitting a whistle and causing only slight rippling on the surface of the lake....

Two days later, a group of fishermen saw a large number of **dead fish floating on the lake's surface** and washed up onto the shore close to La Islita....

The studies carried out to determine why the fish died established that "all the corpses examined showed the cause of death to be a rupture of the swim bladder." The biologists' investigation also indicated that "the cause of the rupture was probably a powerful shockwave."

A high-speed municipal motorboat explored the lake, particularly the zone of the La Islita, searching for any clues. **This inspection revealed a large quantity of seaweed and dead fish floating on the lake's surface 300 m from where the UFO incident was purported to have taken place.** [432]

This is the one case where an analysis was carried out on the fish. If you go to the case on my website, there is an explanation of what the "swim bladder" is.

04-15-1985 *Unnamed pond in Granville Summit, PA, USA*

In **mid-March 1985 there was a large fish kill in a private pond** at a friend's summer home. She asked me to stay in the cabin at the site to protect the property...On or about April 15, 1985...[b]efore I could run, a flat thing, like a coffee cup saucer, rose up out of the pond. It hovered briefly and then turned sideways so that all I could see was a thin edge. As it did so, I could hear water running back into the pond. [433]

While it is a stretch of the imagination to assume the fish kill in March was caused by a UFO just because one was seen at the same place a month later, it is something to consider.

06-09-2004   *Tasman Bay near Nelson, New Zealand*

I saw a teardrop-shaped craft with spinning red lights beaming light into the water...Flat batteries in cars have occurred everywhere at the McKees Domain, and **dead fish are being washed up on the beach in Ruby Bay.** [434]

9.4 b. Dead Ducks and Seagulls

12-31-1997   *Pond in Rockland, Massachusetts, USA*

At 9:15 a.m., Lynn drove back to Reed's Pond. "To my amazement, a black ball hovered in the sky above it, as big as a silver dollar, but nothing was astir in the pond. **I noticed dead ducks and seagulls floating on the surface.**" [435]

9.4 c. Dead Professional Divers

07-02-1974   *Navegantes Beach, Santa Catarina, Brazil*

Two days after the crash, **two fishermen who were the best divers in the area were found dead and naked on the nearby rocks.** They were the only two who had enough courage to dive where the buoy was. They reported to their colleagues that they had seen a disc-shaped object half buried underwater. The ocean water around the disc was terribly hot, so they could not touch it. [436]

Heat, though possibly not directly related to the death of the divers, should be kept in mind in light of our discussion in Chapter 4 and the fact that the water in this account was hot. I believe the corpses were naked as a result of the bodies bloating after death and, with time, splitting the garments they were wearing. For future ufologists this should be a wakeup call for other cases in which a UFO sighting and death are linked—have the specimens analyzed and perform autopsies on the corpses. While there is no scientific proof linking deaths to UFOs, would a court of law consider it to be circumstantial evidence?

Chapter 10

Negative Aspects of Water Ufology

■

*** MISREPRESENTATION ***
Eltanin—"Thing" on the Ocean Floor

■
■

*** ANOMALY ATTACHMENT ***
Water Wheels

■
■

*** GROSS SUPPOSITIONS ***
Missing Ships

■
■

*** HONEST MISTAKES ***
Misidentifications

■
■

*** MALICIOUS INTENT ***
Hoaxes

■
■

■■■■■■■■■■■■■■■■■■■■■

10.1 General Introduction

In the preceding chapters, I have presented cases where researchers have been unable to identify various UFOs as some tangible, earthly object, even after periods of study and in light of today's knowledge of the subject. So in the absence of concrete evidence, they remain unidentified, and ufologists believe they could possibly be extraterrestrial craft. In this chapter, though, I would like to take a look at some situations which I would call negative because, having been interpreted to be something other than what they were originally reported to be, they have had a negative impact on UFO researchers, causing us to spend time and resources on things that have had no real attachment to UFOs.

10.2 Misrepresentation

"Thing" on the Ocean Floor

08-29-1964
A History of the *Eltanin's* Photo of a Deep-Sea Object

Here is a supposedly mysterious event, of which I am not very fond, that sneaks into ufology via the back door because at some point, some tried to link it to something unearthly. The truth as to what this "thing" is lies at the end of an account about the research ship *Eltanin* and happened on 08-29-1964. Keep in mind that the actual date of the photograph is in question as researcher Tom DeMary found a discrepancy between the trip dates and the news article. Since the discovery of what this is makes a change of date inconsequential, I feel the date should remain as originally cited.

10.2 a. This Is the Start of the Whole Thing

In the following news clip, the author of the article starts the ball rolling with the statement: "shows something like a complex radio aerial."

Puzzle Picture from Seabed

The American research ship *Eltanin* sailed into Auckland yesterday with a mysterious photo taken at 2250 fathoms 1000 miles west of Cape Horn.

The photograph, which to a layman **shows something like a complex radio aerial** jutting out of the mud bottom, was taken on August 29 by a submarine camera.

The camera is housed in a metal cylinder pulled along by a cable from the ship. It bounces along the seabed taking pictures at regular intervals.

Dr. Thomas Hopkins, senior marine biologist onboard, who specialises in plankton studies, says the object could hardly be a plant.

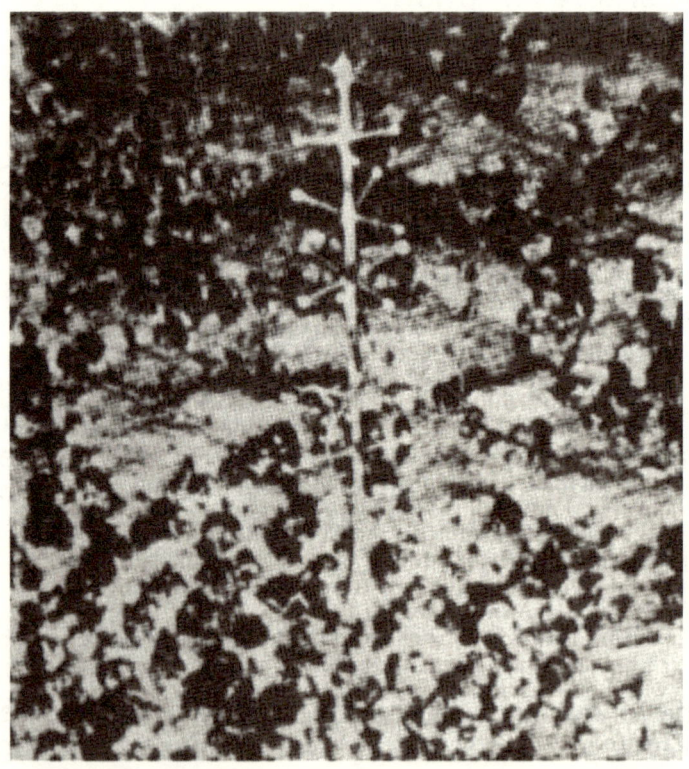

"At that depth there is no light, so photosynthesis could not take place and plants could not live.

"If it is some strange coral formation, then no one onboard has ever heard about it before."

Dr. Hopkins, a graduate of the University of Southern California, said the ship's photographer had been thoroughly questioned on how he had developed the photograph. However, everyone was certain the picture was not faked.

"I wouldn't like to say the thing is man-made because this brings up the problem of how one would get it there," he says.

"But it's fairly symmetrical and the offshoots are all 90 degrees apart. This is why it has been argued over for so long."

It has been estimated the object is about two feet high.

The photograph is to be sent to several United States research foundations for analysis, and Dr. Hopkins will take prints back to the University of Southern California.

The *Eltanin*, owned by the Military Sea Transportation Service, is being used in part of the United States National Science Foundation Antarctic research programme.

It arrived in Auckland yesterday and will be here for six weeks for a refit. [437]

That is as far as the case went in print. Then it was rediscovered in 1968 and used in a magazine article relating various mysteries of the sea, implying through the article's title, *Unidentified Underwater Saucers*, that it was connected to UFOs. The following is some partial text from this article which applies to the *Eltanin*:

The American ship *Eltanin* managed to get a **picture of something that did not belong under the sea**, but marine experts are not certain exactly what that weird photograph might be able to tell our science....

On August 29, 1964, 1000 miles west of Cape Horn and at a depth of 13,500 feet, the camera photographed an astonishing piece of **machinery with a complicated series of projecting rods that made the object look very much like the cross between a TV antenna and a telemetry antenna.** The device, whatever it was, looked as out of place on the ocean floor as a giant squid would look nestled amongst the computers at IBM....

The *Eltanin*'s undersea photo equipment took picture of still unexplained machine-like object.

There are other problems, which the placement of the unknown device leave eerily unresolved. The most immediate one is that our science does not yet possess underwater vehicles capable of descending to such depths; therefore, no one on Earth that we are aware of could have placed the device off Cape Horn. [438]

Coincidently, sometime in 1968, Captain Bruce Cathie, who previously had a sighting of a strange submerged object, published a book entitled *Harmonic 33*. The text of the book includes various theories that Captain Cathie posed in regard to the *Eltanin* photograph:

A surprise was in store when these photographs were developed. On one of the prints, in marvellous detail, was an **"aerial"-type object** sticking up from an otherwise featureless seabed....

As this bit of ironmongery [British for hardware] was situated at a depth of 13,500 ft. below the surface, **I was certain no human engineers had placed it there.**...

It would be interesting to know what the Americans have made of that picture and whether any attempt has been made to salvage the strange object they photographed by accident. In view of my earlier sighting in the Kaipara Harbour, I was willing to accept that the aerial had been placed there by an **Unidentified Submarine Object, or USO**. Can you offer a better explanation? [439]

With the passage of time, another approach was taken by the International Fortean Organization's publication, *INFO Journal*. It was the start of a more reasonable approach to answering this puzzle, but it was not yet the solution:

13,500 feet down—What is it?
[Published 1970]

The photo on the cover [same as the photo in the news clip] was taken by a deep-sea camera at 13,500 feet by the oceanographic ship *Eltanin*, Aug. 29, 1964, 1,000 miles west of Cape Horn. There has been speculation on what the thing is, including suggestions that it is some sort of machine or electronic device—artificial, but maybe not man-made.

Interesting as this theory is, we wonder if maybe the answer is that it's just a "new" type of animal. Note the picture of the *Umbellula,* a long-stemmed polyp about three feet high with a cluster of hydra-like tentacles. This photo was made 350 miles west of the Cape of Good Hope by the oceanographic ship *Kane* at a depth of 15,900 feet. The *Umbellula* was known before the photo, having been dredged up before when it was probably luminescent.

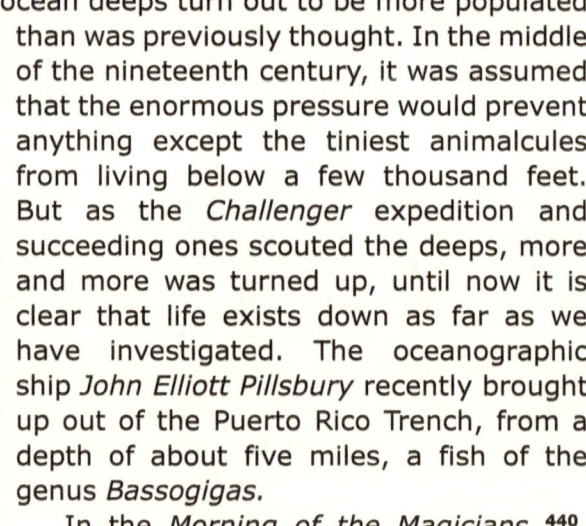

The ocean deeps turn out to be more populated than was previously thought. In the middle of the nineteenth century, it was assumed that the enormous pressure would prevent anything except the tiniest animalcules from living below a few thousand feet. But as the *Challenger* expedition and succeeding ones scouted the deeps, more and more was turned up, until now it is clear that life exists down as far as we have investigated. The oceanographic ship *John Elliott Pillsbury* recently brought up out of the Puerto Rico Trench, from a depth of about five miles, a fish of the genus *Bassogigas.*

In the *Morning of the Magicians* [440], Pauwels and Bergier note a strange track found in the mud by oceanographic probes at 15,000 feet. We obtained copies of this photo, but additional photos have shown the creature making the track—a type of acorn worm.

So maybe our outlandish thing down there in the South Pacific is some sort of animal like the *Umbellula*. But then again—maybe not. There are some other strange stories of tracks way down there. In the Arctic Ocean, 400 miles from the pole, Dr. Kenneth Hunkins dropped a camera through the ice to a depth of 7,000 feet and got photos of "chicken tracks," 2½ inches long, half an inch wide (*N.Y. Times*, Feb. 24, 1958). And in the Kermatek Trough, north of New Zealand, Nikita Zenkevitch got photos of "a big, unknown sea animal" at a depth of six miles (*Manchester Guardian*, Mar. 19, 1958).

These two cases suggest that some rather strange things must be living down there. The thing on our cover may be an animal as strange as we know how to imagine. [441]

10.2 b. *As Promised: The Truth of the Matter after Years of Speculation*

I would like to express my gratitude to Larry Hatch for being the focal point for the researchers who worked to resolve this matter: Murray Bott, Peter Hassall, Tom DeMary, Larry himself, and, of course, me. My thanks to Larry for permission to use his text here.

The *Eltanin* "Antenna" Identified
22 October 2003

The USNS *Eltanin*, an ice breaking cargo ship, was launched and acquired by the U.S. Navy in 1957. Reclassified as an oceanographic research vessel for the National Science Foundation in 1962, it became the world's first Antarctic research ship.

On 29 August 1964 while coring and photographing the deep-sea bottom West of Cape Horn, the *Eltanin* took the photo shown here [same as the photo in the news clip]. The position was 59:07′ S by 105:03′ W, depth 3904 meters.

The "*Eltanin* antenna" is the sponge *Cladorhiza*.

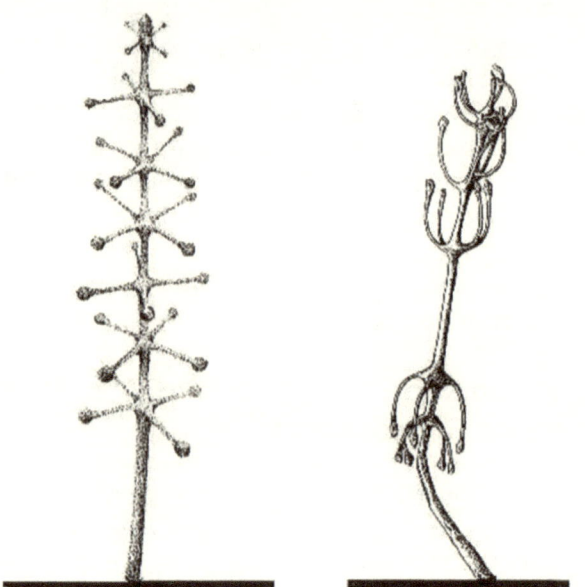

2.15 *Cladorhiza.* R, with clubs drooping as figured by early expeditions. L., straight and erect, from life (see 2.16).

A marine biologist pointed Tom DeMary to the book *The Face of the Deep* (1971) by Heezen and Hollister (see below) which reproduces the photo taken by the USNS *Eltanin* and a redrawing of an earlier drawing by Alexander Agassiz in *Three Cruises of the Blake* (1888).

An extract from *The Face of the Deep*: "*Cladorhiza*, a particularly dramatic [sponge] which somewhat resembles a space-age microwave antenna, was not uncommon in the early dredge hauls of *Challenger* and *Blake* (2.15). Agassiz observed that 'they are sponges with a long stem ending in ramifying roots, sunk deeply into the mud. The stem has nodes with four to six club-like appendages. They evidently cover, like bushes, extensive tracts of the bottom.'"

The familiar *Eltanin* photograph shows only a solitary specimen. Otherwise it would be more apparent that it is a marine animal and not a manufactured object.

Tom DeMary scanned the figures (2.15) & (2.16) from: *The Face of the Deep* by Bruce C. Heezen and Charles D. Hollister, New York, Oxford University Press, 1971 LOC Card Catalog Number: 77-83038 ASIN: 0195012771, pp. 39 & 40.

The drawing was re-drawn from a figure in *Three Cruises of the Blake* by Alexander Agassiz, Riverside Press, Cambridge, MA, 1888, #2: Fig. 541. (*Three Cruises* was 2 volumes, now both out of print. It may have been reprinted in the 1960s. It is packed with photos of life on the sea bottom.)

Tom DeMary deserves major credit for rediscovering the true nature of the "*Eltanin* Antenna." He contacted the marine biologist who pointed out the sources above. All of this was well known to the relevant biologists, but totally ignored it seems by the "new age" people, numerologists, and anomalists of all sorts.

It's a pity that anomalists and scientists don't consult one another. Henk Hinfelaar in New Zealand secured a clipping of the original report from the *NZ Herald*. Murray Bott archived it, and Peter Hassall tediously transcribed it. [See text of the *NZ Herald* article above titled "Puzzle Picture from Seabed."]

This is the original uncropped photo, giving the (often mistaken) location and identification. The source is noted above.

> A Google search for *Eltanin* Antenna brings up about 250 websites. Add the word "sponge" and you might find three or four. [442]

The above article was originally printed in 2003. Now (2009) a Google search for *Eltanin* Antenna brings up about 2,500 websites. Add the word "sponge" and it goes down to 383.

Larry Hatch was the one who was able to gather our information and post it to his website. Unfortunately, his efforts were not well received by the "believers," and Larry received e-mails containing verbally abusive messages and even death threats. It seems that belief in something intangible is, at times, stronger than fact. However, when intelligent investigation or reasoning comes up with a more logical, concrete solution to what has been observed, ufologists do alter their previous positions.

10.3 Anomaly Attachment

Water Wheels

What are water wheels? They have been described as having the appearance of a revolving wheel characterized by spokes, shafts, or lances of light. In some other cases, they are described as compact balls of light that rise from the depths and expand as they reach the surface (perhaps due to a release of water pressure as they rise). Are they real? It would seem so, as they have been reported in nautical journals by many captains of ships and their crews in early and current history.

Water wheels are, in my opinion, near the bottom of the water-UFO barrel. While solid lights have been seen exiting or entering water, I have not found one instance where a water wheel did the same.

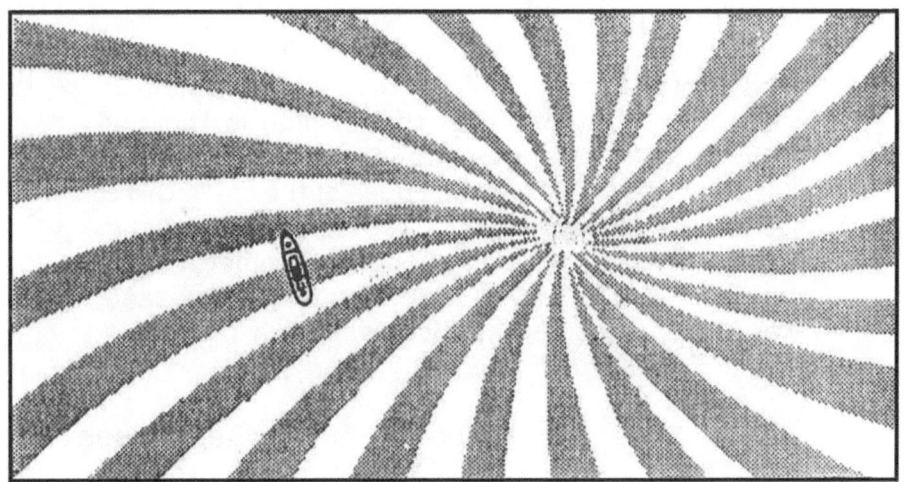

Sketch #46

**General configuration of a typical
phosphorescent wheel** [443]

So how does this anomaly get attached to a UFO? Well, envision that you have just recently been told a story by a trusted friend or associate about a silvery disc that rotated above the water not far from him and suddenly shot away at very high velocity. You want to write about it, but is there a precedent? Thoughts on this did not exist in the early years of ufology, so a little "digging" is in order. Up comes our original researcher, Charles Fort, who *has* written about shining lights in the water that were in a circular pattern—with spokes resembling a wheel. We now have a precedent; the two stories kind of match.

Are they a worldwide phenomenon? Here is the strange part: Probably 99.9% of these events occur in the Indian Ocean where geological forces are in constant play, such as earthquakes and tsunamis, whereas UFOs have been seen worldwide.

At the beginning of my collection of water cases, I had included water wheels as a possible physical representation of a UFO underwater. One case made me think that this was definitely a possibility:

> I have also a copy of a photo taken by a British visitor, R. Johnson, which shows, high in the sky over a lake in Ontario, a strange circle with a black dot in the centre. From the edges of the circle there jet out rays of light like the spokes of a wheel and it has a long tail like that of a larg[e] comet. It was seen on September 14, 1950. The time of passage was fifteen seconds, and if this were a meteor, it was a most singular one. [444]

Of course, the above spokes of a wheel made the connection to the water-wheel cases. However, there are not very many of these flying saucers with a similar description. Yet, in ufology several examples of water wheels have been put into books on UFOs, usually as a companion piece for a real water-UFO case, and therefore continue to add the fascination of a "perhaps" factor. Being (on occasion) circular, they fit our idea of a saucer shape. Their speed in rotation and forward motion makes us ask what natural fish or group of fishes or mammals could possibly do the same? However, until one comes out of the water and is recognized as a flying object, I feel that we have to put such reports at arm's length and wait for a more viable answer.

In the following introduction to water wheels in his book *Remarkable Luminous Phenomena in Nature: A Catalog of Geophysical Anomalies*, William R. Corliss has a similar problem in assigning the phenomena to a natural cause:

> *Possible Explanations.* Certainly bioluminescence is the most likely source of light, although observers frequently remark that the ship's wake is not luminous during wheel-type displays. The aerial wheels of luminous mist, if not illusory, would require airborne organisms in cases where no wheel is visible in the water proper. Earthquake tremors may

stimulate bioluminescence, with the interference patterns created by multiple sources accounting for the complex display geometries. The persistence of intricate geometries over many minutes seems to militate against this theory. Again, as in GLW3, the strong similarity of some marine phosphorescent displays to the so-called low-level auroras (GLA4) is striking. Some wheel observers have noted this, and electromagnetic forces should not be dismissed offhand. [445]

I would like to play the devil's advocate here and mention that in much of our UFO literature, electromagnetic forces are the crux of the explanations given for many of the physical properties observed in connection with UFOs. This could then, in our way of thinking, be the very reason for the stimulation of the marine creatures responsible for the phosphorescence. However, I reiterate that no craft has been seen in connection with this phenomenon, either under, over, entering, or exiting the water, so speculation is all we have.

Mr. Corliss's overview of the phenomenon on pages 357-359 in his book, in part described above, is quite interesting and puts the problem of these light sources clearly in the unsolved category. Hopefully, we can share the information gleaned from ancient and current texts to solve this riddle together.

Individually resembling a ball of light more than a wheel, underwater solid lights are clearly different from water wheels in that they do enter and exit water, so for that reason, I have devoted Chapter 17 to a separate discussion of "Underwater 'Solid' Lights." If the field, as discussed in Chapter 2, is totally encompassing, then we would expect it to ionize the water surrounding the craft just as it does the atmosphere, not as spokes or lines, but totally. This is my main reason for rejecting the water wheel as a UFO.

Despite my feelings about their incorrect association to the field of ufology, I will include several examples of water wheels here from observations made by crew members. The value of this first one is that an attempt is made to study the phenomenon:

04-??-1875   *Veracruz, Mexico*

Upon page 428 of this number of *Nature*, E. L. Moss says that in April 1875, when upon HMS *Bulldog*, a few miles north of Vera Cruz, he had seen a series of swift lines of light. He had dipped up some of the water, finding in it animalcule, which would, however, not account for phenomena of geometric formation and high velocity. If he means Vera Cruz, Mexico, this is the only instance we have out of oriental waters. [446]

In the next account, there are several interesting aspects:

1) The speed, which has been estimated in various cases ranging from next to nothing up to 200 mph, is mentioned here only as "great speed."
2) Origin is described as "somewhere below the surface."
3) There is the appearance of another wheel, but it is rotating in the opposite direction.

05-15-1879   *In the Persian Gulf*

Report to the Admiralty by Capt. Evans, the Hydrographer of the British Navy:

That Commander J. E. Pringle, of HMS *Vulture*, had reported that, at Lat. 26° 26' N., and Long. 53° 11' E.—in the Persian Gulf—May 15, 1879, he had noticed luminous waves or pulsations in the water, moving at great speed. This time we have a definite datum upon origin somewhere below the surface. It is said that these waves of light passed under the *Vulture*. "On looking toward the east, the appearance was that of a revolving wheel with a center on that bearing, and whose spokes were illuminated, and, looking toward the west, a similar wheel appeared to be revolving, but in the opposite direction." Or finally as to submergence—"These waves of light extended from the surface well under the water." It is Commander Pringle's opinion that the shafts constituted one wheel and that doubling was an illusion. He judges the shafts to have been about 25 feet broad, and the spaces about 100. Velocity about 84 miles an hour. Duration about 35 minutes. [447]

In the case which follows we again see the struggle to solve the enigma:

1) Here, the approximate speed is up to 200 mph.
2) Again, water is collected and examined with no specific abnormalities.
3) The travel of another ship through the beams does not disturb the beams.

06-02-1906   *Gulf of Oman (between Oman and Iran)*

Extract from a letter from Mr. Douglas Carnegie, Blackheath, England. Date sometime in 1906:

"This last voyage we witnessed a weird and most extraordinary electric display." In the Gulf of Oman, he saw a bank of apparently quiescent phosphorescence: but, when within twenty yards of it, "shafts of brilliant light came sweeping across the ship's bows at a prodigious speed, which might be put down as anything between 60 and 200 miles an hour. These light bars were about 20 feet apart and most regular." As to phosphorescence—"I collected a bucketful of water and examined it under the microscope, but could not detect anything abnormal." That the shafts of light came up from something beneath the surface—"They first struck us on our broadside, and I noticed that an intervening ship had no effect on the light beams: they started away from the lee side of the ship, just as if they had traveled right through it." [448]

In the following cases, luminous balls are mentioned:

10-14-1923   *Indian Ocean south of Sri Lanka*

"Balls of light" have been seen more frequently—as shown in 23 reports compiled by Professor Dr. Kurt Kalle and published by the German Hydrographic Institute. They describe **explosion-like balls of light** that suddenly appear at the water's surface, then spread quickly to cover 100 square yards or more.

One sighting involved the S.S. *Omar* which was, on Oct. 14, 1923, in the Indian Ocean south of Ceylon. During the night passage the captain recorded:

"These patches of light, whilst expanding, became very brilliant and, after about two minutes, died away. They commenced with a diameter of about one foot and expanded to at least 30 yards." [449]

08-23-1925   *Indian Ocean southwest of Sri Lanka*

Almost two years later, on Aug. 23, 1925, the German passenger liner *Preussen* was steaming through the area west of where the *Omar* had encountered the brilliant bubble patches. An officer of the watch described what he saw:

"From right next to the ship and up to the limit of vision, **luminous balls rose from the lower strata of the ocean**. Their distances from each other varied from 12 to 100 feet. The time interval between appearances varied between 20 and 30 seconds. The balls seemed to rise with a speed of about half a yard per second and had an estimated diameter of six inches. Just below the surface they expanded to a diameter of about half a yard. When reaching the surface, [they] expanded and flattened. It looked as if they were bursting. The area occupied by such a phantom body was ellipsoid in shape with a major axis of 6 to 10 feet. The larger ones retained their intense silvery-greenish luminosity with a shivering motion until we lost sight of them." [450]

10-31-1926   *Indian Ocean southwest of Sri Lanka*

A little more than a year later, on Oct. 31, 1926, the S.S. *Somersetshire* again encountered strange sea luminescence. The ship was in the same general area as that crossed by the *Omar* [See 10-14-1923] and the *Preussen* [See 08-23-1925]. P. H. Potter, the second officer, described this occurrence as "**balls of brilliant light** [that] seemed to shoot from the depth, burst on nearing the surface, irradiate and cover an area, seemingly of a couple of hundred square yards." [451]

Detailed next is yet another feature of both UFOs and this particular mystery—silence and calm seas:

08-25-1925   *Arabian Sea, 150 miles south of Oman*

One manifestation was observed from the S.S. *Somersetshire* on Aug. 25, 1925. It was about 150 miles south of present-day Oman when the second officer, P. Hawkins, observed:
"A white line seemed to be coming toward the ship, at a tremendous speed from the east, which had the appearance of breakers. Very shortly after, the whole sea was quite white, with now and again circular, streaky, black patches, and the whole surrounding area was brilliantly lit up."
Hawkins added this comment which was later found to be a singular condition in many of the luminous appearances:
"During this time (9:20 p.m. until 10:40 p.m.) the **atmospheric conditions were extraordinary. No sound was heard, not even the wind or the breaking of the sea**. No swell was visible and one vessel, which had previously been rolling heavily, had practically no movement. In fact, one could almost have been in dock." [452]

The complete text of the next case is a bit large to print here, but one aspect of it worth noting is the fact that the compass is *not* affected by the anomaly as it usually is in UFO cases:

11-14-1949 *Passed the Strait of Hormuz heading for India*

"At the time of the above, conditions were as follows: Date—14 November 1949; time 1830 GMT; position 26°-17.5′ N, 56°-51′ E. Wind NW′ly force 1. Sea calm with slight surface ripples; no swell. Air 75° (Fahr.), sea 83°. Visibility very good. A clear, bright night with no moon.

Vessel's course 157° T [True]. Speed through the water 11.6 knots. Actual speed over the bottom approximately 9 knots due to strong head current. (Very strong streams are encountered in this area.) **At no time were any unusual deviations of the magnetic compass observed."** [453]

Finally, I will quote one line from another case:

05-30-1962 *Gulf of Siam - SS* Telemachus

The depth finder was turned on but showed the charted depth with no variations. [454]

Assuming the "depth finder" is what we refer to as sonar, the lack of differentiation in charted depth would mean that there was no signal returned from a UFO, only the water's bottom.

10.4 Gross Supposition

Missing Ships

In Chapter Eight, page 115 of his book *Invisible Residents*, Mr. Sanderson delves into the Bermuda Triangle. Between pages 124 and 126 are the dates and names of missing ships, but aside from the fact that many ships do disappear mysteriously, I cannot imagine connecting UFOs to the disappearances. There has not been a single case of which I am aware that had a UFO that was seen visually or on radar "abducting" a ship. Of course, we all have seen the implication of this in the movie *Close Encounters of the Third Kind* (1977), where a missing ship is found in the middle of a desert. Another scene in the same movie deals with the famous missing planes of Flight 19 out of Fort Lauderdale, Florida, which had radio contact with their base for most of their known flight and then, having stated that nothing looked familiar and that their fuel was running out, disappeared forever. This flight never reported a UFO at any point yet this has been a "perhaps" factor in ufology for some time, as is evidenced in the above mentioned movie classic.

A more mundane observation can be found in the next case, where, asked what would have happened if a meteorite had tumbled on the steamship *St. Andrew*, Chief Officer, Mr. Spencer, said:

10-30-1906   *Atlantic, 600 miles N.E. of Cape Race, Canada*

"The ship would have been burned out immediately and every soul on board destroyed. I have no doubt that many of the vessels which have been lost at sea in apparently fine weather have been destroyed by falling meteors." [455]

A few years ago, I posted a note on my website in "Missing Ships and Water UFOs" in regard to meteorites and the sinking of ships in which I asked who shared or disputed Chief Officer Spencer's point of view. I received the following response from researcher Chris Aubeck:

> An observation we made a couple of years ago was that stories of meteoric bodies—almost—hitting vessels at sea were surprisingly common in 19th century newspaper reports. We never quite found a way to test the statistics of it, but I do recall an article from that century in which meteorites were theorized to be a factor behind the mysterious disappearance of ships at sea. This ties in with your [case] from 1906 [10-30-1906]. Check our archives for actual cases.

My humble personal opinion is that any talk of
UFOs abducting ships is nonsense, but sinking or
overturning ships is a different matter. [456]

In addition to meteorites, super waves or waves of great
height, up to as much as 100 feet and also known as rogue
waves, mentioned in the Introduction on p. xxix, have also
been linked to ships going down suddenly. In 2006, the
History Channel had a show, *Rogue Waves*, about the fear of
encountering such waves as well as the anxiety of reporting
them, given the prejudice against their existence. But for
hundreds of years, sailors have reported such waves. In
December 1978, the German vessel MS *München*, the size
of two and a half soccer fields and considered practically
unsinkable, went down in a severe storm in the North
Atlantic, possibly due to a giant wave. In 2001 the *Bremen*,
a cruise ship, sustained damage from a rogue wave in the
South Atlantic. The 680-foot-long (207 meters) cruise ship
Louis Majesty, carrying 1,350 passengers and 580 crew
members, was struck by three rogue waves on March 3,
2010, while sailing in the Mediterranean off the coast of
northeastern Spain, killing two passengers and injuring
more than a dozen others. On February 14, 2014, a rogue
wave struck the cruise ship MS *Marco Polo* as it made its
way into the English Channel with 735 passengers onboard,
killing one and injuring several others. As we recognize that
rogue waves are more common than once thought, we can
hope that science and technology will progress in learning
how to predict rogue waves and that better ship design and
construction will help to save lives and prevent loss of or
damage to ships.

Another History Channel program, *Methane Explosion*,
presented in 2007, demonstrated a scientifically orchestrated
experiment to find if the natural release of large amounts
of methane underwater could cause ships to lose buoyancy
and eventually sink. The result was positive.

For the above reasons, it is not logical to link the UFO
phenomenon to ship disappearances, and any attempt at

such should be regarded only as unnecessary speculation for the purpose of filling books' pages.

10.5 Honest Mistakes

Misidentifications

In many sightings of unidentifiable objects, they are initially thought to be UFOs, but after further investigation, they are recognized as being something other than what was first believed. With the exception of "spectacular" cases, very few of this type are saved for posterity. So why clutter up a book with cases that are known not to be UFOs? I feel that this category is very important from an investigator's point of view as it shows what to look for in new cases—a ruler, if you will, to measure the reality of the current case against the historical records of similar cases. By eliminating cases that have identifiable causes, we know that those which remain are the interesting ones. In short, we need the experience, built on recognition from previous cases, to remove those that will only cause doubt again in future.

In this category, I have collected 63 cases of misidentification. Note that in the following list, there are fixed causes attributed to some of the cases, but there are also other cases therein that are labeled "possible." This again becomes an issue of opinion on the part of both the observer and the investigator. I will not present the text of all 63 cases here, but I have divided them into types and will present one case that exemplifies each type.

I would like to thank Marco Bianchini and his associates at the Italian Center for UFO Studies, or CISU (http://www.cisu.org), for their valuable USOCAT database, from which I was able to learn what to look for in misidentification cases.

All the following cases can be found on my website. Dates in **bold** are the cases which I present following this table:

TYPES	# of Cases

Balloon (of any type) 06
   11-19-1954//11-27-1954//**07-LL-1969**//
   08-01-1986//08-19-1996//03-05-2000
Boat 02
   06-24-1978//10-23-1986
Buoy... 05
   09-21-1968//06-??-1975//12-17-1978//
   05-14-1980//10-25-2002
Distant sighting.. 01
   08-23-1969
Fireworks 01
   07-20-1996
Floating garbage 01
   07-19-1974
Fragment of phosphorus... 02
   08-28-1979//10-??-1999
Glare/Reflection 01
   12-14-1978
Large fish (dolphin, shark, whale).............................. 05
   08-??-1976//07-??-1977//**08-10-1977**//
   10-24-1978//10-27-1978
Meteorite.. 12
   11-17-1960//04-12-1961//05-13-1975//
   08-18~20-1978//08-26-1980//11-11-1980//
   04-21-1990//09-29-1993//10-27-1995//
   04-11-1999//08-01-1999//06-18-2000
Military exercise 02
   12-??-1978//01-01-1995
Parachute.. 01
   12-12-1965
Perception error 07
   08-25-1950//04-21-1959//07-05-1965//
   03-25-1990//09-02-1997//11-13-1997//
   08-29-2002
Remotely Powered Vehicle 02
   01-09-1964//**01-04-1979**

Rocket/Missile... 03
    06-30-1970//**07-??-1970**//12-14-1978
Satellite fragment.. 02
    10-??-1963//**07-20-1977**
Seaweed/Algae 01
    04-16-1980
Signal rocket 05
    04-??-1950//11-30-1957//08-25-1986//
    07-??-1997//**03-07-1998**
Submerged diver... 02
    08-??-1978//**09-16-1978**
Venus 02
    03-MM-1945//**11-25-1989**

Total 63

07-LL-1969   *Morro Bay off Cayucos, California, USA*

Type: Balloon

A large "saucer-like" object, seen landing gently upon the ocean and described by residents in the Cayucos and Morro Bay areas as being over 20 feet long, silver colored and as having a bubble-like covering, has been identified by a Coast Guard official as a weather balloon, possibly from Vandenberg Air Force Base.

Chief Alex Walker, U.S. Coast Guard at Morro Bay, estimated the balloon as 10 feet in diameter and of aluminized material. He said it had a clear plastic covering hood to protect the single frequency radio transmitter used to report upper wind conditions.

Walker said the balloon was deflated and retrieved by fishermen, who delivered it to the Coast Guard in Morro Bay.[457]

06-24-1978   *Adriatic Sea off Civitanova Marche, Italy*

Type: Boat

Around 12:30 a.m., designer Giovanni Salvucci noticed a light that was navigating on the surface of the water and

heading toward the shore. Under it he caught a glimpse of a spherical object approximately the size of the moon. The object seemed to be crushed toward the middle, and at one point, it emitted a flash which upset the witness to such an extent that he ran away.
Evaluation: Likely a boat [458]

Each case in ufology is a serious matter for ufologists. However, I have to include this one even though there are other mistaken buoy cases that are more rational. Enjoy.

09-21-1968   *La Escala (Gerona), Spain*

Type: Buoy

Have the aliens come to the Costa Brava? A sailor from La Escala claims to have been, for several seconds, ten meters away from beings from another planet. The incident put him into a state of great agitation and is the subject of much talk on the coast.

Juan Ballesta, 52, went fishing at daybreak near the small barren island called "Cargol" when he observed a fantastic scene:

"I was with my boat looking for bait near the small barren island when, some twenty meters away, I saw a rounded object that looked like a buoy. I wondered at its size and advanced toward it, but when I was only a few meters away, I saw how the object moved, and two beings emerged from inside. They wore tight-fitting black suits. Their faces were yellowish and presented a terrifying aspect.

"I rowed with all my strength toward the shore which was 100 meters distant and ran like a flash to the tavern where I told what I had just seen," the sailor continued his story.

Two members of the Guardia Civil and several municipal policemen who were in the vicinity quickly went in Ballesta's boat to the spot he indicated. Nevertheless, they did not find any evidence which confirmed his story. [459]

End of investigation:
[Text deleted since subsequent investigation showed it to be] "two people with neoprene suits for underwater fishing and a buoy." [460]

08-23-1969   *Mataró (Barcelona), Spain*

Type: Distant sighting

Two witnesses, Sr. Pedro Queralt and another person who were at the seaside, observed a luminous object that came down to the water level. It was just a light without structure, larger than a star, scintillating, and white. It disappeared, reappeared, rose at increasing speed, and was lost to sight. The sighting lasted about 10 minutes. [461]

End of investigation:
[Text deleted since subsequent investigation leads to the belief that this is] "not a close encounter; distant sighting; closeness to water because of illusion; probably in the horizon. Deleted from the landing catalog." [462]

07-20-1996   *Off Porto Empedocle, Italy*

Type: Fireworks

A telephone call to the Harbor Master's Office at Porto Empedocle reported an observation of a bright object with a blue trail that had been seen falling into the sea. The most credible hypothesis was that of fireworks seen falling into the water because in that stretch of the sea, a procession with boats was taking place for the feast of Our Lady of Mount Carmel. [463]

07-19-1974   *Cannizzaro, Italy*

Type: Floating garbage

An object with iridescent colors and indefinable shape was observed moving on the surface of the water by numerous reputable people on the beach and on the reef. The object

339

came from the north carried by the currents and was heading southward. A curious boy dived, recovering the object, and identified it as a bag of litter. The object came opened and inside a picture postcard depicting the cathedral of Orvieto was also found, addressed to a girl from Messina. [464]

08-28-1979   *Rimini, Italy*

Type: Fragment of phosphorus

A meteorite was seen falling into the sea near the beach about 150 meters from the coast at about 18:30. Immediately after the impact, a tourist seized it, immediately letting go of it because the object began to smoke and was glowing. Its effect ended when it was buried in the sand but returned again when in contact with the air. The object lit up again and became incandescent. This is a classic phenomenon of a so-called fragment of phosphorus. It is probable that the object is some device of a military nature with a component of phosphorus which burns at very high temperatures. The case, which falls more in line with FortCat [Charles Fort - Catalog], was included only because the object was seen falling into the sea and was initially mistaken for a meteorite. [465]

12-14-1978   *Unnamed lake in Avigliana, Italy*

Type: Glare/Reflection

Some people telephoned the police headquarters in Turin to report the sighting of a very bright flying object which landed on the surface of the lake and disappeared afterwards. According to a newspaper of that time, the observation was a reflection on the lake of a fire lit on the shore. According to Ansa [the largest and most prestigious news agency service in Italy], the sighting took place on December 14. [466]

08-10-1977   *Francavilla al Mare, Italy*

Type: Large fish
Another curious sighting that widely hit the press of the period happened on August 10, 1977, in Francavilla al Mare (CH). At the beach some vacationers observed a bluish light which was moving frantically among the waves about a hundred meters from the coast. Immediately rumors circulated about presumed underwater UFOs, so a Roman tourist, who was also a skilled fisherman, rowed in a thin-hulled racing boat and approached the mysterious light. And what happened at that point? Well, the witness, observing the light, did nothing but harpoon it, realizing at that point that all the commotion had been caused by a *Regalecus glesne* or a particular species of large herring. [467]

Though meteors, discussed on pages 333-334, burn after entering the Earth's atmosphere, meteorites may fall on land or in the water. The meteorite is by far the largest category of misidentifications but also one of the weakest categories in the water realm—objects that enter the water. This category is filled with space junk, burning aircraft, missiles, and the like, but it is a legitimate category as meteors are seen glowing brilliantly in our atmosphere, living up to the description of our UFOs.

04-21-1990   *Coast of Lazio, Italy*

Type: Meteorite (comet here)
Over the entire Italian peninsula, reports of sightings and observations followed one after the other of a large green-blue ball crossing the sky at high speed and with a very intense brightness. Even four pilots were able to observe the phenomenon. Some witnesses absolutely believed that they saw this object fall into the waters opposite Lazio so much so that a motor patrol vessel from the Harbor Master's Office carried out reconnaissance for a few hours. The mystery

was later solved with the observed object being part of the fragments left by the comet Thatcher. [468]

12-??-1978 *Venice, Italy*

Type: Military exercise

Some people observed objects that were moving on the sea and immediately thought they were UFOs. In truth, the mystery quickly disappeared when the objects and strange men observed turned out to be Hovercraft involved in a military exercise of amphibious maneuvers by troops of the Italian Army. Naturally the newspapers of the period, in the wake of the Adriatic phenomena, spoke of sightings of UFOs. [469]

12-12-1965 *Tyrrhenian Sea, Isle of Capri, Italy*

Type: Parachute

Some people claimed to have seen an object shaped like a parachute fall into the sea. A photographer, Willy Colombini, managed to take some photos. These were not published unfortunately, but a newspaper faithfully reported the description of what was photographed: clearly noted, above the clouds, an object. They were three black dots in a single vertical position. The center one was biggest and with a hemispheric shape; the two smaller ones seemed tied to a thread. Numerous boats, including the motor-ship *Mergellina*, carried out a search of the sea that did not produce any positive results. From the detailed description of the photographs, which were developed in the presence of Italian military policemen and a marshal of the Italian Air Force, it is probable that the object seen falling was a parachute, perhaps like those used for some meteorological experiment. [470]

08-25-1950 *Cervia, Italy*

Type: Perception error

An airplane was seen falling into the sea approximately five miles from the coast. Bathers observed the airplane falling until it grazed the surface of the sea, and fearing that it was lost, they alerted the authorities. Sources on August 26 clearly indicate that the plane had sunk, while those of the following day reported that the plane, after checks, had landed at the airport of Miramare. [471]

Yet more equipment lost:

01-04-1979   *Favignana, Italy*

Type: Remotely Powered Vehicle
A fisherman noticed an object in the sea and approached cautiously, taking into account its inoffensive nature. At that point, with a rope, he tied the object to the boat, dragging it into the port. Turned over to the military police, it was discovered to be a kind of mini-airplane of orange color with a white nose cone, about two and a half meters long and weighing about two hundred kilograms. The military police then delivered the apparatus to the Harbor Master's Office which said that it was probably a remote-controlled airplane used for oceanographic surveys. [472]

Although the following is listed as a missile case, I wonder about the phrases "remarkable dimensions" and "shaped like a priest's hat," not to mention "flying at a low altitude and nearly touching the surface of the sea," which to me makes this more a UFO than a missile.

07-??-1970   *Off San Benedetto del Tronto, Italy*

Type: Rocket/Missile
A witness, Mario Imperatori, a civil employee of the Prefecture, declared that he saw an object of remarkable dimensions, colored gray-silver, and shaped like a priest's hat rise from the sea approximately three thousand meters from him. Then the object quickly disappeared toward the west, flying at low altitude and nearly touching the surface of the sea. The observation lasted only seconds and the

object took about five seconds to disappear, leaving behind a luminous wake, like a flash of magnesium. The object always remained surrounded by a brilliant white halo. [473]

07-20-1977   *Off Sorso, Sardinia, Italy*

Type: Satellite fragment
While under the surface of the sea, an underwater fisherman, Paolo Pintus, came across an object that was on the bottom, covered with thick marine vegetation. The boy took the object and transported it to shore. Delivered to the military police, it turned out to be an aluminum cylinder with a conical and tapered tip, fifty centimeters long. The case was inserted into USOCAT, and not in Appendix A, since the press at that time clearly spoke about an object falling from a passing UFO. [474]

A seaweed UFO? Well, if you don't know what it actually is:

04-16-1980   *Venice, Italy*

Type: Seaweed/Algae
During the night, the waves of the shoreline suddenly turned from a variable bright color of blue to pink. During the day, the same waves appeared red. In addition, someone who went near the water was able to notice the presence of a glow, probably caused by the decomposition of protozoa that could suddenly have been proliferating in the area. In particular, it could be about the *Noctiluca scintillans* or some particular algae. [475]

03-07-1998   *Pantano Basso - Termoli, Italy*

Type: Signal rocket
A usual report was received of a small airplane that was falling about four miles from the coast, leaving a long bright trail and smoke after impact with the water. The search, carried out with a helicopter from the police in Pescara,

did not produce any results, but some soldiers found four red casings of signal rockets on the beach. Therefore, it is probable that the sighting could be explained as an observation of one of these rockets, probably used by smugglers or illegal immigrants about to come ashore. [476]

09-16-1978   *Off Naples, Italy*

Type: Submerged diver

Around the hours 21:30/22:00, two fishermen intent on fishing from the pier of the port of Naples saw the sea boil about three meters from them. On the surface numerous air bubbles appeared within a diameter of over one meter, and at the same time, they observed a beam of light coming from beyond the boats docked in front of them. The light was similar to that emitted by a flashlight and had a diameter of about 50-60 centimeters and colored the water a bottle-green color. Both the bubbles and the beam of light moved toward a nearby reef until they both disappeared. [477]

Next we get to admire the ever-popular Venus, probably correctly identified in this case.

11-25-1989   *Mediterranean Sea off Licata, Italy*

Type: Venus

Many people saw a very intense light, with colors changing from white to yellow, suddenly appear in the starry sky. The object, which would remain stationary in the sky for more than two hours, dropped slowly until it disappeared in the sea. It had a spherical shape and dimensions like that of a tennis ball. The newspapers of the period gave the observation of the planet Venus and its subsequent setting as a possible explanation. [478]

10.6 Malicious Intent

Hoaxes

Why are hoaxes so malicious and detrimental to ufologists and the work they do? Case investigations are usually financed out of a researcher's own pocket. In addition, they use their "spare time" to investigate cases. Therefore, if a report is a perpetrated hoax, it is indeed a waste of the investigator's limited time and funds. These malicious acts or hoaxes are usually intended to feed off the public trust and satisfy some desire for one-upmanship or self-gratification. What is sometimes done as a harmless practical joke can cause quite a media frenzy with the hoaxers watching the news to see the reports of gullible and baffled eyewitnesses. While only a small percentage of UFO reports turn out to be hoaxes, they challenge the credibility of all sightings. All they accomplish, in reality, is to derail or cast aspersions on legitimate investigations, of both past cases and those in progress.

The following is one of those cases that refuses to die, even though it has been assigned to the grave by many outstanding ufologists. Its most recent reincarnation was done by the History Channel in a series called the *UFO Hunters* in which **no conclusion was reached**.

06-21-1947 Puget Sound off Maury Island, Washington, USA

Capt. Edward J. Ruppelt, onetime head of Project Blue Book, called it the "dirtiest hoax in UFO history" because two U.S. Army Air Force officers died in the course of their investigation of it (Ruppelt, 1956). Kenneth Arnold, who also investigated it, called it "one of the weirdest things I have ever encountered" (Arnold, 1980). Echoes of the incident, which spawned the legend of the "men in black" (see Bender Mystery), resound even today (Rojcewicz, 1987). [479]

The case mentioned above is very long and much has been written about it. Given its length, it has no value here or anywhere else as far as most ufologists are concerned.

03-31-1950   *Riccione, Italy*

The crew of the motor fishing vessel *Maria Grazia* observed a flying object coming from the Carpegna Mountains which was initially confused with an airplane. The strange device, which flew without making any noise, made a sudden nose-dive and headed for the sea. The object fell violently onto the beach and, after a fearful bounce, it sank, lifting a gigantic wave to about two thousand meters from the stern of the boat. The object was later described as having a crushed egg shape of clear color. Naturally, numerous boats headed toward the place of the alleged impact for rescue operations. The mystery was revealed on April 2nd when it was confirmed by a daily newspaper that it was simply an April Fools' joke. [480]

I must admit that I have "feelings" for this next case. It is an early one (the official starting date of ufology was June 24, 1947), but as I have seen in many accounts, the witness's original story, which appeared in the *Steep Rock Echo*, has references to wind, which may have been caused by a UFO's rotating field producing a vortex. It also mentions taking on water, which is virtually the beginning of reports on this aspect. Was the employee's confession perhaps a means of getting the curiosity seekers and the UFO investigators to leave him alone?

07-02-1950   *Steep Rock Lake, Canada*

In a detailed letter to the editor, an anonymous correspondent claimed that on July 2, while picnicking on the shore of a cove in Sawbill Bay, he and his wife saw a flying saucer on the water:

The top had what looked like hatch covers open and moving around over its surface were about 10 queer-looking little figures....These figures I estimated to be roughly three feet six inches to four feet high and all were the same size...The most noticeable thing was that they looked like automatons and did not turn around...[T]hey just changed the direction of their feet.

...No one bothered to investigate the story, however, until the mid-1970s when Robert Badgley, a Scarborough, Ontario, member of the Tucson-based Aerial Phenomena Research Organization (APRO), found that Steep Rock employee Gordon Edwards had written the tale, which was entirely fictitious, to entertain readers of the magazine and to satirize belief in little green men from flying saucers. [481]

04-??-1961   *Montesilvano, Italy*

Probably this is about the most bizarre sighting of the entire catalogue. A few newspaper lines told how aeronautical journalist Bruno Ghibaudi observed a group of UFOs coming from the sky that were intercepted and attacked by other UFOs that had emerged from the waters of the sea. There is a high probability that this was a false account, told in order to enrich aeronautical stories of the time period. [482]

Of course, in this previous account, we have no confession of a hoax, only the supposition. The following case, however, *does* have a confession:

11-09-1974   *A pond in Carbondale, Pennsylvania, USA*

The water's mysterious green-tinged glow, which lasted nine hours, apparently came from a sealed beam lantern tossed into the water by a then-14-year-old boy.

"I created the story," said Robert J. Gillette Jr., who is now 39 and still living in Carbondale.

He said he tossed the flashlight into the water that day in an effort to frighten his sister, Maria, and her friends.

"There was nothing to do in Carbondale...so we had them come over and said there was a monster in the water or something bizarre," he recalled. "The flashlight burned out in the meantime and we had to get another one and that happened to be the lantern...."

When police arrived, Mr. Gillette and his friends...spun a tale of a whirring, sparking, red ball flying over Salem Mountain and plunging into the pond....

"Once the cops came up, I said, 'Oh, oh, we're in trouble,'" Mr. Gillette explained. "I was thinking Boys' Town or reform school. So I made the story up and they bought it...."

"When they threw the lantern in, it flipped over the right way and landed (with the light pointing) straight up in the air," he said. "The pond is green from all of the sulfur and stuff. It looked like the son-of-a-gun pond was green and glowing. To add to it, the warm water was getting cooled off and it was steaming. People thought it was boiling." [483]

In the following case, the joke is on the newspaper, even though it does not set out to deliberately mislead people. Here a university research group is involved in a legitimate experiment trying to gather data for its particular discipline. However, the method used to collect information is objectionable because, while the anthropological field may get its data at no cost to its own reputation, false reports harm the reputation of ufological studies.

07-09-1978   *Lake Paluda, Jesolo, Italy*

The Venetian newspaper *Il Gazzettino* published numerous letters relative to many sightings that happened in the area of Jesolo. One in particular was the story of two students who had observed something round, bright, and green that had landed in the lagoon. Only after a while was it discovered that all the letters had been sent by one research group, headed by Professor Franco Cagnetta, psychoanalyst and teacher of cultural anthropology at the University of Paris, which was carrying out a study on the creation of the myth. [484]

??-??-1979   *Off Roseto degli Abbruzzi, Italy*

A sailor, Renato di Carlo, while he was on board his boat, observed a sphere of dark red color that dissolved while plunging into the water. The same witness managed to take five photographs. The witness of the action later confirmed that it was a joke created with some friends in the climate of the ufological psychosis of those years. [485]

The following was not intended as a hoax but simply as another April Fools' Day joke:

04-01-2000   *Off the coast of Rimini, Italy*

On the wave of an April Fools' joke thought up by Radio Gamma, a listener called the radio station saying that he had seen the water ripple while he was fishing at sea; subsequently another radio listener stated that he had seen a violet light. [486]

04-11-2001   *Torremezzo, Italy*

An anonymous person has circulated a message via the **Internet** in which he said he saw a film of a UFO that emerged from the waters of the sea at 5:00 during a television broadcast by a private Calabrian network, led by the father of soccer player Mark Juliano. Ufologist Pietro Torre, who has been interested in the phenomenon, said that he telephoned the television network in question, TeleItalia, and the owner said that such a film never existed and they never put it on the air. (The case is probably from the second half of the '90s, as this is the period in which this athlete played in the Italian soccer championship.) [487]

Oh, yes. Give them a new medium to play with and play they will.

Chapter 11

UFOs—Taking on Water by Conventional Means

11.1 For What Purpose?

In this chapter we find our friends taking water into the craft using various means other than the one mentioned in Chapter 2 ("Physical Influences..."). On my website, there are currently 36 such cases. Sometimes they involve stories with aliens arriving like the next-door neighbor coming to borrow a cup of water.

The need that all earthly life forms have for water is quite evident, and I am sure we can extend this need to other life forms as well. But, for what purpose or purposes would extraterrestrials need water? There has been speculation that aliens may be testing our water for pollutants or simply for drinking. Of course, this planet could be considered the local "gas station" if their power is derived from H_2O, which I highly doubt, as that is more our own blast-furnace mentality of propulsion. Certainly, the need to water any vegetation they may extract from our soil might be another explanation. At any rate, in our solar system Earth does have the monopoly on *liquid* water.

In this chapter are cases of UFOs that hover over water, and yet in most of these instances, no effects are seen on the body of water over which they are hovering. In all likelihood, the reason for this is because the field is *off* in order to lower the apparatus needed to extract water from below the field. In most of these cases, observations are made of water being pumped or sucked up via hoses, pipes or tubes to the vehicle from a body of water. This makes us think, in our own narrow conception of possible purposes, why we would perform the same operation. In the following accounts, we will examine some of the reasons that the aliens give us for needing water. For starters, here is a simple taking-on-water case:

??-??-198. *Lake Constance off Stein am Rhein, Switzerland*

At Stein am Rhein, Lake Constance narrows until it becomes, after it flows around some tiny islands, the Rhine River and flows towards Basel. In the 1980s, Mrs. H. was in Kaltenbach near Stein am Rhein across from the palace Hohenklingen. From there was a panoramic view of the lake and its transition into a river. It was around 6 a.m. The sun rose, and from her position, Mrs. H. saw "a large, strange flying object," a classic flying saucer, which had a wide disk with a cupola above and a curved out cupola-like form below. **The disk-like UFO hovered over the river. Then the underside of the UFO opened. Out of the opened hatch came something like a long, very thin tube. It went into the Rhine. Nothing happened for ten minutes. Then the hose was drawn in again and the hatch closed.**

The flying saucer shot up vertically with enormous speed. It revolved, sped towards Germany, and vanished. [488]

Again, there are many short and long cases with just such a scenario, all of which display the aliens' need for water. So, in order to make this more interesting, I will present accounts which reveal more than simply taking on water because they contain additional information about both the need for water and demonstrate other physical effects noted in ufology as well. I begin with some of my favorites from years gone by, even well before the supposed beginning date of UFOs in 1947, thus showing they were around even before then. The first two cases are from what is known in ufology as a wave—multiple sightings that progress over a large expanse of geography. The year 1897 was such a period because there were many sightings that went from the west coast to the east coast of the USA.

11.2 Encounters with Aliens Fetching Water

04-22-1897   *Josserand, Texas, USA*

Considerable excitement prevails at this writing in this usually quiet village of Josserand, caused by a visit of the noted airship, which has been at so many points of late. Mr. Frank Nichols, a prominent farmer living about two miles east of here and a man of unquestioned veracity, was awakened night before last near the hour of twelve by a whirring noise similar to that made by machinery. Upon looking out he was startled upon beholding brilliant lights streaming from a ponderous vessel of strange proportions which rested upon the ground in his cornfield.

Having read the despatches [sic] published in the *Post* of the noted aerial navigators, the truth at once flashed over him that he was one of the fortunate ones, and with all the bravery of Priam at the siege of Troy, Mr. Nichols started out to investigate. Before reaching the strange midnight visitor, **he was accosted by two men with buckets who asked permission to draw water from his well.** Thinking he might be entertaining heavenly visitors instead of earthly mortals, permission was readily granted. Mr. Nichols was kindly invited to accompany them to the ship. He conversed freely with the crew, composed of six or eight individuals, about the ship. The machinery was so complicated that in his short interview he could gain no knowledge of its workings. However, one of the crew told him the problem of aerial navigation had been solved. The ship or car is built of a newly discovered material that has the property of self-sustenance in the air, and **the motive power is highly condensed electricity**. [Here we have a possible connection to the oft-used term "electromagnetic" which is quite frequently found in UFO literature.] He was informed that five of these ships were built at a small town in Iowa. Soon the invention will be given to the public. An immense stock company is now being formed, and within the next year the machines will be in general use. Mr. Nichols lives at Josserand, Trinity County, Texas, and will convince any incredulous one by showing the place where the ship rested. [489]

And two weeks later:

05-06-1897   *Hot Springs, Arkansas, USA*

We next look at the testimony of Constable Sumpter and Deputy Sheriff McLemore of Hot Springs, Arkansas:

While riding northwest from this city on the night of May 6, 1897, we noticed a brilliant light high in the heavens. Suddenly it disappeared and we said nothing about it, as we were looking for parties and did not want to make any noise. After riding four or five miles around through the hills, we again saw the light, which now appeared to be much nearer the earth. We stopped our horses and watched it coming down until, all at once, it disappeared behind another hill. We rode on about half a mile further when **our horses refused to go further** [yet another example of animal reaction to a UFO]. About a hundred yards distant we saw two persons moving around with lights. Drawing our Winchesters—for we were now thoroughly aroused to the importance of the situation—we demanded: "Who is that, and what are you doing?"

A man with a long, dark beard came forth with a lantern in his hand and, on being informed who we were, proceeded to tell us that he and the others—a young man and a woman—were travelling through the country in an airship. We could plainly distinguish the outlines of the vessel, which was cigar-shaped and about sixty feet long and looking just like the cuts that have appeared in the papers recently. **It was dark and raining and the young man was filling a big sack with water about thirty yards away**, and the woman was particular to keep back in the dark. She was holding an umbrella over her head. The man with the whiskers invited us to take a ride, saying that he could take us where it was not raining. We told him we believed we preferred to get wet.

Asking the man why the brilliant light was turned on and off so much, he replied that the light was so powerful that it consumed a great deal of his motive power. He said he would like to stop off in Hot Springs for a few days and take the hot baths, but his time was limited and he could not. He said they were going to wind up at Nashville, Tenn., after thoroughly seeing the country. Being in a hurry we left and upon our return, about forty minutes later, nothing was to be seen. We did not hear or see the airship when it departed. [490]

??-??-1906   Mitchell, South Dakota, USA

Herbert V. Demott, 10, saw a "craft" come down near the well. "As I approached it, a door rolled back and I was welcomed inside." The two occupants, who looked like ordinary men, "sat on camp stools"; they conversed with him but did not divulge their origin. "The outer shell of the craft was filled with helium gas, and when the lever was moved, the magnetism from the earth was cut off, allowing the craft to rise." **The pilots drew water from the horse trough "to be used in making electricity** [again referring to electricity]." [491]

In the following case, the reason the aliens need water is not spelled out; however, one of the aliens does give the witness a clue to the way the craft rises, not using water but "a very heavy fluid." Note that in Chapter 4, I spoke of the possibility of a conduit running around the UFO where the temperature was at or near absolute zero and where electrons could travel at extreme speeds because of the lack of resistance. This "fluid" might be the conductor of the electrical transmission:

SP-??-1951   *Paarl, South Africa*

H. M., a British instrumentation engineer, had driven part way up a mountain called the Drakensteen, 7 miles from Paarl, and was just turning back when a man appeared, waving his arm to stop the car and asked, "Have you any water?" H. M. had none. "You see, we need water!" he said. H. M. offered to take him to a stream; the man asked if it was far away, and on hearing it was only ¼ mile, got into the car. **H. M. offered an oilcan as [a] container for the water.** "That will be all right," said the man. When they got back, the man pointed toward the mountain into the area darkened by its shadow, saying, "There, please, there!" So, H. M. drove into the shadows and discovered, about 100 yards from the road, a lenticular object with squarish windows around its circumference, standing on 2 broad feet. It was 35-50 ft. in diameter [and] about 13 ft. high. Between the support legs, a flight of steps led up to a lighted opening. The man then invited the dumbfounded H. M. to enter the UFO, and with a "friendly gesture," persuaded him to do so. Standing in the doorway, he saw that a circular bench, or couch, ran all around the circumference; in the center was an array of about eight 3 ft. hand levers projecting from a rectangular aperture in the floor, each ending in a sort of "fork," like that of an old hand brake, and beyond these was an object shaped something like an upright piano, perhaps an "instrument panel," though no dials or instruments were visible. The white light came from no visible source. In the UFO were 4 more men, one of whom was lying on the couch; H. M.'s companion explained that he "had got burnt in a slight accident." H. M. was not permitted to approach him. All the men were short, 5-5'3" tall, and wore belted, beige laboratory gowns extending below the knees, with trousers and shoes visible below. They totally ignored the witness. H. M.'s companion gave them water; on being asked if they needed a doctor, he said no. The[n] he asked if H. M. had any questions about the craft. H. M. asked where the engines where [sic-were]; he replied, **"We don't have any** engines." Pointing at the levers, he said, "We nullify gravity; that is how we rise." Asked how this was done, he answered that they used "a very heavy fluid" which circulated in a tube at near light velocity, creating a magnet. H. M. said such a speed was impossible. "No, it is simple," he replied. "When the fluid is leaving the tube, it is already entering at the other end, thus its relative speed is infinite." H. M. asked where they came from; he pointed at the sky and said, "From there"; to more specific questions he would not reply. After 15-20 minutes of conversation, the man pointed toward the door and invited H. M. to leave, which he did. About 45 minutes had elapsed since the man had first accosted him. The next day he went back to the spot and found some "very strange marks." H. M.'s interlocutor seemed older than the others, perhaps over 40. The men's bodies and faces were normal (perhaps with somewhat pronounced foreheads), their hair short and chestnut colored, none showing any traces of beard growth. Their hands seemed delicate, like women's. The witness told this story to the investigator in 1977 while in Spain doing work for a Spanish firm. He had never divulged it before. [492]

In the following case, besides the extraction of water, there are two instances of disturbance of the water's surface, as expounded in Chapter 2, while still hovering above the water:

03-23-1978 *Atlantic Ocean off Guaiúba, Brazil*

Sixty-seven-year-old Joao Inacio Ribeiro was walking along the oceanfront when he saw a circular object about ten meters in diameter approach from the sea. It had brilliant blue-, yellow-, green-, and lilac-colored lights. On its upper section it had a sort of cabin or cupola with three circular windows. **The craft approached, stopped, and hovered above the waters, causing the area directly below the object to become agitated and choppy.** Ribeiro then saw two short humanoids appear on top of the craft. The figures were about one meter in height, and wore gray-colored, tightfitting outfits. The beings were baldheaded, with small noses and mouths and huge black eyes. One carried something resembling a walking cane; the other had what appeared to be a vacuum cleaner-like object. **Both beings then began operating the vacuum-like object, seemingly extracting water from the ocean. After ten minutes they re-entered the object. Moments later the object descended to ten meters above the waters causing them to become even more turbulent, creating large waves and foam.** Afraid, Ribeiro ran home. [493]

Although very leery of stories of beings seen without a craft, I feel that the following contains many of the observations from other cases where a craft is seen. In many such reports, the alien is seen exiting or entering the field surrounding the UFO, whereas a human could not do this. It insinuates that the alien is wearing a suit that has properties similar to the field itself. Perhaps the suit has a reversed field which counteracts the craft's field, affording the alien immunity to its effects. The parting of the vegetation in front of the alien implies that this field is doing the parting.

02-13-1980   *Stream near Rafaela, Argentina*

The witness, C.R.S., along with two other men had gone to inspect a newly acquired property and lot on the north side of a local stream called "Las Prusianas." After a while the three men split up and the witness, feeling tired, decided to rest and wait for the other men. As he sat drinking some water, he suddenly noticed, at about 25 meters away, a human-like figure walking on the stream bed. Curious, the witness stood up and advanced several steps in order to obtain a good look at the figure. Suddenly, filled with an unearthly peace and harmony, the witness stopped and stared at the totally human-like figure, seemingly losing all notion of time. The figure was wearing a metallic, white-colored, one-piece suit, very tightfitting, that covered its hands, feet, and head, except for the face. And the areas around its shoulders, knees, neck, elbows, and waist had what appeared to be 3 "rings" which apparently facilitated its movements. On its back it had a large, square "backpack" similar to the ones used by modern-day "astronauts," and on the right side of the backpack, **the witness saw a long, hose-like protrusion shaped like a golf stick which the entity seemed to introduce into the stream's waters, presumably to absorb water.** As the entity performed this operation, it bent over slightly. Once this was done, the hose-like implement slid back into the backpack. Moments later in a "mechanical"-type movement, the entity turned sharply to its right facing the witness. The entity immediately then started moving towards the witness, stopping at about 10 meters away. **The stunned witness noticed that the figure moved in strange, fluid movements as if stepping on an "air cushion." He observed as the grass seemed to flatten and part in front of the entity.** Despite the fact that the figure was about 10 meters from him, the witness, for some reason, was unable to distinguish the entity's facial features and could not see if it was wearing gloves. The entity seemed totally indifferent to the presence of the witness. **The entity then turned and walked into the nearby thick brush which also seemed to part as he advanced through it, and it also closed behind him.** The witness pointed out that the thick brush in the area was very rough and spiny, but it appeared not to affect the entity. At no time did the witness see any anomalous lights, [hear any] noises, or see any unidentified crafts. Later when the two other men returned, the witness did not tell them what he had seen but did ask them if they had seen anything strange; apparently, they had not. [494]

Here are some interesting excerpts from other taking-on-water cases:

05-EE-1940 *Stream, Broadwater County, Montana, USA*

[Letter written to Senator John Glenn:]
He asked me if it would be alright if they took some of the water. I could not see why not; I said sure. **He then gave a signal and a hose or pipe was let down.**....

He said as you noticed we are floating above the ground, and though the ground slopes, the ship is level.

There are in the outside rim of the ship two flywheels, one turning one way and the other the opposite direction. He explained that this gives the ship its own gravitation, or rather overcomes the gravitational pull of the earth, other planets, or the sun or stars. And though this pull is light, we use this gravitational pull of the stars and planets to ride on.

He went into somewhat greater detail on the power development by these two flywheels. He mentioned something about them developing an electromagnetic force. As this was quite new to me and he realized that, but he saw I had gotten the picture, so he stopped.

I asked him where he got the energy to run the ship. He said from the sun and stars, and he would store this in batteries, though this was for emergency use. [495]

Again the rotation of the disc is noticed:

08-??-1969 *Kama River near Voronov, Russia*

The stranger wore a grayish-metallic overall with thick-soled boots. The man did not wear a hat and had a short haircut. In his left hand **he held a hose-like implement, which was apparently extracting water from the Kama River.**.... At the end of the encounter, the stranger asked Ivanovich to stay back from the object and to just observe it take off. [This is of interest as we now know that the field is dangerous and not easily seen like a propeller.] The object had a semi-transparent, green-colored, globular dome on top. As the stranger approached the object, a door appeared on the dome and he entered it. The door then shut. The external disc immediately began to revolve and its outer rim became invisible. The object then rose into the sky and instantly disappeared into thin air. [496]

Besides drawing water from streams, rivers, lakes, and oceans, UFOs also have an interest in water from other sources—storage facilities. Let us examine some enigmatic disappearances in the next section.

11.3 Missing Water

Another mystery of water-UFOs is the repeated depletion of water from water storage tanks, local farm reservoirs, swimming pools, and in one case a water tanker truck. Water has mysteriously disappeared in locations all over the world.

With no evidence of leaks, spills, or flooding in the immediate vicinities, reports have been made of large volumes of water rapidly and mysteriously disappearing, but instead of taking many hours or even days to drain, swimming pools and huge water tanks are emptied within a very short period of time.

In the case of large amounts of water being drained by a craft measuring 30 to 100 feet in diameter, the question of storage is also a puzzle because storage tanks on land are sometimes as large as (or even larger than) the craft itself.

Occurrences involving the mysterious disappearance of water may be accompanied by sightings of UFOs, strange lights or noises, or EM effects, but this is not always the case. In fact, there are very few accounts connecting UFOs to this type of water drainage.

One of these is a report dated 09-30-1980 which occurred in Rosedale, Victoria, Australia, where a UFO was observed and approximately 10,000 gallons of water disappeared overnight from a water tank. The closest we come to an association with UFOs in other cases is that they were mentioned as being seen in the general vicinity of the missing water. One should exercise caution when evaluating cases of this type because even when a UFO is seen, it would not necessarily imply that it is responsible for the loss of the water.

Prior to receiving the 09-30-1980 case, I was a skeptic (as uncomfortable as that word is to us ufologists) on the subject of disappearing water cases. For many years I dismissed these types of missing water stories as they were relatively rare and it seemed that a natural explanation for them would eventually surface, though to date, there has been no satisfactory clarification for these strange occurrences. Consequently, the accounts in my collection are fairly recent additions. I want to thank the editor of my book who kept insisting that there was something to this aspect of the water-UFO mystery. I am now interested in continuing to collect cases of these strange depletions.

Yes, water is taken on board the UFO, as we have seen before, and in the next case, we can see how it was done—differently. It is the ONE case thus far that bridges taking on water conventionally using pipes, tubes, and hoses with the cases that follow it. Note that this is a long case and can be read in full on my website, so I will only present small segments of the text for your immediate attention:

09-30-1980 *Water tank on farm near Rosedale, Australia*

The percipient, a 54-year-old farmhand and caretaker of the "White-Acres" property, 19 km west of Sale, Victoria, went to bed at 2230 hrs on September 29, 1980, and experienced no trouble in sleeping. At 0100 hrs, September 30 (daylight saving), he awoke and immediately heard a noise like a screeching whistle, unlike the noise of anything on the nearby road. It should be said that the Princess [sic-Princes] Highway, a main road Melbourne to Sydney, carrying interstate traffic, runs right alongside the house where he lay sleeping.

There were also noises of cattle bellowing and a horse running around in apparent panic....

Immediately he observed, out of the corner of his eye, an object moving from his right to left. Its estimated height was 8-10 feet above the ground....

The object itself seemed to consist of two sections—a white dome on top and a larger orange section underneath. Around this bottom section there appeared to be circular windows or lights. Suddenly, the noise level increased to "an awful scream," **something like a black tube appeared around the base of the UFO, and this seemed to inflate to a tremendous size, just beyond the diameter of the object. There was a tremendous "bang," and a blast of air and heat came from it**. This blast almost knocked him off his bike. The UFO lifted up and gradually rolled off the spot going eastwards....

The witness relates that he had noted that **the 12-foot-high, 8-foot-wide, concrete water tank had been full the day before** (September 29, 1980), and held about 10,000 gallons of water which is pumped by a wind pump through an incoming one and a half inch pipe. Once full, it is used as a summer reservoir. A valve is closed and water bypasses the tank and is pumped out by 1 inch pipe directly to the area around the house.

When it was inspected on the day of the event, he states that a) about 10,000 gallons of water had gone, and b) in the centre of the tank the muddy residue on the bottom was piled up (the outlet is on the side, not in the centre), to a height of perhaps two feet, and c) there was algae, which had been floating on the top of the water, stuck on the walls near the top hanging upwards. It was estimated that it would take 72 hours or so to empty the tank by normal pipe means. [497]

Back in Chapter 2, I said that using the vortex to import water into the craft would be time consuming due to the minuscule amount of fluid water reaching the bottom of the craft. But, as with other "one-witness" cases, I noted some interesting details in this account. The observation by the witness of the swirled and upward-stacked muddy deposits shows that this upward suction of the field is at work again, due to the craft's proximity to the water in the tank, as in the mounding scenario. However, something else has evidently been added to the mix. The question here is how does the water end up in the large storage tank attached to the craft? Is it possible that the "black tube [which] appeared around the base of the UFO" and "seemed to inflate to a tremendous size, just beyond the diameter of the object" could be used as the UFO's storage area for water that is taken for possible transport to a waiting mother ship? In reference to the "bang" and the "blast of air and heat [which] came from it", I believe that this was caused by the field which was then switched from off to on in order for the vacuum to pull up the water into the tank.

As a side note, there is something else that occurred to me. Our alien friends may have **specialized craft** to accomplish the task of gathering water for a "mother ship." This idea is not far-fetched considering the fact that we have specialized variations of aircraft in our own inventory. For example, we have passenger aircraft, cargo aircraft, helicopters, military aircraft, seaplanes, gliders, and so on. Even with a specific type of aircraft, there may be multiple uses for it, for example, the DC-10 (Douglas Commercial-10) built as a passenger craft is also used as a cargo carrier. Now in its golden years, it has been equipped with tanks which hold a large amount of a chemical fluid for firefighting.

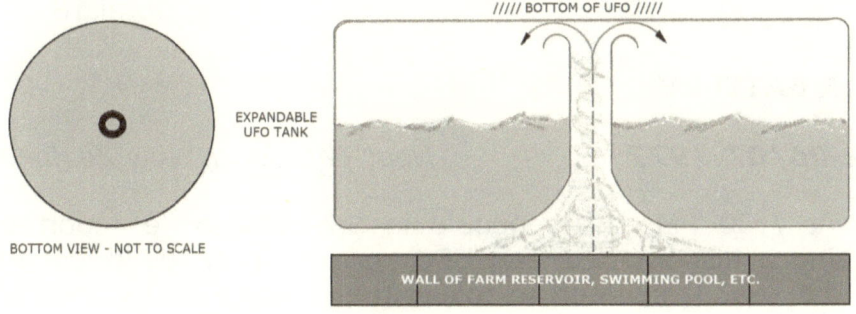

Sketch #47
The expandable UFO exterior tank

While the witness in this Rosedale account did see a UFO, in many cases involving disappearances of water, a UFO is not seen. It is in these situations that it would be foolhardy to initially assume that a UFO is responsible for the loss of water.

The table which follows includes the missing water cases which I currently have in my collection. All of them can be found on my website. Dates in bold are the cases which are presented in this chapter:

TYPES	# of Cases
Car battery	1

12-06/07-1977

Farm dam	1

04-26-1985

Pond	1

08-19-1973

Swimming pool	6

07-11-1981//11-LL-1991//06-23-2002//
12-LL-2005//10-15-2005//11-??-2010

Tanker truck	1

06-12-1981

Water tanks	8

06-24-1979//**09-30-1980**//**11-24-1999**//
04-??-2002//01-05-2005//10-31-2005//
07-02-2006//**01-09-2011**

Total 18

CAR BATTERY

12-06/07-1977   *Near Tatapouri, New Zealand*

At 7:30 PM, three people in an automobile reported seeing a disc-shaped, red UFO coming from the direction of the sea. It seemed to follow their car for a distance of about one mile until the driver stopped the car so that they could take a better look at it. At that instant, the UFO veered off into the hills towards the Waimata Valley (an area of increased UFO sightings during that month). The witnesses noted that their car lights were much dimmer than usual and on **arriving at their destination, discovered that the car battery had no water in it despite the fact that it was nearly new and had been serviced two days before the sighting.** [498]

In the above case, I sincerely doubt that water was taken from the battery deliberately because it is such a small amount as to make the effort more than it is worth. Perhaps, it is more explainable as a result of the heat of the field and the extraction of electrons from the battery in one automatic process.

FARM DAM

04-26-1985   *Gomaub, Namibia, Africa*

Case No. 28 (C) UFO AFRINEWS 4 Namibia
Our Namibian correspondent, R. M. Roeis, has been in contact with the daughter of Mr. V., the witness mentioned in this case. Mr. V. himself died in 1987, two years after the phenomenon was reported in the newspapers. The girl, whose name is Hannetjie, is married and lives in Mariental, and is not connected with the farm where the incident occurred. At that time, she was 12 years old, which makes her about 18-19 years old now.
Hannetjie says: "It was about midnight. I was feeling very feverish. I and my mother got up and I had to drink some medicine. It was then that we saw the strange phenomenon. It was a round ball, bigger than the headlight of a motorbike. It appeared as if the ball had several little stars that seemed to blink. It was not a blue-white colour but it glittered like the stars do. **The next day was a Monday and that afternoon at about 1400 hours, one of father's farmhands, who was guarding the sheep, arrived at the homestead and said that a red ball had 'drunk' [the water from] the Lower Dam [or reservoir].**

"My mother described the light as blinking, but it appeared [to me] like little stars that grow light and then darken.

"The ball of light that I saw arrived from the south and then went north. It seemed to me like a ball with a thousand little stars in it.

"The weather was definitely dry, the wind still, and there were no clouds. It was a very quiet happening; there was no sound. It was one of those nights when our dogs were at their most quiet. But **the man guarding the cattle said that the sheep refused to come close to the camp where he had seen the red light ball.**..."

I have always maintained that no case should be validated when only one investigation has been made. In my experience, the more one probes or discusses the case with witnesses, the more material comes to light and the more accurate one's final report will be!

In the first reports received of this case, it said that Mrs. V. had heard her child crying in its cot—and one was under the impression that the "child" was a small baby who could not give any evidence. But, the child was not "in her cot" as previously stated; she was 12 years old and remembers the event quite vividly.

The other interesting fact is that another witness has appeared on the scene: the farmhand who was guarding the sheep. **It was he who saw the ball of light "drinking" the water. Previously we had only heard that one witness, Mr. V., had found his dam dry.** [499]

POND

08-19-1973   *Yellville, Arkansas, USA*

My mom, dad, niece, and I were in our backyard after making ice cream. We noticed, hovering over the tree-filled field across Highway 62, what looked like a long flat array of lights. They were flashing in a sequence: green, red, green, red, blue, green, red, white then they would sequence again. I couldn't tell if there were any windows or exactly what shape it was, it was high up, maybe 300 yards....Field 1 was completely normal, as was cow pond 1. We wore our knee waders because the mud around the ponds was thick and pretty deep if you got within 3 feet of the water. It was also, as usual, surrounded by frogs. It took us a while to get the dog out of the water and the mud, but we continued on. Field 2 had strips of dry, flattened grass in it starting about halfway through it, but none of the strips connected. At the edge of field 2 for about 5 to 10 feet, the grass was all flat and dry. **The dog wouldn't go anywhere near the flat grass or the pond that abutted it.**

Pond 2 was WEIRD. The usual mud around the water still held the impressions of the cows that had last drank [sic] from the pond. The deep holes where they had pulled their legs out of the mud were completely dry. The mud had dried and cracked...it crumbled easily in our fingers when we picked it up, but on the ground it was solit [sic-solid] and hard...sll [sic-all] the way to the **water's edge which was down from about a diameter of about 40 feet to only about 10 feet. We circled the water three or four times, but as far as we could see there were no frogs or even bugs in, on, or near cow pond two.** We tossed chunks of the dirt into the water to try to stir anything, but nothing was there. Less than a week before, the pond had been full of frogs and bugs and even crayfish. The only thing we got on our waders was dust. We went into field 3, but once we were past the 5-10 feet of flat, dry grass there were bugs and normal field flora and fauna...no frogs though. **500**

If nothing else, the following account introduces an aspect that I, for one, had not expected...the purification of the water *before* taking it on board. Of course, what the aliens did with the extracted chlorine when finished is yet another major question that has not been answered.

SWIMMING POOL

07-11-1981   *Port Byron, Illinois, USA*

The Girl Scouts have left camp for the day. It is late evening, and as the day camp supervisor, you are alone in the camp headquarters which also serve as your home. Your wife and two older children are away.

Suddenly your dog sets up a warning barking. Intruders, no doubt. Through the trees that screen the camp swimming pool, you see bright lights. Unusual. Someone slipping in for a late night swim would hardly turn on the pool floodlights!

You slip through the trees and into the clearing and an unobstructed view of the pool; the pool lights are not on. Instead, directly above the pool, and higher than the regular lights, is a brilliant light, bright enough to obscure its source. Suddenly, the lights go out. You hear a whirring noise that rises in pitch as it apparently rises into the sky.

Gathering your wits, you turn on the pool lights. **The water level is down by some 3 feet—30,000 gallons of water are gone!** You quickly trace a possible leak. All natural means of egress for the water are dry. Now you notice that the 24-hour pump is not running. Normally, it runs constantly unless manually cut off. **Later you discover that there is no chlorine in the remaining water.**

Reader: What is the solution? Hoax? Hallucination? A real event? Investigation so far indicates a real event. The witness did not report the incident until a month had elapsed and then only because he was encouraged by a televised account of the Muscatine, Iowa sighting (included in this issue).

Those acquainted with the UFO phenomenon will readily note that water and UFOs have more than a passing association. **There appears to be a correlation between small bodies of water (reservoirs, small lakes) and the sighting of UFOs. In addition, UFOs have been reported lowering a tube into a body of water and sucking up a quantity of water.** [501]

TANKER TRUCK

06-12-1981   *Alice, Texas, USA*

Mr. Robert Gomez, a vacuum truck driver for a firm in Alice, Texas, was headed west on FR 665 toward Alice on June 12, 1981. He had just finished a job, **leaving approximately 165 gallons of water in the tank on the back of his truck** and zero pressure in the tank. The sky was cloudy and it had rained earlier. He turned the radio on.

About 2:10 p.m. he noticed a bright object that he first thought was an airplane, until it increased in brilliance and stopped in midair. The object was disc-shaped and brilliant white, with a dark ring around the outer perimeter and a dark inner ring around its center. It also had a slight "cap" on top.

Gomez then **noticed that the truck started to slow down, so he pressed the accelerator trying to pick up speed. The exhaust stacks were blowing smoke but the truck seemed to be about a foot off the road.** His AM radio had stopped playing but his CB was working, so he called his dispatcher to report the incident. Afterwards, the UFO went back into the clouds and disappeared.

When he arrived back at his shop, he noticed that the AM radio was playing again. When he got out of the truck, the maintenance man told him that it looked like the back of the truck was smoking at the valve. Mr. Gomez noticed that there was now 55 pounds of pressure indicated on the gauge, so he opened up the valve to drain the remaining water, but there was no water. All that came out was steam. [502]

WATER TANKS

The following is an excerpt from another interesting case which occurred on a military base. Nowhere in the account from which the passage is taken does it state the amount of missing water in liters or gallons:

11-24-1999   *Santa Fe, Argentina*

A Search for Truth

"Upon arrival, we were greeted by two soldiers. We told them that we were researchers of the subject and wanted to carry out some studies. They whisked us off to a superior officer, to whom we had to explain matters in greater depth, why we were doing it and what we would do with our conclusions. Permission was granted," states the researcher.

He later explained that "the soldiers themselves led us to the impression, located in the remotest area of the base, which covers six blocks. **We started by looking at the imprint and other facilities, such as sheds, the water tank, trees, wires and the cistern, which supplies water and has a depth of 15 meters and a diameter of 2.5 meters.**"

The first clue emerges here. According to the story that Sgt. Carlos Villano told the researchers, "The cistern was practically drained the night of the event." **The officer also stated that the morning following the event, the principal of the primary school that operates on the base's premises called him to report a lack of water in the school's rest rooms. "I was sent to the cistern to see what was going on, thinking that the fuses had blown, but they were fine," said the officer. At the same time, he found that the cistern and tank were both out of water, while noticing the circular impression for the first time.**

Direct Witnesses

Another witness was Claudio Chena, the water basin's custodian, who was 25 at the time. "He said that he was a professional photographer and showed us pictures taken of the event. He also explained that he was playing soccer with his cousin in the base's soccer pitch, which is in the front of the base, facing Avenida Freyre," comments Brussa.

"Claudio said that after 9:00 p.m., **they saw a bright light toward the water tank, and that the lights went out throughout the district, but not the streetlamps, which remained on.** They paid little mind to the situation, thinking the lights were from a truck driving along Calle San José, behind the district," states Brussa in his account. [503]

And finally, we have another description of a huge quantity of water mysteriously disappearing:

01-09-2011   *Salitral de la Vidriera, Argentina*

"We were having lunch on Sunday with a cousin who owns a field in the vicinity of the 'La Vidriera' salt flats. In order to spend time with us, he left another relative alone in the field in charge of the animals. Just as we were done eating, **the relative phoned him in a state of near derangement. The previous day he had filled up the water tank** [and] **shut the windmill down because the tank was full. It has a capacity of 60,000 liters (16,000 gallons) of water more or less.**

"**But there was not a drop to be found on Sunday morning.** No sign of puddles or spills around the tank. No cracks on the tank's floor either. He told me, furthermore, that it has a 6-inch flood cap and when they drain it, it takes more than half a day to empty it out completely.

"The temperatures were in excess of 36 centigrade at the time, and I put off further research until the next day (Monday morning). But by then, the other relative had filled up the tank, because the animals had been left without any drinking water. Therefore, I no longer wanted to go in person.

"**This is the second time that this has happened.** The first time was some 15 years ago. A bit of information I forgot to mention is that there were no missing animals.

It should be ascertained if any neighbors were missing livestock." [504]

While the connection between UFOs and their need for water has yet to be resolved, we see many cases of UFOs "taking on water" where pipes, hoses, or tubes are used. There are also instances of mysterious disappearances of water, which may not necessarily be linked to UFOs. In any event, after examining numerous reports of taking on water in this chapter, we are still faced with the question of why the aliens need this water.

11.4 Taking on Water for Camouflage?

Earlier, we took a look at some of the reasons why they may take on water, including the need to reduce the heat generated by the crafts' engines, similar to what automobiles require so they do not overheat, and also the need to water any vegetation that they might extract from our soil. Now I would like to explore another possible purpose for the water they take: camouflage.

Back in Chapter 4, I presented definitions for the terms cloud, fog, haze, mist, smoke, steam, and vapor and went on to relate them to heat and the relationship of water to the UFO's field. However, there are many accounts in which UFOs emerge from clouds or a witness drives a car into fog and is suddenly faced with an alien craft. This also occurred in Chapter 5 when Anna Jamerson and her mother had a fog bank come onto the beach that engulfed them, and then they were both abducted.

Many of the cases that I have collected indicate a steam-related event, resulting from the heat generated by the UFO's field touching the surface of the water. But there is compelling evidence that UFOs cloak themselves with clouds, fog, etc., which they actually produce. Some examples of this camouflage follow with the various meteorological terms used interchangeably by the witnesses to indicate what they saw coming out of the craft:

02-11-1937   *Near Kvalsvik off the coast of Norway*

Red and green lights were visible on the machine but as the boat approached, the lights suddenly went out and **the object was enveloped in a cloud of smoke** and disappeared almost magically. [505]

09-14-1954   *Several towns southwest of Paris, France*

"All at once (by now we had been watching for several minutes) **white smoke exactly like a vapor trail came from the lower end of the cloud**. At first it pointed toward the ground, as if spun from an invisible shuttle falling free, then it gradually slowed down while turning around, and finally rose up to describe around the vertical object an ascending spiral which wound it up in its coils." [506]

12-07-1954   *Isla del Francés, Uruguay*

Around the perimeter [of a metallic-appearing elliptical object] was a row of **"exhaust ports"** from which a bluish luminous **haze was emitted**. [507]

09-12-1963   *Atlantic Ocean off Cape Cod, MA, USA*

The men on the tower, including foreman Patrick Loreno, who was in charge of the crew dismantling the tower, said the object, whatever it was, had had a controlled light and that **smoke or steam had appeared on its surface.** [508]

05-??-1965   *Tasman Sea, near Sydney, NSW, Australia*

When she again looked at 7 p.m., she was amazed to see it moving slowly towards the rocks at the base of the cliff. As the "cloud" came nearer, the observer was able to look down into the cloud and was astounded to see "a magnificent snow-white flying saucer". **The cloud was formed by a grey-colored steam which soon turned a pink after**

issuing from vents around the outer edge of the saucer. [509]

06-24-1967   *Waters off Trenton, Maine, USA*

Two persons sighted a silver-gray, hat-shaped object hovering about 500 feet from the shoreline. **It emitted vapor from its base.** The object ascended into a fog bank and descended again at a greater distance before moving away. [510]

01-21-1971   *Brush Mountain, Idaho, USA*

The light was said to last several seconds moving neither right nor left at eye level and changing color from red to orange before becoming **engulfed in a cloud of "steam"** which apparently extinguished it. [511]

09-30-1973   *Giles County, Tennessee, USA*

After the craft left, Swanner observed a **fog** of whitish vapor where it had rested. He crawled down from his perch and went to the landing site. Breathing his [sic-this] **fog**, he had a smothering sensation and felt as if his lungs were being pumped full of air to the bursting point. He ran from out of the **fog** and the fresh air made the feeling disappear. [512]

05-02-1978   *Escazu, Costa Rica*

...he suddenly noticed a hovering, silvery, disc-shaped object that emitted a strong yellow light.... **A gray cloud-like mist emerged from the object and descended towards the ground**. Stunned, Enrique watched as the **mist-like vapor** assumed the shape of a human-like figure that remained suspended above the ground. [513]

04-01-1982   *Petrolia, Pennsylvania, USA*

It was a very large triangular-shaped construction, a dull gunmetal gray color and was **surrounded by a luminescent mist which seemed to originate from the back of the main body**. [514]

05-14-1991   *Lancaster, California, USA*

Red lights were revolving above her and a thick **fog was emanating from ports** set around the large craft hovering above. [515]

04-12-1992   *Near Deming, New Mexico, USA*

Suddenly what looked like a huge **luminous cloud of smoke** began stretching across the highway ahead; it formed a solid curtain across the road. The witness drove into the **grayish-white vapor** for an hour. [516]

07-07-2000   *Red Rock Canyon, Nevada, USA*

The devise [sic-device] / saucer seemed to be **creating some kind of fog**, but it just kind of radiated from the craft. It didn't blast away from the craft, and the light **left no trail in the fog it was creating**.... in the blink of the eye, the craft and light were gone. **But the fog stayed. The fog didn't sweep up with the leaving of the craft.** It just hung there. [517]

11-10-2005   *Ligurian Sea off Genoa, Italy*

The object was elliptical**, had four portholes on its side and had smoke puffing out of one end**. It seemed to be made from solid light and have a black shadow all around it. [518]

Since the extraterrestrials have not declared war on us, I view this camouflage of the craft as a technique the aliens use to avoid contact with us. In this way, they would be able to observe us, as the Trekkies did when visiting another species:

> In the fictitious universe of *Star Trek*, the **Prime Directive** is the guiding principle of the United Federation of Planets. The Prime Directive, used in four out of five *Star Trek*-based series, prohibits Starfleet personnel from interfering with the internal development of alien civilizations. This conceptual law applies particularly to civilizations which are below a certain threshold of development, preventing starship crews from using their superior technology to impose their own values or ideals on them. [519]

In the next chapter, we will see the beam of light, which has been seen often in land-based cases. Is it any different in water?

Chapter 12

Beam of Light... Into a Body of Water

In eyewitness accounts involving beams of light, there are no indications of physical changes to the water, but beams of light have been observed many times, so they must have some function. In abductions, which I discussed in Chapter 5, they are used to transport people from one place to another as in "Beam me up, Scotty." Could the beam also be a tool for measuring heat, pollutants, or density of life forms in a particular body of water? Could it be a means of communication with another craft still underwater? Another possibility might be that the beam is used to mark a destination to which the aliens would want to return in the future. Something which affects how a beam may be described is the visual distance between the witness and the craft so that if a UFO is up close, the witness is possibly going to see more than someone who is much further away and can only describe the beam from a distance. Although we still have more questions than answers for the moment, the accounts which follow deal with the beam and may leave some clues for future researchers.

??-??-1958   *Ocean off the west coast of South Korea*

Residual Illumination
A somewhat similar report was given to me by an Indianapolis computer expert who had served in the U.S. Air Force. He had been a sergeant in charge of a fire truck crash crew at an air base on the west coast of South Korea overlooking the China Sea. Late one night in 1958, two American jet fighters flying in from Japan to another base in Korea requested permission to make an emergency landing because one of the planes was low on fuel. As is normal in such emergencies, the fire truck and its crash

crew were positioned near the runway in case the plane ran into difficulty.

As the men waited for the planes to arrive, they saw a bright light approaching from across the China Sea. The light grew bigger and bigger and, when it got within several hundred yards of the shore, it stopped and hovered a few hundred feet above the water. The sergeant radioed the control tower and asked what it was. The controllers didn't know. They were examining the light through binoculars but couldn't determine anything.

Suddenly, the object shined a beam of light straight down on the water. A short time later the light went off, but the water remained luminescent for a while before fading out. The object again shined a light down on the water and turned it off again a minute or so later. Once again, the water remained luminescent for several minutes.

By this time the jet fighter that was low on fuel was landing, and the control tower asked the pilot of the second plane to check out the light, which was still sitting in the same spot. The pilot of the second jet circled around, and as he approached the object, it instantly shot back toward China and disappeared in seconds. [520]

01-??-1980   *Tapajós River, Amazon Basin, Brazil*

Again Residual Illumination

Late one Saturday night in January 1980, a Brazilian electronics businessman and nine other people **saw a disc-shaped object shine a light down on the water** near the eastern shore of the Tapajós River in the Amazon. The man, his family and some relatives were camping at a beach thirty miles south of Santarém and appeared around eleven o'clock.

Several teenagers were still awake, lying in hammocks and talking, when they saw the UFO come toward them from across the river. **It stopped about thirty yards from the beach and shined a light down on the water**

twenty yards below. The kids awakened the others. The UFO hovered briefly and then began moving north, or to the campers' right, parallel to the shore. It traveled more than half a mile, with the spotlight still shining straight down, before it disappeared.

The curious thing about all this was that as the beam of light moved along the river, it left behind a trail of luminescence on the surface for several hundred yards. The trail gradually faded out, south to north, at about the same speed as the light itself had moved. [521]

??-??-1967   *Gibsons & Keats Island, BC, Canada*

Bent Beam

The following year [Previous article was in 1966.] Miss E. R. East, another resident of Gibsons, saw more strange action in precisely the same area, which she described in a letter:

"I live alone in the upper half of a house that overlooks the channel between Gibsons and Keats Island. Closing Gibsons Bay is Shoal Point, with a beacon on the end of the shoal which has a blue-white flasher. My kitchen windows are of casement type, and as the ground is built up at the back of the house to allow access, the window sill is only about three feet above ground level. I was awakened in the night by a banging noise and thought someone was trying to open these windows, so I got up to investigate, but when I parted the curtains and looked out, I saw a ball of brilliant orange-red light soaring above the hills back of the village. As I watched, the color changed to a glowing white, and I rushed to my side windows and watched the object sail over the hill towards Georgia Strait. I was bemused and went back to my room and sat on the edge of my bed looking out at the water and thinking of what I had seen. **Suddenly, from high above the village, a beam of green light shot down to the water** between the shore and the beacon, and **as it struck the surface, it bent and lay flat on the water. It reached right across the channel and lit** up the wharf on Keats Island. **The beam seemed to be made of many small beams, for I could see dark streaks between the green, and it must have been at least ten feet wide, the same width all along the length I could see; I mean, it did not ray out like a flashlight beam would but was compressed into a pencil beam.** As I stared at it, my eyes began to sting, and I fell back on my bed, almost in shock. When I roused and looked out, the beam had gone, and I was too stunned to think of looking to see if the UFO was still in sight.

"On thinking the episode over, I reasoned that the occupants of the UFO had been attracted by the flasher light of the beacon and were trying to find out what it was. Except for the initial bangs, I heard no sound throughout this incident."

(Miss East's sighting was an example of one of the most fascinating UFO tricks, the manipulation of light. While bent and/or compressed beams are occasionally reported, the water-surface beam she saw was rare, if not unique. The banging noise seems to be another UFO trick.) [522]

11-05-1970   *Cholla Bay, Sonora, Mexico*

Change in Angular Width

Albert Formiller, Phoenix, Ariz., was fishing for black sea bass in Cholla Bay about 9:00 p.m. when he observed a light in the sky coming from a circular saucer-shaped object. The sea was quiet. The object appeared to stop and hover about 200 to 300 feet above the surface of the water. **A light appeared from the bottom of the object and illuminated a broad spot of water, which Formiller estimated to be perhaps one-half mile wide.**

Formiller described the light as appearing to come from within a tube and changing from a broad floodlight to sharp spot on the water surface, apparently as the light was raised or lowered. There was no sound.

After a few minutes a cloud seemed to form around the vehicle. Formiller does not believe this was caused by exhaust gases because there was no apparent "blowing" of the cloud from an exhaust. About five minutes after illuminating the water, the searchlight was turned off and a similar light was turned on atop the vehicle, illuminating the upper part of the milky cloud.

The vehicle then began to move in a westerly direction. It was visible for about 20 minutes in all. Other lights on the strange vehicle gave the cloud a greenish cast. Similar reports came to the *Phoenix Weekly American News* from other Phoenix residents. [523]

SM-??-2002   *Mosinee, Wisconsin, USA*

No Change in Angular Width
At Mosinee when I saw it, it came from way up! And it almost landed in the water, but came to a sudden stop like nothing moving at that speed could. It hovered and my friend and I could actually feel it watching or observing us. We got the hell out of there. One thing in Mosinee though: **it shined a light on the water like a searchlight, but the beam did not spread out; it was straight up and down, like [a] 5-feet-wide beam. And as it went higher above the water, the beam did not change size or shape.** [524]

05-16-1975   *Stephenfield Lake, Manitoba, Canada*

Communication?
Three men had wandered away from a party on the shore of a lake and were standing by a dock when they saw a "moon-shaped" object, hovering over a dam near the far side of the lake. **As they watched, a "solid" beam of light shot from the object to the surface of the lake. Underneath the surface of the water, a glowing object appeared.** This submerged object then began moving towards the witnesses. When it was about 20 feet away from them, one of the men threw a rock at the object. It then appeared to break into several pieces and return to its original location. The "beam" went out, the hovering object

split into two pieces, which flew off in different directions, and the baffled men went to tell their friends. [525]

04-06-2007   *North Park Lake near Pittsburgh, PA, USA*

Illumination and Fish

I want to keep this relatively short, but I had quite the experience tonight, one that I will remember forever. I was driving by North Park with friends and **a strange light was shining directly on the lake**. We looked up but couldn't see the source, which seemed very strange at the time. **We got out and moved towards the lake, while trying to keep our eyes on the light in the water. The water was strangely illuminated. Visibility was about 4 feet deep** I'd estimate. This was very clear visibility, from about 100 yards away, probably a little more. But the sight was vivid; **I actually thought I saw fish beneath the surface.** This was not long ago in pure darkness, from far away. I shouldn't have been able to see what I saw. Moving on though, the light that had been causing this, we noticed either moved from the lake to above the trees (and was the source the whole time) or was silently waiting several hundred feet above and was never noticed. Either way I don't know. I do know that the last time I saw the light/object, it was slightly above the trees, then below them and disappearing, then back above and a dark amber color, then below again, then back up, then completely gone and not to be seen again. It didn't noticeably fly into the distance or leave in a spark of glory. That was it. We just lost sight and it was gone.

The strange thing to me was the size of the object. It couldn't have been large enough for even the smallest human to fit inside. This thing was literally a floating street light with intelligence. I've heard of these sized objects possibly being involved in crop circles. I didn't see anything remotely similar in this case. Please reply if you have relevant information. [526]

10-10-2007   *Lake Easton, Washington, USA*

Beaming up fish?

Lake Easton—I saw a silver/gray disc about 35 feet radius hovering over Lake Easton for 3-5 minutes on October 10, 2007, around 10:40 p.m. **I saw a beam of very bright greenish light which illuminated the lake. Then we saw small objects floating up to the craft. My friend said they were fish.**

Investigator's Note: Reports of unknown objects gathering material are rare. [527]

To summarize, in the 18 cases that I have where a "beam of light" is seen, there are 11 with no description of the color of the beam. I would assume that this would indicate what we would consider a normal beam of white light, such as a flashlight or searchlight. In two cases, witnesses were specific about the beam being white. Two other cases described it as green or greenish, a separate one depicted it as amber, yet another as red, and one more as light blue. All we learn here is that there are color variations to their beams, which *might* serve as a visual indication of their various functions.

Other effects can include residual illumination in the water after the craft has departed, as seen in two of the above cases. This could also be due to *Noctiluca scintillans*, a bioluminescent ocean species also known as sea sparkle. *Noctiluca scintillans* may also form red tides. [528]

Additional effects we have seen in these reports include two mentions of a bent beam—variation of the beam on water from vertical to horizontal at the same time; one instance of possible communication with an underwater UFO via the beam; and three cases involving fish—one involving their death, another dealing with illumination, and one with fish being beamed up. All other accounts only mention a beam going into the water along with other normal UFO observations.

The composition of the beam and its various uses remain a mystery, hopefully however, not an unfathomable one. Bring on the scientists.

Chapter 13

Burning Objects or Water-UFOs?

One of the worst types of case scenarios involves "burning aircraft" going into the water. In these cases, the local rescue services are notified, and an air-sea search is started. The end result is that nothing is found floating on the water—no oil slick, no cushions, and no aircraft or pieces thereof. The authorities check with airway controllers and it is usually found that no aircraft were scheduled in the zone where the event occurred, and here the mystery begins.

If you wanted to make an argument against the possibility of UFOs, you could say that an unscheduled aircraft might be carrying persons like smugglers or others up to no good who, for the sake of covertness, do not file a flight plan as all aircraft are required to do. In the event that their aircraft caught fire and went down, there would be little hope of rescue, with no air-sea search ever being conducted, as there would be no "burning aircraft" to identify. But having played devil's advocate, let us now examine the "burning aircraft" from the ufologist's viewpoint where a fiery red object might conceivably be a UFO.

On the pro-UFO side of this type of sighting, I will recount an overland case to show that a fiery aircraft can have unique characteristics. The following is from John Magor's book *Our UFO Visitors*:

> One day in February [1974], Gordon Matlock, a member of the Civil Air Patrol at Sullivan, Missouri, and his wife were startled to see what looked like **an airplane on fire** above their house. **It was a ball of flame in the sky**—which seemed to be burning out of control. **But strangely, the spectacle moved slowly ahead without losing altitude** and then lowered or dropped a section

of "bright intensity" which abruptly disappeared. After a moment the blazing object moved out of sight. [529]

There are many more of these mistaken "burning aircraft" cases over land, but there are also those that make it into our water files. Not seen as UFOs but normally reported as "aircraft," some of these painfully short accounts follow to illustrate the uselessness of such sightings for ufologists.

13.1 Very Simple Reports

10-14-1948 *Into Lake Ontario, Bear Creek Harbor, NY, USA*

General Description: "Fiery ball"; "burning plane"
Behavior: Going at terrific speed; moving up and down
Witnesses: Two adults [530]

01-27-1964   *Into Lake Winnebago, Oshkosh, WI, USA*

General Description: Fiery object; possible burning plane
Behavior: "Fell" into lake
Witnesses: Police and others [531]

09-09-1964   *Into New River, Radford, VA, USA*

General Description: "Plane on fire"
Behavior: Fell, trailing flame and smoke 30-40 feet; appeared
          that something flew off the side before it crashed.
Witnesses: Several adults [532]

In the preceding reports, we see a woeful lack of detail about the sighting and any follow-up after it. There is very little correlation to a UFO by maneuvers, speed, shape, or anything else other than the fiery colors. We continue with this same type of report but with added text that brings it more into the UFO realm.

The following presents a slightly better report, with more witnesses and a relationship to the earlier chapter on heat.

01-17-1956 *Pacific off Newport & Laguna Beaches, CA, USA*

A fiery finger which streaked across ocean skies near Newport and Laguna last night plunged into the sea and caused a widespread search by a Coast Guard cutter and three Marine helicopters.

The unidentified flying object appeared shortly after dark, setting off a series of telephone calls to Laguna Beach police. Residents placed the time of the sighting at 5:45 p.m. **Marine Corps officials said the burning object was about the size of an airplane** and was seen by coastal residents to plunge into the ocean **causing the ocean to boil and steam to erupt** at the point of impact. [533]

13.2 The Similarity between a Fiery, Glowing UFO and a Burning Aircraft

10-13-1958 *Lake Ontario off Sea Breeze, New York, USA*

A Bear Creek Harbor resident and his wife said **they saw what looked like a burning airplane apparently crashing into Lake Ontario** about 13 miles east of Sea Breeze about 1:15 this morning.

Charles Powell of East Lake Road, Ontario, said **he saw "an orange-red fiery ball moving up and down like a burning plane trying to land on the lake in front of his house."**

"It was going at a tremendous speed," he said, "faster than any boat could have gone." Powell's call to State Police started a Coast Guard search of the area. A Civil Air Patrol spokesman this morning said no missing planes had been reported.

A trooper arrived at the Powell home about 2:15 a.m. and watched the ball of fire with Powell and Mrs. Powell until it burned itself out about 3 a.m.

A 7½ hour Coast Guard search of the area by boat was unsuccessful in locating any clue as to the location or identification of the object. A Coast Guard spokesman said the search would not be continued unless a plane was reported missing. [534]

03-14-1965   *Everglades Swamp, Florida, USA*

On the night of March 14, 1965, James W. Flynn, who is a rancher and hunting dog trainer, was camped out for the night in the Everglades. Just as he was settling down for the night, his dogs became restive and upset [another case of animal reaction]. He looked around expecting visitors, but instead **he sighted a bright light silently and slowly descending about a mile away**.

Thinking that perhaps an airplane had gotten into trouble and gone down, he prepared his swamp buggy for the journey and set out to render aid if needed. He was guided directly to the spot by the glow which continued unabated. This worried Flynn as **he expected to find a burning plane** and probably injuries or fatalities. He found neither.

Pulsating Glow

When about a quarter mile away, he grounded his swamp buggy and continued on foot. Soon he found himself in a large clearing and he wasn't alone. Some twenty yards away **he saw a circular, cone-shaped object with a pulsating glow**. It hovered just above the ground with a slightly perceptible wobbling motion. He detected, after a bit of study, a sound he could only describe as a hum. He estimated the size at well over seventy-five feet in diameter and twenty-five to thirty feet thick. There were four rows of ports or windows encircling the craft, each emitting a yellow light unlike the color of the craft's overall glow. A partition immediately behind the windows prevented him from seeing any internal details or occupants.

For many minutes Flynn just stood there, amazed. He had heard of such things but, until now, had never really taken them seriously. Overactive imagination, he'd thought. But this was not imagination. It was real.

Curiosity overcame fear. He started to approach the craft to get a better look. He never made it. A pencil thin blue light shot out from "somewhere" on the craft hitting Flynn on the forehead "right between the eyes." He was unconscious before he hit the ground. [This is very similar to what happened to Travis Walton in his book *Fire in the Sky*. [535]]

When he regained consciousness, he was partially blind, sluggish, had a terrific headache and a large, sore bruise on his forehead. The craft was gone. Somehow he got his swamp buggy going and got to his hometown of Ft. Meyers, Florida. He was rushed to a hospital for examination and treatment.

Trees Burned

After he told his story, investigators went back to the spot. They found a large circular spot in the clearing where the ground and grass were charred. The tops of some nearby trees were severely burned. The trunks and limbs of some of the trees were scarred.

The Air Force, at that time, normally debunked and belittled UFO sightings and the people who reported them out of hand, but in this instance, they did not reckon with Flynn's standing in his community. On this occasion the Air Force had to partially back off. They only took one shot, that I can find, in that they labeled it by innuendo to be a hoax. Question is, how did Flynn fake the charred circle of ground, how did he burn and scar the trees and, most importantly, how did he self-inflict a bruise of such shape and intensity?

One final piece of evidence helped exonerate James Flynn. One of his physical injuries was atrophy of internal muscles. Medical science tells us that one can't be faked, period!

It happened, probably exactly as Flynn described it. The trouble is he dosen't [sic] know what it was nor where it came from. So, even today, it's still carried in the annals of the UFO as unexplained! [536]

In the next case, at least we have a possible UFO because burning aircraft are not known to come to a stop in midair.

05-01-1974   *Cowichan Lake, British Columbia, Canada*

Momentarily he froze. High in the sky through an opening of the cloud layer, **he saw what he had seen many times during the war—a burning aircraft**. "My gosh, one of the large airliners on its way to Alaska or Japan is on fire!" flashed through his mind.

For a few seconds he watched the spectacle, and as the fiery, tumbling object, emitting a flaming substance, zeroed in on his property, he ran to the trailer to get his wife and sons out. Seconds later the whole family stood and stared.

The object was still tumbling towards them, growing bigger as it plunged towards the ground, when suddenly beneath the layer of broken clouds which touched the 3,000-foot mountains behind the motel, **it stopped over the lake.**

There were three very bright masses of orange lights spaced in a triangular shape, Meuser recalled, and his son, Ralph, remarked, "You could not have missed them even if you were blind." Tilted at a slight angle, the lights rotated and although no forms could be distinguished—"not even with binoculars"—it was something very large to which those lights were attached. [537]

13.3 The Similarity between a Fiery, Glowing UFO and a Burning Boat

12-05-1954   *Lake Erie, towns in PA, NY, & Ohio, USA*

The object resembled [a] fish tug which looked to be much nearer the beach than one normally drifts, closer even than private fishing boats anchor. **Because of the brilliant**

orange light radiating from it, the lady thought it was on fire.

The glow of a cloud-hidden moon on the water, however, **revealed the thing to be suspended six or eight feet above the lake** off Orchard Beach. The light it made remained steady and blindingly bright.

"If it had been a fire," the lady said, "such intensity would have been white—not orange."

During the eight or nine minutes that she watched it, the ship skittered east, then west—"so fast that by comparison a jet would be standing still." [538]

09-01-1957   *Ocean off Porthcawl, Wales, UK*

Chief-Inspector Reginald Jones, of "D" Division, Glamorgan Police, told *Flying Saucer Review* that the two policemen **thought at first that they were seeing a ship on fire** on the horizon towards Ilfracombe. But then it rose out of the water like a blood-red sun, a good deal larger than a full-sized harvest moon.

While the two police officers watched, two more streaks appeared above and below. It remained at sea level, then suddenly took off at a fantastic speed towards the Atlantic. [539]

08-28-1960   *Oyster Bay Harbor, Long Island, NY, USA*

Seeing is believing and John Krebs of Plainview believes he saw a boat explode in Oyster Bay Harbor at 7:45 last night.

Krebs, who lives at 20 Edison Ave., told police he was about three miles offshore in a boat when he saw a flash to the right of him. **The flash grew bigger, and he thought he could make pit [sic—out] the outline of a boat.** Then there was a curl of black smoke, he said.

Although he didn't hear an explosion, he said **he was sure it was an engine on fire.**

A speedboat pulled alongside of him, Krebs said, asked "if I saw the flash." We both headed for the area but **found no trace of a boat or debris**.

Krebs then headed for shore and called police.

Police and Coast Guard boats searched the area but reported they could find no trace of an explosion. [540]

09-10-1965   *Mediterranean Sea off Le Brusc, France*

"I was out alone in my boat; it was a mild night, with a starry night, and the sea calm.

"It had gone [past?] midnight when I saw a great big light developing very rapidly at sea level. It rose up very high into the sky. It was red, and when I saw it, **at first my thought had been that it was a ship on fire**, but I saw no flames or smoke from it. It was more like an immense flare with a little vapour around it. It changed colour, passing first from red to orange, and then green, and then blue, and then orange again. I heard no sound of any explosion; it all happened in absolute silence. Then this light or glow split up into several parts, and then gradually they faded away bit by bit till the whole sky was black again. The sighting had lasted 15 minutes. I have never found out what the glow could have been or from where it could have come. I made enquiries to find out if a boat had been on fire or had exploded in the area, but I was told that there had not been, nor had there been any flares or fireworks. I never found out what it was." [541]

12-20-1978   *The Adriatic Sea off Bellaria, Italy*

Thousands of people on the shore saw a large, blinding object situated in the water about a mile from the coast. **Everyone compared the sight to a ship on fire.** Photographer, Elia Faccin, took some photographs, which were published in the newspaper of that time. Strangely enough, while taking the first picture, the shutter on his Olympus camera jammed. Roberto Mantovani and Sandro Manaresi observed the object with a telescope and described it as an overturned dish with a tower on top showing an opening. The object was lit by greenish and yellowish-orange lights, which became dimmer towards the outside of the dish. **It was moving very slowly and leaving behind it a**

small wake. Other witnesses described the object as having twelve lights with a much brighter one in the center. [542]

13.4 A Meteor?

08-01-1975   *Trinity Lake, New York, USA*

At about 8 p.m., the sun had sunk below some trees on the far side of Trinity Lake, and shortly thereafter, at 8:25 p.m., a movement in the sky caught Condon's attention, and looking up, he noted an orange-glowing spherical object with the approximate size of a marble held at arm's length. **Condon's first thought was that it was a meteor**, and he noted the time on his wristwatch so he could later check observatories to see if they had recorded it. Then he realized it was not a meteor; it left no trail and was moving too slowly. It did not appear to be fiery or burning as would be the case if it had been a meteor entering the atmosphere. He fully realized that he was watching something unusual when it changed direction and began circling the lake. [543]

13.5 Descriptions as Resembling "Fire"

09-08-1767   *River Isla, near Coupar Angus, Scotland, UK*

"We hear from Perthsire [sic-Perthshire] that an uncommon phaenomenon was observed on the water of Isla, near Cupor [sic] Angus, preceded by a thick, dark smoke, which soon dispelled, and **discovered a large, luminous body, like a house on fire**, but presently after took a form something pyramidal, and rolled forwards with impetuosity till it came to the water of Erick, up which river it took its direction, with great rapidity, and disappeared a little above Blairgowrie. The effects were as extraordinary as the appearance." [544]

01-15-1956 *Pusan, South Korea*

Just a week later another incident occurred about 1,000 miles west of Komatsu. This one is unique in U.F.O. sightings for two reasons:

One: It was seen by a large number of witnesses including civilian and American personnel.

Two: The object was under direct and relatively close observation for about 90 minutes!

The incident began about 8 p.m. at Pusan, Korea, on January 15. The object was described to military authorities as being "about the size of a large washtub and emitting a blue-gray glow. It was seen falling into the water about 50 yards offshore near Heunde [sic-Haeundae]."

It was early enough in the evening to attract the attention of a large number of Korean townsfolk. They reported that the glow continued for about an hour and a half before the object "apparently sank into the sea."

By this time Korean National Police arrived at the scene and they, in turn, alerted U.S. Military Police. Cpl. Ben Elliot, an M.P. on patrol duty that night, was on the scene quickly enough to observe the **object floating in the water for almost an hour.**

He described its glow as being similar to the flames from burning alcohol or benzene. The glow, he said, appeared to be about the size of a large washtub, but the object itself could not be seen on the surface of the water.

None of the witnesses expressed any desire to row out to the object for a closer look. As a result, it eventually sank out of sight into the sea's depths without being inspected.

At this writing, no further reports concerning the object have been made. It was thought that Pusan University staff members might arrange for divers to attempt to recover remains of the object. If they did, their findings remain as much a mystery as the object itself. [545]

In summary, sightings of what are initially thought to be burning aircraft, boats, or other objects are slightly better than those found in the misidentification section which appears in Chapter 10. The reason for this is that as the sighting progresses, the witness continues to view the burning object, and he or she realizes that it must be something other than what he or she originally thought it was. The fiery glow is well known in ufology as an evening event because the rotating field of the UFO is ionizing the air in contact with it, unintentionally producing the image of a flaming body as a byproduct of its operation.

Nevertheless, this is the weakest segment in the case files, as usually no remnants of the object are found by the rescue crews. Then again, what remains do we have in the best of cases?

Chapter 14

Sea Serpents or Water-UFOs?

On various maps of the ancient world, there exists the phrase "Here there be sea serpents and dragons," alluding to unknown, uncharted areas of the oceans. Newspapers of long ago, and even those of not so long ago, regularly featured articles about these creatures. Of course, as we have found out in the UFO arena, many UFOs are attributable to natural phenomenon or manmade objects. In like manner, back then, many of these observations were of the same nature. However, what if some of the observations were of something else? What if these descriptions, given the knowledge of the times, were quite different than what was actually observed? Our problem here is that the witnesses are probably deceased, the newspapers that published any of these articles no longer exist, and who knows what the articles that were published represent? How can we discern whether the content was filler or fact?

In doing some reading about sea serpents, I was "nudged" uncomfortably by sentences that brought to mind, not an animal or fish, but what I had come to see as physical observations of a craft. One has only to understand that the technology of then and now is quite different and this forces us *now* to think along the lines of *then*. We try to comprehend the event as witnesses saw it but with an up-to-date perspective using our current knowledge.

The cases which follow most closely resemble the occurrence of mistaken identity as I see it. Within the first two texts, I have inserted some of my personal observations on various aspects of water-UFOs with explanations of why I think these accounts might belong to the water-UFO domain.

I found this first article, which might be viewed as possible newspaper filler or fiction, to be disturbing, as several points in the text were not known to the people of the period as a

"UFO." The first part of the story is mostly introductory and then it proceeds with the sighting. I have never heard of a "sea creature" of this size possessing an electrical current.

Here There Be Sea Serpents and Dragons
Caution ye who tread here.

07-02-1893   *Puget Sound, Washington, USA*

AN ELECTRIC MONSTER

Flashes of Light and Terrible Sounds Emitted by One in the Bay

W. L. McDonald Struck Senseless in Attempting to Rescue a Shocked Comrade

Nearly 150 Feet Long and Covered With Coarse Hair—A Fishing Party's Trip Cut Short

A party of Tacoma gentlemen have good reason to remember the morning of the 2nd of July as long as life remains in their bodies—and to quote the exact words of one of the party, "There are denziens [sic] of the ocean that man never, in his most horrible and fantastic nightmare, even saw the likes of."

On Saturday morning a party, composed of the following well-known gentlemen, set sail on the sloop *Marion* from the boat house at the end of the wharf for a three days' fishing and hunting excursion on the Sound. The party consisted of Auctioneer William Fitzhenry, H. L. Beal, W. L. McDonald, J. K. Bell, Henry Blackwood, and two eastern gentlemen who are visiting the coast, and it is from the lips of one of these gentlemen, who declines to allow his name to be used, as he says that shortly before he left the East he took the Keeley cure[13], and he fears that if his name was used in connection with this article, his eastern friends might think he had "gone back" and got 'em again.

The party were well supplied with all the necessaries of life, as well as an abundance of its luxuries, though it must not be inferred from this fact that the luxuries played any part in creating the sights seen on that memorable morning. Of course, as a person having much respect for truth, I merely chronicle the story as told me and leave each reader of this remarkable yarn to judge for themselves the necessary amount of credence to give it.

"We left Tacoma," said the eastern man, "about 4:30 p.m., Saturday, July 1st, and as the wind was from the southeast, we shaped our course for Point Defiance, intending to anchor off that point and try our luck with rod and line. We cast anchor about 6 o'clock, the wind having died out, and had fair success fishing. The wind coming up again pretty strong, Mr. McDonald suggested getting under way for Black Fish Bay, Henderson Island, as he knew of a fine trout stream running into the bay and also an excellent camping place near the fishing ground. So about 8 o'clock we weighed anchor and shaped our course for Black Fish Bay, which place we reached about 9:30. We landed and made everything snug about the boat and made a nice camp on shore, and as it was by this time 11 o'clock, we all turned in to get a little sleep as it was agreed upon that at the first streak of daylight we should all get up. About 100 yards from our camp was the camp of a surveying party, but as it was so late, we decided that we would not disturb them but that we would call upon them the following morning, and would probably get some valuable pointers as to the best places to fish and hunt on the island. After a few jokes had been cracked, the boys laid down and in a short time everything

---

13 Keeley cure: A proprietary system for the treatment of drug addiction and alcoholism by the administration of gold chloride. [after Leslie J. Keeley, 1832-1900, American physician] 008

about camp became as still as death. It was, I guess, about midnight before I fell asleep, but exactly how long I slept, I cannot say, for when I woke, it was with such startling suddenness that it never entered my mind to look at my watch, and when after a while I did look at my watch, as well as every watch belonging to the party, it was stopped."

[In ufology, the presence of a UFO usually affects instruments subject to electromagnetic interference. As an example, the watch being stopped would indicate the possibility of a UFO or at least an object of strong electromagnetic force. Continuing with the case:]

"I am afraid, sir, that you will fail to comprehend how suddenly that camp was awoke.

"Since the creation of the world, I doubt if sounds and sights more horrible were ever seen or heard by mortal man. I was in the midst of a pleasant dream when, in an instant, a most horrible noise rang out in the clear morning air, and instantly the whole air was filled with a strong current of electricity that caused every nerve in the body to sting with pain, and a light as bright as that created by the concentration of many arc lights kept constantly flashing. [The unpleasant sound cited is infrequent in a ufological sense, but the sound of an electric generator, which I would think might be uncommon at that period, might be a plausible cause as it is heard in some up-close UFO cases. Also, pertaining to what was just said, there are many cases that mimic this scenario. Similar comments referring to hair standing on end because of the electrical effect are well documented. The event, taking place at night, again introduces the ionized air effect that causes UFOs to glow in various colors due to power settings. (See *Unconventional Flying Objects* by NASA scientist Paul Hill, p. 13.) The account continues:] At first I thought it was a thunderstorm, but as no rain accompanied it, and as both light and sound came from off the bay, I turned my head in that direction, and if it is possible for fright to turn one's hair white, then mine ought to be snow white, for right before my eyes was a most horrible looking monster." By this time every man in our camp, as well as the men from the camp of the surveyors, were gathered on the bank of the stream, and as soon as we could gather our wits together, we began to question if what we were looking at was not the creation of the mind, but we were soon disburdened of this idea, for the monster slowly drew in toward the shore, and as it approached, from its head poured out a stream of water that looked like blue fire. [I have at least three cases where water was pushed forward, not by the UFO, per se, but by its surrounding field. It led to boats being almost swamped by water and in another case by two people on the beach being almost sucked into the sea by the undertow caused by the water forced towards them. The coloration of the water, which "looked like blue fire," might be explained either by the coloration of the UFO at the time of the sighting or its described "arc lights." Let us continue with the story:] All the while the air seemed to be filled with electricity, and the sensation experienced was as if each man had on a suit of clothes formed of the fine points of needles. [This is indicative of static or other electrical effects as previously mentioned.] One of the men from the surveyor's camp incautiously took a few steps in the direction of the water, and as he did so, the monster darted towards the shore and threw a stream of water that reached the man, and he instantly fell to the ground and lay as though dead.

"Mr. McDonald attempted to reach the man's body to pull it back into a place of safety, but he was struck with some of the water that the monster was throwing and fell senseless to the earth. By this time every man in both parties was panic-stricken, and we rushed to the woods for a place of safety, leaving the fallen men lying on the beach. [As will be seen later, the men were not harmed but evidently suffered from an electric shock. Again, the field of the UFO is evidently an electric configuration and affects all things that come into contact with it and, in this case, water is a good conductor. Moving on:]

"As we reached the woods, the 'demon of the deep' sent out flashes of light that illuminated the surrounding country for miles [illumination that is noted in many UFO cases], and his roar—which sounded like the roar of thunder—became terrific. When we reached the woods, we looked around and saw the monster making off in the direction of the Sound, and in an instant, it disappeared beneath the waters of the bay, but for some time we were able to trace its course by a bright luminous light that was on the surface of the water. [The "monster" disappeared "beneath the waters" and was traceable by a bright light that was "on the surface of the water." Phrases like these are similar to those in several other cases where a UFO either enters or exits the water. A solid light (not of the water wheel category) is seen over or under the sea and is referred to as being seen "through" the surface. Continuing on:] As the fish disappeared, total darkness surrounded us, and it took us some time to find our way back to the beach where our comrades lay, and we were unable to tell the time, as the powerful electric force had stopped our watches. We eventually found McDonald and the other man and were greatly relieved to find that they were alive, though unconscious. So we sat down to await the coming of daylight. It came, I should judge, in about half an hour, and by this time, by constant work on the two men, both were able to stand, and both agree that the moment the water the monster threw touched them, they became immediately unconscious." On being asked to give some description of the fish, for it was, he said, "an electrical fish," the eastern man said:

"This monster fish, or whatever you may call it, was fully 150 feet long, and at its thickest part, I should judge about thirty feet in circumference. Its shape was somewhat out of the ordinary in so far that the body was neither round nor flat but oval, and from what we could see, the upper part of the body was covered with a very coarse hair. The head was shaped very much like the head of a walrus, though, of course, very much larger. Its eyes, of which it apparently had six, were as large around as a dinner plate and were exceedingly dull, and it was about the only spot on the monster that at one time or another was not illuminated. [I would think he meant diameter when the witness mentioned the "circumference" of the creature. After all, he could not have measured around it. However, the dimensions he gives are quite common in ufology, and, if the shape he mentions is that of the craft's length, it then fits quite nicely into the "cigar shape" so familiar to ufologists. As to the six eyes on the nautical creature, could those possibly be windows? Still, the case goes on:] At intervals of about every eight feet from its head to its tail, a substance that had the appearance of a copper band encircled its body, and it was from these many bands that the powerful electric current appeared to come. The bands nearest the head seemed to have the strongest electric force, and it was from the first six bands that the most brilliant lights were emitted. Near the center of its head were two large horn-like substances, though they could not have been horns for it was through them that the electrically charged water was thrown.

"Its tail, from what I could see of it, was shaped like a propeller and seemed to revolve, and it may be possible that the strange monster pushes himself through the water by means of this propeller-like tail. [This fits neither a nautical species nor a UFO.]

"At will this strange monstrosity seemed to be able to emit strong waves of electric current, giving off an electromotive force which causes any person coming within the radius of this force to receive an electro tonus[14]. This fish probably receives its power from some submarine cavern of volcanic origin which owing to its peculiar construction, and having an extra-large deposit of copper, it charges the fish that inhabit that region with a strong electric force that is displayed by this peculiar specimen. The peculiar-shaped, copper-like bands may be caused by the strong magnetic force of the fish and the copper deposits of the ocean, as the strong current would form the copper into a solution, whilst the strong attraction of the fish would naturally form an electric battery, drawing towards it this solution, thus forming deposits on the fish,

---

14 tonus: The ability of a muscle to contract in response to a stimulus. [008]

so that in reality the electric fish is completely encompassed in copper, and its rapid movement through the water is constantly generating frictional electricity, which, I should judge, would in a measure account for the fish being so constantly and powerfully charged with electricity, though far from its original source of supply. One of the strange characteristics of this fish, and one by which it undoubtedly obtains its food, is its high electric control of dense and foggy atmosphere surrounding it, which amalgamates with the electrification of the fish, making a potential which causes any living creature, such as birds or insects, flying through the air to fall dead into the water. Of course, that is merely a theory, and I may be mistaken as to its origin or where it goes to, but one thing I do know, that I would not encounter the same monster again for the universe, and you can ask the rest of the party and you will find that they all agree with me, that to be within so short a distance of such a terrible monster and yet live to tell the story is something that only happens once in 1000 years. I hardly need to tell you that we were not long in getting underway for Tacoma, and I can assure you that I have no further desire to fish any more in the waters of this bay. There are too many peculiar inhabitants in them. I am going to send a full account of our encounter to the Smithsonian Institute, and I doubt not but what they will send out some scientific chaps to investigate.

"Now I must be going, as I have to leave on tonight's train, but if you need any further particulars, you can obtain them from any of the party. No, I do not know who composed the survey party; all I know about them is that they are from Olympia and that they were on the island running farm lines on some disputed land." [546]

03-12-1908   *Atlantic Ocean heading north from Jamaica*

Monster was "All Lit Up"

"March 27—At 8 p.m. Thursday, March 12, latitude 22.06 north, longitude 72.21 west. One-half mile off starboard bow sighted strange marine monster. It approached and followed ship all night. Friday at 9 a.m. monster crossed our bows. Passengers in a panic. Reduced speed to five knots."

—Extract from the log of the *Admiral Farragut*.

Sitting in the doorway of his cabin, collarless and in his shirt sleeves, Captain Mader of the fruit steamer *Admiral Farragut*, which arrived recently from Port Antonio (Jamaica), told the story:

"We picked the sea serpent up—or rather the monster picked us up—late one night," said he. "I was on the bridge when one of the passengers, an elderly man, rushed up and excitedly called my attention to a phosphorescent light several miles astern. At first I thought it was a new submarine boat. As it came nearer, we played the searchlight on it and could see that it was some strange sea monster.

"It seemed about 120 feet long and threshed its way through the rough sea at a fearful speed. All night long it followed the vessel, and during that time most of the passengers and crew remained on deck. The phosphorescent glow of the monster lighted up the sea within a radius of fifty feet. [What a "fearful speed" might be is unknown to me, but I would assume that it would be much faster than the speed of the ship. The fact that the sea was lit up to a radius of 50 feet indicates that either the plankton was disturbed beyond the "sea serpent" by some means or that it possessed an extremely bright, artificial light source. Moving on:]

"On the next morning about breakfast time, the serpent swam within thirty feet of the starboard side. The creature resembled a huge boa constrictor, with the exception that its body was green. From its side streamed seaweed and other marine growth. [Green is a color commonly found in water-UFO cases. Snake-like movements could be attributed to a cylindrical object, with the water's movement distorting its shape into that of a snake. The case goes on to say:]

"The monster raised its head several feet above the water. It had huge eyes projecting from the top of its head and two green horns that projected upward nearly five feet.

The horns resembled large antennae and moved about continually. [Five-foot horns that resemble antennae are one thing, but to have them "moved about continually" seems more like a mechanical operation.]

"Three times the serpent crossed our bows and fearing to run it down, I signaled the engineer to reduce speed to five knots an hour.

"Some of the persons onboard thought the monster was hungry, and we threw over several sacks of peanuts and a few bunches of bananas. When we arrived off Cape Hatteras late that day, the monster circled around and swam south." [547]

08-16-1934   *Atlantic Ocean near Bermuda*

Deep-Sea Monster with Headlights
Seen on Record Descent in Ocean

A huge deep-sea fish, possibly unknown to man, was one of the curious sights which greeted Dr. William Beebe, American scientist, in a daring record-breaking descent toward the bottom of the ocean.

The underseas explorer and his associate, Otis Barton, were unable to identify the monster they sighted from their "bathysphere" yesterday.

Sealed in the two-ton iron ball, Beebe and Barton were lowered to a depth of 3028 feet, more than a half mile under the surface. The descent exceeded their record of last Saturday (August 11) by 518 feet and surpassed the earlier mark of half a mile by 388 feet.

A large gray "shadow" at 2750 feet was the first appearance of the unknown fish. The object seemed to be illuminated by scores of tiny lights, glittering like a diamond necklace, Dr. Beebe said. He estimated its length at 20 feet. Phosphorescent parasites are believed to have given off the lights.

Beebe described the monster as the largest he had ever seen in a deep-sea dive. Barton attempted to photograph it, but his results were uncertain as underwater creatures flee when a searchlight is turned on from the bathysphere. [548]

Could these so-called sea serpents be yet another class of "misidentifications"?

Chapter 15

Physical Influences of a UFO on a Boat's Motion

Besides being similar to the title of Chapter 2, "Physical Influences of a UFO on Water," the content of that chapter also has a bearing on this one. In Chapter 2, the vortex under the UFO lifts the water up unintentionally and only as a byproduct of the field's operation. This lifting has been known for many years in ufology, but, in many respects, I believe the cause was considered something akin to the *Star Trek* tractor beam or some similar unknown force.

Regarding cases where cars are lifted off the road by this force, the terrified driver—aware or unaware that there is a UFO hovering over the top of the car—suddenly loses operational control of the vehicle. The driver hits the brakes, but the car keeps moving; guns the engine to escape, but the car continues at some speed not governed by the driver; turns the steering wheel to avoid a ditch or other obstacle, but the car continues in the direction it seems it wants to go. Why is this? The car, and as we will also see with boats in the water, is being lifted off the surface. Some cars have been elevated to great heights, probably due to the lifting power of a large UFO's field. In one case, a car was picked up and carried across another country's adjacent border. The occupants, who had not intended an international trip, had a difficult time trying to explain how they got into the country without having the same documents necessary to leave it. The smaller variety of UFOs might not have the power to raise that much weight, but if they lift a vehicle so much as a half inch vertically, all steering, braking, and acceleration are nullified.

I think that you have already visualized what could happen if a UFO did the same thing over a boat in the water, hovering above it and influencing its movement as well. All this means is that what is observed on land is in everyway the same as on water. In short, the same physical principles are at work, wherever the craft may be or in what area it is seen. Keep in mind that the following two cases illustrating this movement are 339 years apart.

03-??-1638   *Muddy River, Massachusetts, USA*

Puritan James Everell and two others were stunned as they saw a luminous mass that hovered and returned over a three-hour period. **Their boat was pulled upstream by the phenomenon.**

The settling of the first Puritan colony in Boston was chronicled by Governor John Winthrop, who arrived in Massachusetts Bay in 1630 with one thousand English emigrants. A historian himself, Winthrop kept a record of the colony's first years in the New World. His journal is far from being a mere collection of unlikely anecdotes or village gossip. It is quite significant, therefore, that he regarded two spectacular sightings of unexplained phenomena as being sufficiently important to be recorded for posterity.

The first sighting took place in March 1638. A member of the Puritan Church, James Everell, "a sober, discreet man," was crossing the Muddy River one evening in a small boat with two companions. **Suddenly a great luminous mass appeared in the sky above the river. It seemed to dart back and forth over the water.** When it remained motionless, it "flamed up" and seemed to measure three yards square. When it moved, it "contracted into the figure of a swine" and flew away towards Charlton.

It did this repeatedly over a period of two or three hours, always returning briefly to the same spot above the water before shooting off again.

When the light had finally vanished, Everell and his friends stood up and were surprised to learn that the boat was now further upstream than it should have been, as if it had been pushed, pulled or carried

by an unknown force. In fact, they had been carried against the tide to their original starting point, one mile away. [549]

01-21-1977   *Canal in St. Bernard Parish, Louisiana, USA*

UFO Induces Abnormal Silence, Halts Motion of Boat

January 21, 1977: St. Bernard Parish, LA, 8:45 p.m. While hunting for nutria along a canal, a man's boat was engulfed in brilliant light. **The man** (name on file) **felt heat, and all sounds of nature ceased.** A glowing object then sped away, and everything returned to normal. After a while he picked up a friend and resumed his work. **Then a light again approached and hovered over the boat. Once again silence prevailed, and his hair stood on end "like wire." Everything felt warm. The engine continued to run, but the boat stood still as if being held by "strong gravity forces."** (Later investigation showed that there were no impediments in the water, and the channel had been dredged recently.)

The object appeared round, about 20 feet (6 meters) in diameter, with a textured surface showing a pattern of connecting diamonds or squares.

When the object departed, the boat suddenly lurched forward throwing both men down, as if the force gripping the boat had suddenly been turned off. They saw the UFO retreat and hover near an oil refinery, where a company guard also saw it. Then it moved and hovered again in a new spot, where it emitted a light beam downward. After about 30 minutes it disappeared. The principal witness reported a period of about 20 minutes during the sighting that he could not account for. [550]

In the last case, other effects are noted besides the halting of the boat's travel. The heat and utter silence experienced by the witness, I believe, are caused by being enclosed within the UFO's field. If this field is rotating at high speed,

411

no molecules of air are being transmitted to the interior of the field, and sound travels on these molecules. Evidence that it is an EM (electromagnetic) field is verbalized in the statement "his hair stood on end 'like wire.'"

Chapter 16

Totally Submerged

The only proper place for the term "USO" (Unidentified Submerged Object)

While it is obvious that UFOs which enter or exit bodies of water must also navigate beneath it, logic does not follow that an object, known only to be totally submerged and given the label of "mysterious, unknown, underwater object," is necessarily what we commonly refer to as a UFO. In these situations, if these craft are not seen visually or by sonar (or hopefully both) and do not perform beyond the capabilities of modern craft, they should only be regarded as mistaken objects that probably have a rational explanation. No mere "bump" in the water should be automatically described as an extraterrestrial craft.

16.1 Hooked onto Something

The following newspaper article illustrates the possibility of attributing too much to the mysterious:

04-22-1956 (Not on my website)

Trawler Snags Atom Sub, Is Nearly Pulled to Bottom

Groton, Conn. May 2. The $55,000,000 atom-powered submarine *Nautilus* narrowly escaped tragedy off the New Jersey coast April 22 when it tangled with a net and nearly dragged a little fishing trawler and its crew to the bottom of the Atlantic, the Navy disclosed today....

413

The captain of the trawler estimated his damage also at $1,300—in fish and gear that was towed away by the submarine, which apparently never surfaced after the mishap and continued on its course to Groton with its crew unaware of the near tragedy.

Thought It Was a Sea Serpent

Tonnes Anderson, Point Pleasant Beach, N.J., skipper of the 61-foot trawler *Jennie*, made his estimate last Friday in telling of his strange encounter with what then was to him a mysterious, undersea object. **One of his crew members thought it was a sea serpent,** but another said he believed it was a submarine. The size of his crew was not disclosed....

Engineer Shirley Friend said the *Jennie* suddenly lurched to a halt and then was drawn backward. The winch cable to the fishing net became taut, he said, and the stern of the vessel began to sink. The deck was level with the water, he said, when the cable suddenly snapped.

Naval officials investigated the story and the mystery finally was solved today by Commander John Dudley, flag secretary for the commander of submarine forces of the Atlantic Fleet. [551]

Just as the crew member in the above story thought that his ship was hooked onto a sea serpent, the additional examples which follow continue the trend of being helplessly pulled by some unidentifiable "thing."

07-EE-1961   *Port Aransas, TX, USA*

A 67-foot shrimp boat sank under **mysterious** circumstances in the Gulf of Mexico off Port Aransas the first week in July. **The boat simply hooked into "something" that ripped the vessel's stern right off.**

Ira Pete, the owner of the *Ruby E*, said, "I hooked into something and pulled the stern off." He said he had no idea what the "something" was. He and two crewmen were taken aboard another fishing vessel which was close by and no one was hurt.

The world seems full of these **mysterious** "something's." **552**

04-08-1963   *Montauk Point, NY, USA*

On the 8th of April 1963, the New Bedford dragger *Sunapee* was fishing sixty miles south of Montauk Point, L.I., when **she started to move backward**. Captain Nelson Ostman, skipper of the 74-footer, said **the crew had seen a submarine in the vicinity earlier and realized immediately that they had caught onto the sub.** He tried to swing the vessel around so that the net would pull free, but the maneuvering was unsuccessful and the *Sunapee* lost her nets, lines, and drags, valued at $3,000. The Navy at Boston said it knew nothing of the incident and was investigating. **553**

08-09-1963   *Georges Bank off Portland, Maine, USA*

The 90-foot dragger *Resolute* limped into Portland Harbor today and the crew told the Coast Guard **the vessel had been towed backward and nearly submersed by an unidentified object.**

Capt. John Larson, the boat's skipper, said the vessel was nearly capsized yesterday when the object became tangled in its nets. Larson said his boat was moving so fast backward that it began to heel over and take on water.

Larson said he and his six-man crew cut thousands of dollars worth of nets to save the *Resolute*. The boat also sustained damaged radio aerials and loosened planks.

The crew said the object was either a submarine or a huge whale. The Navy said it had no submarines operating off the Georges Banks [sic–Bank] where the incident occurred. [554]

10-11-1969 *Mediterranean off Cabo Cope (Murcia), Spain*

The fishing boat *Agustin Rojas*, owned by Sr. Francisco Simo, was about 5 km from Cabo Cope when **its nets caught something from which they could not be freed**. The Spanish Navy was alerted and issued a statement on October 13 noting that a buoy had been used to mark the spot and that Counter-Admiral Pery, aboard the helicopter carrier *Dedalo*, was in command of the operation which involved the destroyers *Jorge Juan* and *Valdes*, the submarine S13, and another ship. The object was never found. [555]

16.2 Submarines?

These were the years of the Cold War, and many countries had subs of one type or another in the oceans. Explanations of how we can be led astray by speculation of "mysterious" or "possible" underwater UFOs are illustrated in the following cases:

02-16-1955 *Atlantic Ocean off Fort Pierce, Florida, USA*

The Coast Guard sighted an unidentified submarine only 30 miles from the Air Force's huge guided missile center at Cocoa, Fla., today. The Navy threw an anti-submarine squadron into a search for the mystery vessel.

The Coast Guard said one of its cutters sighted the submarine about 10 miles off Fort Pierce on the Florida east coast. The submarine submerged when a 30-foot Coast Guard boat approached within a mile of it, the officer said.

The submarine was first seen around 5 a.m., the Coast Guard said. **Officers noted a "red glow" in its conning tower and a light in the superstructure.** [This is interesting from a UFO point of view.]

An officer at the Jacksonville Naval Air Station said the Navy had sent an anti-submarine squadron into a search for the sub. But he would not reveal the number of planes engaged. [556]

11-14-1961 *Off Sydney Heads, Australia*

Here's the kind of **mystery** that has turned up several times in recent years. The report just reached us although the incident occurred Tuesday, November 14, 1961.

Australian and New Zealand warships were conducting Navy exercises off Sydney Heads when they detected and pursued **a large, underwater object that interrupted the maneuvers.**

There was no visual sighting of the object, but Senator Gorton, the Minister for the Navy, said it could have been an ocean-going submarine. If so, "the speed and ease with which the mystery craft eluded the fleet suggests it was nuclear-powered," the *Sydney Sun* stated.

But suppose it wasn't a nuclear-powered submarine? All the Navy would say officially is that it was an "**unidentified object**." Contact was soon lost. Neither U.S. nor British submarines were in the area at the time. [557]

I guess that the British and Americans are the only countries with subs. Did someone forget the Russians?

08-24-1962 *Ocean off Gottland, Sweden*

For those readers who like to be kept informed of mystery objects beneath the waters as well as those seen in the sky, we reprint the following from the London *Daily Mail* of August 24:

"Ships of the Swedish Navy exploded depth charges and shells near a **mystery submarine** off the island of Gottland tonight after it ignored orders to surface. The submarine, detected by echo-sounded apparatus, was first ordered to surface through hydrophone contact.

"Then a charge of gunpowder was detonated underwater—the international order to surface. Again the order was ignored, so the Swedish warships dropped two depth charges 'a safe distance from the submarine' and fired anti-submarine shells.

"The Swedish Board of Defense, announcing this tonight, said the submarine then left the area." [558]

09-13-1969   *Waters off Norrtälje, Sweden*

"An **unidentified submarine** has been reported Saturday night (September 13, 1969) inside the prohibited military sector at Björkö-Arholma, north of Stockholm.

"Navy helicopters and a Coast Guard cruiser were sent immediately to the area to search for the intruder. It was first sighted by a Finnish ferry outside Norrtälje. A man stationed on the island of Hämtan said he saw the submarine from his motorboat near Havssvalget. He picked the object up with his searchlight and watched it from a distance of only a few yards. **He was unable to identify it**. After he radioed his report to the Coast Guard, an extensive search was held Saturday evening by aircraft, ships, and military troops. No trace of the object was ever found." [559]

12-07-1972   *Disko Bugt (Bay), Greenland*

Copenhagen, Denmark, Dec. 12—The Danish Defense Ministry said today that an **unidentified submarine** was believed to be operating deep inside Greenland's fiords under cover of the long polar night.

The report—following **a protracted, fruitless hunt last month for what was finally said to be an unidentified foreign submarine** in a Norwegian fiord—was based on observations by Greenland fishermen in the Disko Bay area on Dec. 7 and 9. The bay is on the west coast of the outlying Danish territory.

The Danish Defense Minister, Kjeld Olesen, informed the Government of the observations at a Cabinet meeting. There was no indication that any major search for the submarine was underway. [560]

I have almost given up on these types of reports because they are so widespread and, at the same time, have no explanation as to whether they are definitely submarines or could possibly be UFOs. Only the instances that show a definite anomaly allow us to add those cases to our shopping cart. However, the "what-ifs" do not stop there. We do but turn a page.

16.3 Hit by Unseen Metal Craft (UFO or Submarine?)

02-05-1964   *Off Cape Mendocino, California, USA*

The yacht *Hattie D*, a converted Navy search and rescue craft, was struck and sunk on Wednesday, 5 February 1964 by an **unknown object** about 25 miles off the rugged coast of Cape Mendocino, California. A Coast Guard helicopter, dipping between 30-foot-high waves, pulled a German shepherd puppy, nine crewmen, the captain, and his wife from the pitching deck of the yacht.

The *Hattie D* set out from Seattle on January 24, then after various stops, left Neah Bay, Washington, on February 2 for California.

All eleven survivors insisted the yacht struck or was rammed by a "metal object." "I don't care how deep it was," said crewman Carl Jensen. "**What holed us was steel and a long piece**. There was no give to it at all." Jensen was referring to the 7,800-foot depth at the point where the sinking occurred. [561]

Subs are made of steel, so why attribute this incident to a UFO when we do not know the material composition of such a craft?

10-06-1969   *South China Sea*

"In a report dated October 10, 1969, it was suggested that a submarine may have sunk the Swedish tanker *Seven Skies*. During the preliminary investigation in Singapore, Commander Otto Ferdinand Henning said that the ship behaved strangely just before the explosion. **The tanker rose upwards and then rolled over as if it had struck a reef or collided with some huge underwater object.**

"The explanation was given that the *Seven Skies* collided with an unknown submarine running at periscope depth. Commander Henning had no memory of any explosion, he said, but other crew members stated that the ship was shaken by a series of violent explosions before it sank. The accident took four lives: three Swedes and one Indonesian." [562]

11-19-1969   *Between Domsjö and Bureå, Sweden*

"At 4:00 p.m. on Wednesday afternoon, there was another mysterious collision. The German **ship** *Insulanur*, en route from Domsjö to Bureå, **collided with an unknown object** at Sydostbrotten. The weather was clear and quiet and the collision was quite violent, but **the crew failed to see the cause**. The *Insulanur* is now being examined in the harbor at Skellefteå." [563]

11-20-1969   *Waters off Hälsingland, Sweden*

"Is there a **strange submarine** in the waters outside of Hälsingland? A collision occurred yesterday, Thursday, November 20, 1969, 16 nautical miles from Galström, southeast [of] Sundsvall. The helmsman of the trawler *Silverö* told the press that **a light appeared just before the collision and then disappeared. It was visible for about ten minutes.** [At least here we have a slight anomaly inasmuch as a sub would not advertise its presence with lights on.]

"After the collision, the trawler heaved to and remained stationary. **There was no sign of the object which**

caused the collision. The revenue cutter TV 245 searched the area but found nothing.

"The press reported that no Swedish units were in the vicinity at the time. **The collision was severe and must have damaged the object**. The trawler is now undergoing repairs at Ljusne. The damage was apparently caused by a very sharp object. No traces of stones or rubble were found in the ruptures, but two reddish-brown spots have been discovered on the damaged hull. A planking about one metre below the waterline was badly damaged as were three planks above the waterline." [564]

16.4 The Reverse—Ship Hits Something Unknown

10-05-1959   *Atlantic Ocean off Portsmouth, NH, USA*

Collision of Nuclear Vessel Causes Some Damage

The nuclear-powered submarine ***Seadragon*** **hit a whale during a test run last night**.

The *Seadragon*, second nuclear submarine built at the naval shipyard here, left yesterday for four days of builders' sea trials. The boat came back to port today with a report that she struck a whale while running on the surface last night.

Some damage was reported to a propeller and shaft.

Aboard the submarine in charge of the test was Vice Admiral Hyman G. Rickover. The *Seadragon*'s skipper is Lieut. Comdr. George P. Steele. [565]

02-03-1965   *Sea of the Hebrides off Barra, Scotland, UK*

The **trawler** *Star of Freedom* (70 tons) from Fleetwood, Lancashire, **struck an unidentified object** in the early hours of the morning of February 3. She was steaming at the time at nine knots in 80 fathoms, 15 miles E.S.E. of Barra. The crash lifted her bows from the water. A distress call was sent out, and the crew manned the pumps. Eventually the badly holed trawler was beached in Castlebay harbour.

Skipper George Wood is convinced that he hit a surfacing submarine, but both British and American naval authorities denied that any of their submarines was responsible and refused to comment on the possibility that a foreign vessel had been involved. [566]

In the following case, witnesses said that an underwater light actually hit a ship. I have not corrected the English used in this account so that the reader can get a feel for the story as it was originally verbalized:

01-15-2000   *Waters off Indonesia*

My cousin told me that he never saw anything like it. But his friend did see this things just one month earlier. The location is nearby, eastward from the first location. His friend was during the trip toward city of Manokwari. He saw the same lights under the sea but in different formation. These ones form a long line formation just ahead the boat. When the boat he was in trying to turn back to avoid crash with these things, suddenly the same lights formed the same formation just behind the boat. **The boat was lifted up a bit when hit these lights.** Next thing the boat turn right and escape from both the lights formations. [The report says that the boat hit the lights. As I stated previously, a submarine wanting to travel covertly underwater would not want to advertise its presence by having lights on. Also, it is mentioned that "the boat was lifted up a bit," which, in my opinion, more clearly represents what would happen if a UFO actually did come too close to a ship. Rather than ramming it, I believe the field would be moving whatever it touches upward as a result of the rotating field. The text continues:]

The last sighting was when the KM *Lintas Samudera* (a passenger ship) was having an accident in the Sea of Banda on the Moluccas. Some people were killed during the accident (I don't recall the number of casualties but you can check it out on the local and national newspaper on 22-30 January 2000. One of them that I read was *Jateng Pos* dating Friday, 28/01/00). **Some survivors told that before the ship was sunk, there was two shape of lights from under the water moving fast toward the ship and crash the ship.** The ship was later found in the beach of nearby island. [567]

16.5 Submarines That Could Not Have Been Subs

Submarines do not operate in shallow waters due to the danger of damage to the ship, however, this is not the case with UFOs. There are many cases of UFOs submerging in shallow waters. Keep in mind that a UFO has a protective field that would prevent damage to the craft.

02-22-1955   *Atlantic Ocean off North Carolina, USA*

Hunt Fails to Find Two Strange
Subs Reported off N.C.

The Eastern Sea Frontier Command said tonight **two submarines were reported sighted off the North Carolina coast, but a search failed to find any trace of them**....There are no American submarines in the area, the spokesman said. He said **the water at that portion of the coast was shallow far out to sea and the supposed submarines would have to have been operating in only 50 feet of water, "which doesn't make sense."**

The spokesman said numerous fishing boats that operate in the area could have been mistaken for submarines at that distance. Civilians on the Nags Head beach said the atmosphere was hazy and visibility poor.

Alpheus Drinkwater, veteran telegrapher and news correspondent at Manteo, N.C., said, "There very definitely is something out there." He said several Nags Head beach residents reported seeing what appeared to be two unidentified subs.

Atop 40-Foot Pole

"If they were submarines, they were unmarked," Drinkwater said. D. A. Sutton, a telephone construction company employee, said he observed the strange objects from atop 40-foot telephone poles.

"I definitely saw something and it looked like two subs," he said. **He was unable to identify the objects definitely as subs because of poor visibility**.

Last Wednesday two Coast Guardsmen reported sighting a submarine off Fort Pierce, Fla. Navy planes searched the area but failed to find it. [568]

03-10-1965 *St. Lawrence River, Baie-St.-Paul, QC, Canada*

Another "impossible" USO was observed in the St. Lawrence River near the city of Quebec, Canada, in March 1965. For four or five minutes, Captain Claude Laurin of Quebecair and his co-pilot **could see the "submarine" lying below the surface over 200 miles (320 kilometres) from the open sea, a situation that would be extremely hazardous for any ordinary submersible**. [569]

16.6 Actually Seen Underwater

Included in this section are those cases that support the submerged UFO sightings. Observed attributes include qualities such as shape, lighting configuration, and speed.

??-??-1929   *Tarpon Springs, Florida, USA*

I thought you might like to hear a story my father told me recently (and having been in the military was even afraid to tell me) about a sponge diver in his hometown of Tarpon Springs, Florida. It seems that an old Greek sponge diver back in 1929 was underwater doing his sponging in his old-type sponge diver's suit and helmet, and as he looked up from what he was doing, **he saw a saucer-shaped silver disk fly by him underwater with lights on it, and then**

it flew by him again, this time in the other direction as if to watch what he was doing. The man thought it must be something the Americans must have made or maybe there is another ship in the area. When he came out of the water, he told the other spongers what he saw underwater. No one believed his story. They told him no such thing could be true, and there were no other boats in the area. For years he stuck to his story and told everyone who would listen to him. [570]

??-??-1943   *Unknown position in the Persian Gulf*

Seaman Matthew Mangle, from the bow of his ship, **sighted a huge disc beneath the surface of the water. The object, glowing with a soft, greenish light**, paced the ship at about 12 knots before speeding up and moving out of sight. (Similar reports are on record.) [571]

06-??-1959   *Buenos Aires, Argentina*

The object described above [See case dated 07-28-1962 on my website] sounds strangely like the fast, submarine-like object which the Argentine Navy had bottled up in the harbor at Buenos Aires in June 1959. Our latest information on this particular incident indicates that the object was shaped generally like a huge fish, was silver in color, and sported a very large tail which generally resembled the vertical stabilizer on a B-17. **Divers got a good look at it and could not identify it as a submarine. This object also was extremely fast and maneuverable.** [572]

03-12-1965   *Kaipara Harbour, New Zealand*

It was on 12 March 1965, eight years later, that I had my next stroke of luck. This sighting was the best and most interesting of them all, and thenceforward my investigations were pressed on with all speed until they culminated in my present findings....I suppose we were about a third of the way across the harbour when I spotted what I took to be a stranded grey-white whale....As far as I can remember, I gave no indication of surprise and I said nothing as I looked down. **My "whale" was definitely a metal fish. I could see it very clearly**, and I quote from the notes I made later:

(a) The object was perfectly streamlined and symmetrical in shape.
(b) It had no external control surfaces or protrusions.
(c) It appeared metallic, and there was a suggestion of a hatch on top, streamlined in shape. It was not quite halfway along the body as measured from the nose.
(d) It was resting on the bottom of the estuary and headed towards the south, as suggested by the streamlined shape.
(e) The shape was not that of a normal submarine, and there was no superstructure.
(f) I estimated the length as 100 ft., with a diameter of 15 ft. at the widest part.
(g) The object rested in no more than 30 ft. of clear water. The bottom of the harbour was visible, and the craft was sharply defined....

Inquiries made from the Navy confirmed that it would not have been possible for a normal submarine to be in this particular position, due to the configuration of the harbour and coastline. I added as much information as I could to my growing file of notes and tucked them away for future reference. [573]

04-??-1973   Gulf Stream off Bimini and Miama, FL, USA

"In both cases **an object**, gray, light, smooth, and tapering a little, like large cigars with rounded ends," he said, "**passed at great speed across his bows, underwater**. Delmonico estimated the size of the machine to be between about 120 and 160 feet and its speed about at least sixty to seventy knots. When the captain first saw it moving under the water, he thought that a collision was inevitable, as the object appeared to be surfacing just in front of him. Apparently becoming aware of the presence of the boat, the machine dived and disappeared, passing under the keel. **No turbulence or wash appeared to accompany the apparition.** No rudder, fin, or any other projection broke its smooth, polished surface, which contained no portholes." [574]

07-??-1992   *Ligurian Sea off island of Gorgona, Italy*

Three friends who were on board a deep-sea fishing boat had an interesting experience. One of them became aware that at some distance from the boat, the sea had a strange color, like there was a sandbank, even though nautical maps did not show any. They approached, thinking that there was a large school of fish, but when they reached the location, they saw that **under the water, at an estimated depth of thirty meters, there was a circular structure about 200 meters in diameter.** The structure, which was stationary, emitted a weak light of a silver color. Unfortunately, there are no other details or investigations; this is a pity since the story is undoubtedly interesting and worthy of further investigation. [575]

So what do we have in this chapter? Hooked on "something," "unidentified" submarines, hit by an "unseen" craft, sub hits "something unknown." In short...speculation.

I find myself in the strange position of being on the side of both scientists and skeptics for whom it must also be strange to try and explain away something that they themselves have not witnessed. This confusion is further compounded by the actual witnesses who, through their usually unaccepted anecdotal tales, are at a loss to accurately describe in understandable terms what it was that they saw to critics and scientists.

Give up this category? As much as I would like to, the remainder of these cases do give us pause to review their value. For instance, shallow water would rule out almost all submersibles, yet UFOs have been seen flying into shallow water. I know this for a fact because I did an interview with a friend of my family's who was a missionary on a remote island in the Pacific. The UFO that she observed in this case entered water that was only 5 feet deep, and the witness was very familiar with the water as she had skin dived in this reef (See my website for case **Undated-4**, 1975~1979). No remains of a crashed vehicle were found by island natives who went to the site spurred on by thoughts of reward.

Can a UFO actually be seen underwater? Yes, as case dated 07-??-1989 in Chapter 3 illustrates. However, as with many UFO cases, "seeing is believing." A note of caution should be made here though. A case on my website, dated **07-05-1965,** was originally reported by a scientist, Dr. Dmitri Rebikoff, as an unknown object that he had seen while underwater in the Gulf Stream off the coast of Florida. He said that it was first thought to be a type of shark, but that **"its direction and speed were too constant.** It may have been **running on robot pilot**. We received **no signal** (from it) and therefore do not know what it was." [576] Fortunately, the object was photographed and was later identified as a species of stingray. Obviously, scientists can make mistakes, also.

Another category involving totally submerged objects is the underwater solid lights, which is the topic of the next chapter. Since UFOs have been seen entering and exiting water, with the objects glowing both in the air and underwater or even solely as totally submerged, I *reluctantly* advise you to keep the door open for other totally submerged cases in the future.

Chapter 17

Underwater "Solid" Lights

In this collection of cases, there is a solid light seen going into water and continuing to glow underwater or seen initially underwater and then exiting water. This is not to be confused with what has become known as "water wheels," which I discussed in Chapter 10, that have a different configuration than the underwater solid lights. The confusion, I believe, was due to early UFO authors who had an underwater account and tried to compare it to some previous unknown water case from the past. The trouble here is that water wheels and underwater solid lights are distinctly different. The water wheels are like pinwheels or spokes on a bicycle wheel: lines of light off a center and not joined except at the hub. The solid light, in contrast, is more like the UFO field with which we are so familiar—a glowing ball of ionized light (in this case ionized water) created by the solid field that surrounds the UFO.

Another mystery between the two groups is that, while in-and-out-of-water balls of light appear all over the world, water wheels are 99.9% confined to the seismically unstable area of the Indian Ocean, thus making what is observed there suspect to some other geological source.

So it is here that we begin the attempt to find a correlation between the "flying" balls of light and their connection to water.

17.1 Glowing into Water and Continuing to Glow Underwater

In this section, a glowing UFO is seen descending and entering the water without the glow being extinguished. This would be quite a feat if this were conceived of as a meteor. For one thing, a meteor would slow in its descent due to the denser air and would cool due to the lowering friction of the atmosphere. However, if we still assume that it is a fiery meteor (hot, yes, but glowing, no), its descent into the water and the fact that it is still glowing would seem to definitely rule out a meteor as the glow would be extinguished by its submergence into the cold and suffocating water. We are therefore left with the enigma of what this object is that continues to glow beneath the water's surface. Examples follow:

03-24-1955   *Rhoslefain, Wales, UK*

As they watched, it exploded and, still in the shape of an orange ball, plunged into the sea. **The strange thing was that they could still see it glowing *beneath the surface of the water*, and this continued for upwards of an hour after the object finally struck.** [577]

12-02-1965   *Pacific Ocean off San Pedro, CA, USA*

[She] said her son, Martin, took pictures of **a glowing orange and red object as it sank into the sea** late in the afternoon....Mrs. Cohen said she and Martin were in their yard and looking out to sea when **they spotted something glowing in the ocean**.
  "**You could see steam rising from the object**," she said. [578]

12-22-1977   *Atlantic Ocean near Boston, MA, USA*

"Now comes the wacky part of the whole thing. While we were watching the UFO, suddenly this other glowing thing drops out from underneath it. The damned thing drops out from underneath it. The damned thing **looked like a neon-green smoke ring. It dropped away from the larger UFO down toward the water...and submerged! We saw the glowing green circle of water where it went in, and then the glow disappeared!**" [579]

10-15-1983   *Sizewell, N. of Orford Ness, England, UK*

One of these men had a second encounter at 05:30 on 15 October when **a green blob of light appeared in the sky and fell slowly towards the surface of the sea—plunging into the waves and disappearing beneath the surface, still glowing.** [580]

09-24-1998   *East River, New York City, NY, USA*

A red, glowing object hovering above the East River. It then **dropped into the river, and its glow could still be seen after the object was under the water.** [581]

17.2 Glowing Underwater and Continuing to Glow out of the Water

Well, so much for supposed meteors. Next we will try our luck with glowing flying fish or glowing missiles coming out of the water and flying away.

09-LL-1954   *Off the coast of Georgia, USA*

Looking in the direction of the thrashing, I perceived a **great glow**, like a big school of herring **coming near the surface** and disturbing luminous bacteria...only it was not the right color. It was more deeply orange. In color it was like the annelids which are common off Bermuda in certain seasons, but **this was one continuous glow and not a collection of nickel-sized blobs of light. The glow seemed to come from a plateau of water which was causing solid water to flow over my decks**, hence impeding my forward progress....Even as I realized my "windjammer" was something else, the splashing and sloshing of water stopped, and **this vessel, or whatever it was, rose out of the water**... [582]

10-18-1978   *Off Hardys Bay, New South Wales, Australia*

He suddenly noticed a "bright green glow" emanating from below an escarpment [a steep slope]. **The glow was lighting up the underside of the sea mist from the ocean.**... Then a "huge, bright, silver object, tubular in shape with bright green glowing ends came up the escarpment. It was so large it filled the entire windscreen," he reported....

That same night a professional fisherman from Hardys Bay was 20 km out to sea off Maitland Wreck. At about 10:05 p.m., he said he heard a strange noise like "water boiling and hissing." **He then noticed a bright green light traveling under the water about 200 meters in front of his trawler** but lost sight of it through the sea mist. [583]

04-02-2008   *Lake Ontario off Hamilton, Ontario, Canada*

The object looked like a big full moon, **a really bright yellow glowing ball that came out of the lake.** [584]

17.3   In and Out or Out and In

I love this game of hide and seek, except we cannot play it in the same way.

08-MM-1964 *Upper Nemahbin Lake off Delafield, WI, USA*

He and his brother were out on a lake at four in the morning heading for a fishing spot when the craft he described descended suddenly and entered the water very slowly, remained underwater for a while (**he could see the glow in the depths of the lake**), then emerged very slowly and, once it was free of the water, accelerated rapidly and had the appearance of a star once again in a matter of seconds. [585]

LL-??-1972   *Unknown waters off Puerto Rico*

The most notable part of the observation was that **the objects had a glowing, greenish-blue field around them**, and they entered the water at considerable speed without any splash. The water just parted ahead of the visible field, and the objects flew into the hole, and it closed up behind them with only a flurry of ripples. Other disc-shaped objects came out of the water the same way and flew northeast over the island. **After the UFOs entered the water, there was a greenish glow for a few seconds that faded as the objects presumably went deeper. Conversely, when they came out of the water, they were preceded by the greenish glow** and then the water would part and the ship would emerge—again with no splash and no water dripping from the object. [586]

EE-??-1974   *Waters near Vietnam*

We were watching the *Mason* in front of us and the glowing trail of ocean it was kicking up from the phosphorus algae in the water when **the ocean in front of us lit up, started glowing. It got brighter and brighter and then this really bright orange-yellow ball came out of the water** on the right-starboard side of the destroyer. **It flew over the top of the destroyer and went back in the ocean** on the port side, **with the same glowing ocean water** and then disappeared. [587]

At this point I think that it is obvious that the protective, ionizing field of this craft functions very well and glows when the field is on at nighttime or in the darkness of the water. However, if it were a submarine, built with an emphasis on stealth, whose primary defense lies in its ability to remain concealed in the depths of the ocean, it certainly would not want to have a glowing appearance which would advertise its presence. I am sure the UFO's occupants are not afraid of being conspicuous in a glowing craft because they are confident that the technology which they possess can protect them from any possible threats that we may pose to them.

17.4 Totally Submerged

Now we can include that elusive group of cases where there is only a bright glowing light underwater, exhibiting its electromagnetic (EM) effects along with speeds which would not be within a normal range for our terrestrial ships.

04-07~18-1951   *Red Sea near Djibouti*

The night fell softly dark and *Sheila* sailed herself gently across a warm breeze. Mike and I were talking in the cockpit when we noticed a light far out to the southeast. As we watched, it grew more vivid and was seen to be sweeping towards us; **it seemed like the beam of a very powerful lighthouse, pivoted in the south and sweeping from one horizon to the other—but** *under* **the water.** It rapidly came closer, relentless and inexplicable, until **it lit up the sails with a greenish light quite bright enough to read by**. I watched the defined beam as it passed under *Sheila*, throwing the dark shadow of her hull momentarily over the sails, and then it fled to the western horizon.

It left us speechless, but **another great beam appeared in the east, swung towards us, underneath us, and silently fled into the western darkness**. This happened about five times, always the same, at the same regular intervals, in complete silence and with no change in the wind or sea. [588]

01-23-1964   *Northeast point of Groote Eylandt, Australia*

"Seen at sea by crew of a vessel NE point of Groote Eylandt, WA. **Large lights in water, made compass go 'haywire.' Shadow in centre of lights rotated clockwise, causing lights to pulsate....**"

As soon as he had corrected the vessel as best he could, he switched off the compass light and found the unnatural light was about 6 ft. on the starboard side. The light was in the water. It was described as a ghostly white light; in the centre was a shadow which rotated in a clockwise direction causing the light to pulsate. The light appeared to draw away to the stern. **It is estimated that it was miles across and a few hundred yards through**... [589]

??-??-1966 *Unknown location in the Atlantic Ocean*

An AF active-duty jet pilot with the rank of captain, who is also a NICAP member, came to our offices and reported that a group of service pilots had seen **a large disc-shaped object rotating under the surface of the Atlantic Ocean**. The sighting was at night while the pilots were on a routine mission, and **the UFO was clearly visible because of its brilliant blue-green glow**. [590]

05-23-1968 *Atlantic Ocean near the Azores*

I was aboard the USS *Monrovia* APA-31, which was in route to Norfolk, Va., from a six-month deployment in the Mediterranean. At approximately 9 p.m., **a large submerged object was sighted** on the starboard side, just aft of the stern.

The object was an elongated ovoid in shape, luminescent orange in color, and appeared to have a translucent quality.

The USO matched several course and speed changes and rendered compass, radar, and radio equipment inoperable.

As quickly as the object seemed to have appeared, it disappeared. [591]

??-??-1968 *Atlantic Ocean off Palm Beach, Florida, USA*

I was on the command bridge outside on the port side and looking to the lights of Palm Beach in [the] distance just abeam, and then **I saw the lights under the surface some 30-40 meters away** just abeam **in the depth of some 10-15 meters**. It looked like a big airplane without wings, tail, with all the windows lit on it. There were some 10-15 windows, but I did not count them.

The object was moving with us but some 30 degrees, crossing our course. The speed was about double of ours and it went under the bow of our ship and I could see just a gloom when it came on the other side in front of [the] port bow. Our ship was 250 meters long and her draft was some 9 meters. [592]

??-??-1977   *Unknown location in the Caribbean Sea*

We were watching the phosphorescent glow that trailed the edge of ship, caused by certain plankton that light up when disturbed. This effect flickered like low flame and did not extend far behind the ship before the glow dimmed out. Like an underwater magnetic field, dancing here and there while we watched it.

Then we saw a large, blue glow approaching our ship underwater from the stern. As it caught up to us and began passing underneath, we sat on the fantail and looked right down at it. It was much wider than our ship, and **it was pulsing almost like the way embers flicker**. It was brighter and a deeper hue than the flickering plankton. My first thought was that maybe it was a submarine, disturbing plankton much like our ship did....**It was sort of oval in shape, but being underwater, one could not see a definite shape, except that is** [sic—it] **was about as wide as it was long and that our ship's length was about the size of its diameter.** [593]

02~09-??-1982   *Atlantic off the coast of Virginia, USA*

To the best of my knowledge, **the light was not oval but round and it was solid.** No feathering of the perimeter, no spiral arms. **It was like a beam of light directly under us that was extremely bright**. This was not the result of bioluminescence. Bioluminescence is faint and fades off as the water calms. [594]

06-??-1988   *Pacific Ocean off Antofagasta, Chile*

A pilot onboard a "well-known" domestic airliner reportedly saw on the coast directly across [from] Antofagasta and **deep under the waters a "fleet" of orange lights** under the waters. [595]

07-06-1997   *Ionian Sea off Fiumefreddo di Sicilia, Italy*

At around 1:30 a.m., three friends were out fishing. The sea was calm and **they saw a bright light under the surface**. It was somewhere between blue and white in color, it was oval in shape, and **it moved extremely fast**. [596]

LL-??-1998   *Unknown position in the Pacific Ocean*

I had been watching the foam churn up around the hull of the ship as it passed through the water, and the glow of the phosphorus turned the white wash to a luminescent green. My attention was suddenly turned toward what looked like an approaching surface contact. Within a few moments, **I could tell that the object in question seemed to be below the actual surface of the ocean**. The contact was a mere 100 yards off the starboard side, just aft of our ship at a bearing of approximately 165 degrees. As I examined the approaching contact, it seemed that there were only two white lights, parallel to each other....Keeping in mind that a submarine has no underwater lights attached to its hull and **it was clearly under the surface of the water**, we quickly ruled out the possibility that it was anything mundane. [597]

10-05-2005   *Atlantic Ocean off Myrtle Beach, SC, USA*

About 15-20 minutes later as we were talking and looking out over the ocean, **we both noticed 2 lights in the ocean about 300 feet or so (estimate) from the shoreline**. It could have been a bit further as it was hard to tell the distance from our balcony. **The glowing lights were 2 separate lights under the water. You could tell**

they were under the water because they seemed to be glowing rather than a bright light on the water. They were round, glowing lights that had some size to them, maybe 10 feet or so around, again it was hard to tell how big from the balcony, but they were larger then [sic – than] just a small light. One light moved quickly to our right paralleling the shore then seemed to stop, then the other light quickly followed it like it was chasing it. The first light quickly darted ahead again (maybe 100 feet or more, it was hard to tell); the second light darted after it again to catch up with it. It made one more quick move again at an angle (I believe away from the shore) and the second light followed again. Then it seemed to just disappear. Both of us witnessed it and after it was gone said to each other, "What was that?" [598]

08-09-2006   *Lake Springfield, Springfield, MO, USA*

Two fishermen (names on file) in their early 60's were fishing at night on Lake Springfield (just south of Springfield proper). At about 2140 **they observed a dim, circular light under the water surface** and about 100 feet from them. As they watched, it turned right then left and approached the boat. **At a distance of less than 30 feet, they could see a disturbance of water above and behind the lighted area**. It appeared to be perhaps 8 feet across, and they could see **a series of light clusters making up the light mass. The light varied from yellow to orange** as it dropped down into the water and could no longer be seen. The observation lasted about 3 minutes. [599]

06-03-2007 *Tyrrhenian Sea btwn. Paola & Stromboli, Italy*

At 3:45 p.m., while flying over the Eolie Islands [also known as the Aeolian Islands] on an airplane coming from Milan and heading in the direction of Reggio Calabria, offshore between Paola and the volcanic island of Stromboli, a high school student named Rosalia **noticed under the water a white light of circular shape with a radius of approximately one hundred meters.** The color of the light was white-blue, deeper in the center and less bright towards the edges; however, it always stayed visible, well-defined, and noticeable despite the bright sun. [600]

In and out of water and totally submerged. Sounds like a very famous case from Nova Scotia, Canada: Shag Harbour.

Chapter 18

Shag Harbour: The "Water Roswell"

The UFO crash at Shag Harbour, near the southern tip of Nova Scotia, Canada, is, in many respects, better than the famous incident at Roswell because of its vast amount of documentation. A twelve-year-old boy who witnessed a UFO event in Halifax, Canada, in 1967 became one of the principal investigators of the Shag Harbour case years later. That researcher, Chris Styles, remembers getting a phone call from his grandfather who lived in Shag Harbour, telling him about the UFO sighting there.

Chris, along with his friend and fellow UFO researcher Don Ledger, wrote the book *Dark Object* which details the case and includes many documents and interviews with personnel of both the Canadian and U.S. Navies. Chris told me that he has a four-drawer file cabinet full of Canadian and U.S. documents relating to this case. Can anyone say "paper trail"? Both Don and Chris have interviewed naval personnel that were sent to investigate the strange craft. Chris, along with Graham Simms, co-authored a second book, *Impact to Contact: The Shag Harbour Incident*. Published in 2013, it details their investigation of this event.

18.1 An Early Account of the Sighting

The following is one of the many early articles published on the case. Due to the fact that data has changed during the years of the investigation, Don Ledger has kindly made changes to errors in the document. I show these changes as [Text—DL], signifying Don Ledger. A change was also made by Chris Styles and is noted as [Text—CS]:

10-04-1967

UAO Dives into Sea at Nova Scotia

One of the more detailed and well-documented sightings out of Canada took place on the night of October 4 at about 11:05 p.m. [more like 11:20-25 p.m.—DL]. At that time several individuals observed a row of lights which glided into the water. The first report came from five young people between the ages of 15 and 20 years old who saw three or four distinct lights arranged in a straight row, the total of which appeared to be the size of a large aircraft [perhaps "60 feet in length" instead of "about the size of a large aircraft" since the various eyewitnesses were so consistent with suggesting this number—CS] coming down at an angle of 45 degrees toward the water. The object was lost to sight when it passed behind a low hill, after which time the witnesses observed a single white light apparently resting on the water at a point which they estimated to be about 200 or 300 yards offshore in the Barrington Passage [the Sound, the western approach into Shag Harbour—DL].

The witnesses, who were in a car, immediately notified the Royal Canadian Mounted Police [Corporal Victor Werbicki at the Detachment in Barrington Passage—DL]. When Corporal Ron O'Brian, one of three officers who went to the scene, arrived, a light which he compared to that of a flashlight was apparently on the surface of the water about 300 yards offshore. Two [One—DL] of the officers went for a rowboat [went in search of fishing boats to aid in the search while the other contacted the Rescue Coordination Centre (RCC) in Halifax, N.S.—DL] and one stayed and watched the light which extinguished five minutes later. About ten minutes after the light went out, the two [three—DL] RCMPs [officers—DL] were at the spot where the object was believed to have been seen. **There was nothing to be seen except a patch of yellowish foam made up of 1 to 1½ inch [3-4 inch—DL] bubbles, the total of which was between 30 and 40 yards [80 feet—DL] across.**

Fishermen in the area felt [swore—DL] that this foam was not normal tidal foam.

Other subsequent reports by residents in the general area indicated that at least a dozen people were witnesses to the phenomenon. One observer said that he noticed a brief streak of light coming from a point between the first and third lights and also heard a whistling sound.

A check was made with aeronautical authorities [the Rescue Coordination Centre in Halifax, NS—DL], and it was found that there were no aircraft missing in that area.

The object apparently landed in Cocherwit Passage, which is between Cape Sable Island and the mainland [This should read in the Sound between Outer Island and the mainland.— DL]. Two girls [people, a woman and her daughter—DL] in their mid-twenties [ages incorrect—DL] who were driving on Cape Sable [Island—DL] reported, independent of the other observers, that they had observed three yellow lights which tilted, then descended [down toward Shag Harbour—DL]. After that they observed a yellow light on the water [Not so—DL].

On the following Friday, Navy divers were organized for a search which began on Saturday [Friday—DL] morning. To this date there is no indication that anything was found. It should be noted here that a disc, apparently in trouble, fell into the Peropava River in Brazil in the fall of 1963. Despite extensive operations with diving equipment as well as metal detection equipment, nothing was found in this instance either.

A.P.R.O.'s Nova Scotia members were instrumental in getting a quick report to headquarters after which Mr. Lorenzen [co-founder of A.P.R.O.] notified the University of Colorado Committee so that they could affect immediate investigation. [601]

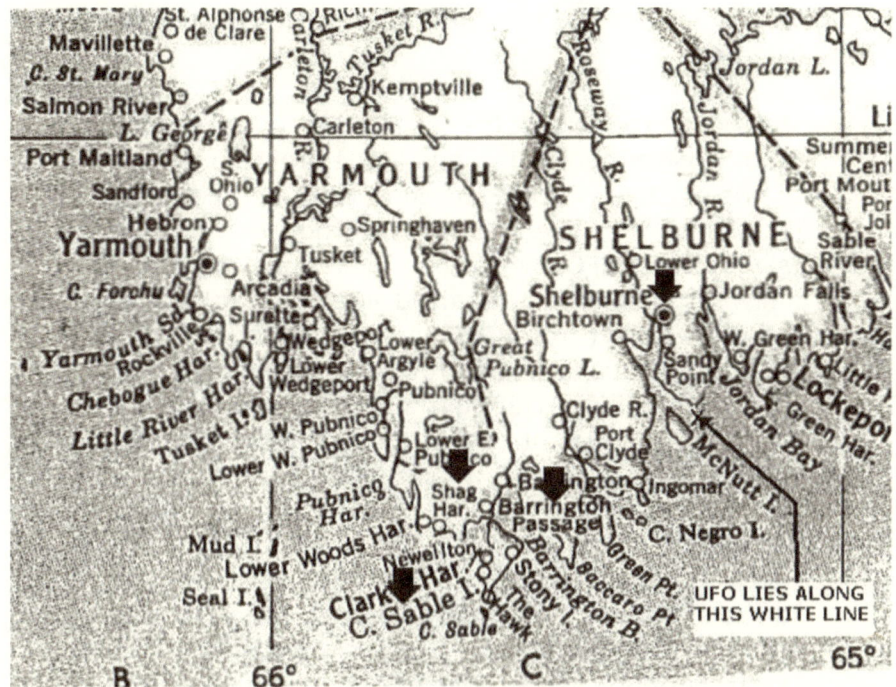

Map showing points mentioned in the text

There is more to the story, although this part is not as well-documented. The object that entered the water off Shag Harbour was tracked traveling underwater for about 25 miles to Government Point near the site of a submarine detection facility, CFS Shelburne. Navy ships were sent and positioned themselves over the UFO where it had stopped. While waiting and planning a possible salvage operation, the Navy detected a second UFO which joined the first one. There was speculation that this second craft came to help the first one. For almost a week the Navy ships remained positioned over the submerged UFOs. Unfortunately, they became distracted by a Russian submarine that came into the vicinity. During this time, both UFOs quickly moved away from the area, exited the water, and flew away. Multiple witnesses reported seeing two UFOs in the area on October 11, 1967.

Chris Styles sent an e-mail to me in reference to the UFOs' location:

> It was told to me that the 2 UFOs sat on the seabed off of Government Point, and if one drew a line from the tip of Government Point that would extend to the location of the Cape Roseway Lighthouse on McNutt Island, the 2 UFOs would be situated exactly at the halfway point of that charted line. In fact, one of the retired staff from CFS Shelburne, who took the [chartered] bus tour that you were on, agreed with that positioning. [During the Halifax Conference, Oct. 11-12, 2003, I traveled round trip from Halifax to Shag Harbour.]

[He continues:]

> To me there are two things that set Shag Harbour apart from so many other UFO sightings and UFO crash scenario claims. They are: 1. No one reported a UFO! It was the authorities who first bestowed and stuck with that designation. 2. The Shag Harbour incident is the world's only UFO crash that is supported in that interpretation by government documents that are freely available and without controversy as to their origin or authenticity (unlike Roswell). [602]

18.2 How Are They Going to Fix It Underwater?

Going back to speculation that a second UFO came to aid the first one, we might wonder how a repair could be made underwater. If I am right, all we need is an elementary physics lesson in magnetism.

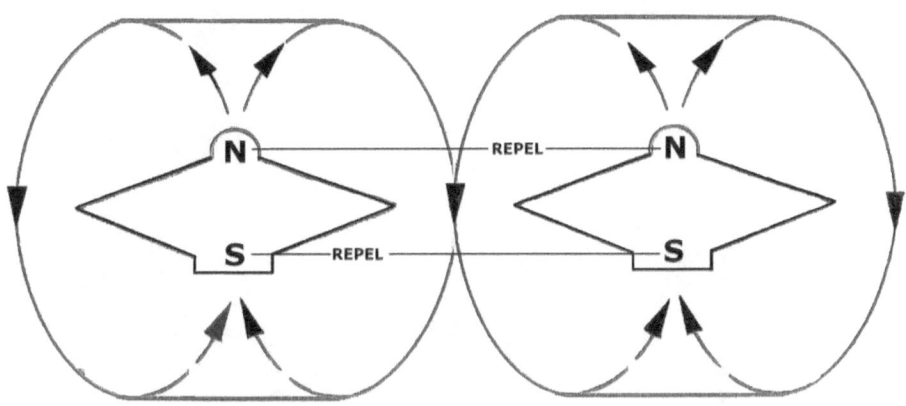

Sketch #49

If we imagine both saucers side by side as in the above sketch, this presents a little problem in that the repair crew would not only have to walk through their craft's extremely hot and powerfully rotating field, but, upon exiting into water, they would also have to reenter the disabled craft through its extremely hot and powerfully rotating field. Somehow I do not think this is a feasible idea, even if one of those fields is reversed, inasmuch as an undesirable attraction is created between the two craft. However, there is another way. In elementary physics we learn that two magnets in close proximity to each other still retain their own fields. So both craft still have their magnetic fields. But if we take the rescue saucer's south pole and place it over the disabled saucer's north pole, how many fields do we have?

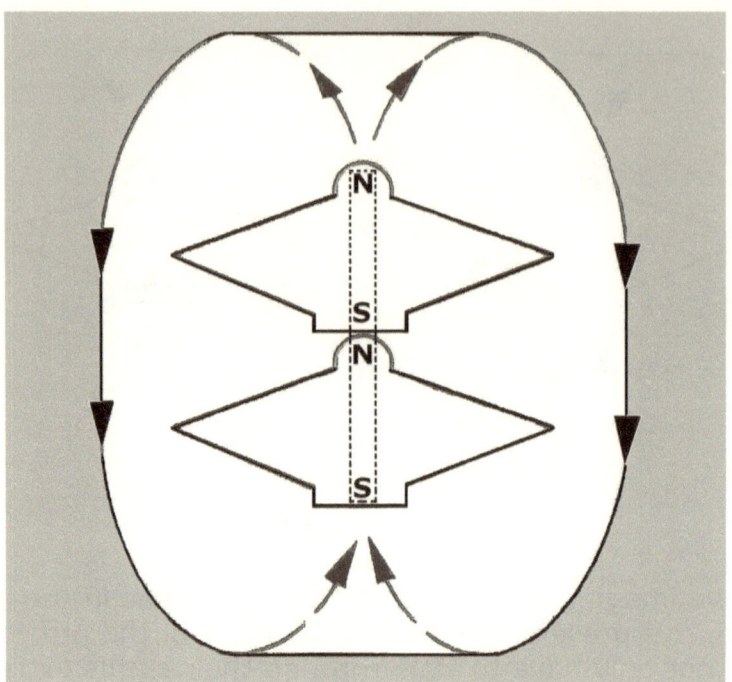

Sketch #50
One Field—For Both UFOs

This combined field would allow the rescue team to open the door of their craft and float down (as aliens are wont to do) to the disabled craft without a drop of water touching either the craft or them. Work, work, work. Fix, fix, fix, and all is well, except the two craft are joined like Siamese twins.

On these craft, I assume that they not only have an on and off switch, but they also have a "reversed field" switch, as explained in Chapter 3. So when the top craft wants to leave, they just flip the switch and they now have their north pole facing the repaired craft's north pole and they can subsequently separate. For those of you who saw the movie *Close Encounters of the Third Kind*, this is essentially the scenario near the beginning of the movie where the police cars are chasing one very bright white UFO. The police come to a screeching halt at the edge of a cliff when the light they are pursuing separates into three UFOs.

Nevertheless, this is all speculation since I have no knowledge that the divers actually saw one saucer on top of the other or that any type of repair operation actually took place. I believe "there was activity" was a phrase used by either Don or Chris in stating what the navy divers had seen underwater. Though several decades have passed, the crash of a UFO into Shag Harbour is still discussed today and the sighting remains unsolved. Hopefully as time goes by, more will be learned about the Shag Harbour incident.

Chapter 19

The Sea Was Like a Sheet of Glass

Another Mystery Solved

While in the final stages of preparing my book for publication, I hit upon a last-minute puzzle piece that had lay undisclosed until I realized what was causing a particular effect seen by some witnesses. The key word in this instance was glass. When doing a keyword search on my website, waterufo.net, the search engine produced three cases (so far) containing the same basic terminology I was looking for—something similar to a sheet of glass—to describe the surface of the water. Although the answer was in front of me the entire time, the cause which had eluded me now seemed simple. These three cases follow:

WW-??-1979   *South China Sea*

I was the Boatswain's Mate of the Watch and it was the mid-watch about 2:30 AM. **The sea was like a sheet of glass and there was only starlight. The lookout stationed up on the signal bridge called down and said he saw something glowing in the water ahead of the ship.** It being the time it was and nothing else going on, the OOD [Officer of the Deck], JOD [Junior Officer of the Deck], and myself all put on the binos [binoculars] for a look.

We all saw it and kind of looked at each other in amazement. There was what turned out to be five large, long, vitamin-shaped objects [Dimensions given below would make them vitamin "capsules."], lined up in a directly opposing course which would take them straight under the length of our keel.

They were approximately 100 feet long and 40 feet across with about 20 yards between them. They glowed the typical bright, bioluminescent green and at first we thought it was some kind of marine life, but as the first one passed beneath the stem of the ship, we lost power, steerage, the gyro and compass went wacky AND the chronometers and wristwatches stopped! This prompted the OOD to have the messenger fetch the captain, whose cabin was steps away.

By the time the skipper got on the bridge, the front portion of the first object was below the bridge and we were hanging over the side for a look. The captain gave the helmsman orders, but the wheel would not respond, and the guys in the engine room could not explain why we'd lost power.

There was no pulsing or shape changes or any appendages nor fins. Just sleek, silent, glowing, lozenge shapes moving effortlessly directly below our hull in the opposite direction.

Somehow the ship just kept gliding slowly over the top of them, and we just stared at them as they passed below while we all voiced our opinions as to what the heck it was.... [603]

In the next case, there is not only the flattened water but also the hole that drains the water downward. This is the water being pulled down into a tube and exiting at the north pole of the cylinder.

06-27-2006   *A pond near Flagler, Oklahoma, USA*

On June 27, 2006, five of our grandchildren were fishing in our pond about 3:00 p.m., when they saw a big splash on one side of the pond. Everything became still, and the **pond became like glass** as the object sank. **Then they saw a whirlpool in the pond** about 8 to 10 feet across that moved very fast across the pond towards them. [The "whirlpool" moved towards them because the craft was moving towards them.] **When it hit the water, there was steam that went up from the object** then the object disappeared to reappear in another area. [604]

09-03-2006   *White Fish Bay, Michigan, USA*

Lake Superior — On Sunday, September 3, 2006, at or near 11:30 PM (2330 hrs.) EDT, while sitting outdoors looking across White Fish Bay toward Canada gazing at the red beacons from many electricity generating windmills on an island in the bay, I suddenly became aware of a brilliant gold-white sphere of light just below the cloud layer to my east. I was in a lawn chair and had binoculars with me. Seeing this light appear below the overcast, my first thought was of an aircraft with its landing light turned on. As an airplane watcher, I kept my eye on the light expecting to see and hear an aircraft approach directly toward me. I said nothing to my wife who was next to me tending our fire. I picked up my binoculars for a better view and noticed while watching the light that I was lowering the glasses to keep it in view as the surface of the lake was becoming visible.

I realized the object was not heading toward me but instead was in a slow, controlled, steady descent. As it neared the lake's surface, its own brilliance formed a reflection on the water. As the object continued its descent, the reflection on the water grew brighter and larger. The object, visible only as a ball of light (which may have had a dark band across the center), lowered itself and the object and its reflection merged at the surface of the lake and the object was gone from view. As I watched the object, it had descended into and below the surface of the lake. **Not a ripple or other disturbance appeared on the surface**. There was no dark outline of a landed craft. There were no sounds. There was never any steady, flashing, or strobing red, green, or white navigation lights as would appear on a conventional air or watercraft. I continued to watch but nothing more was seen, just empty, still water. [605]

In the following sketch, which I originally drew to show how the field in reverse worked underwater, I did not address what the south pole (top of the sketch) was doing as it had no consequence for me at the time. I only realized, after a moment of inspiration on January 14, 2015, that the south pole was also functioning. Though not necessary in function (just in polarity), the inward flow of water does not come from just two directions in the sketch but the full 360-degree flow.

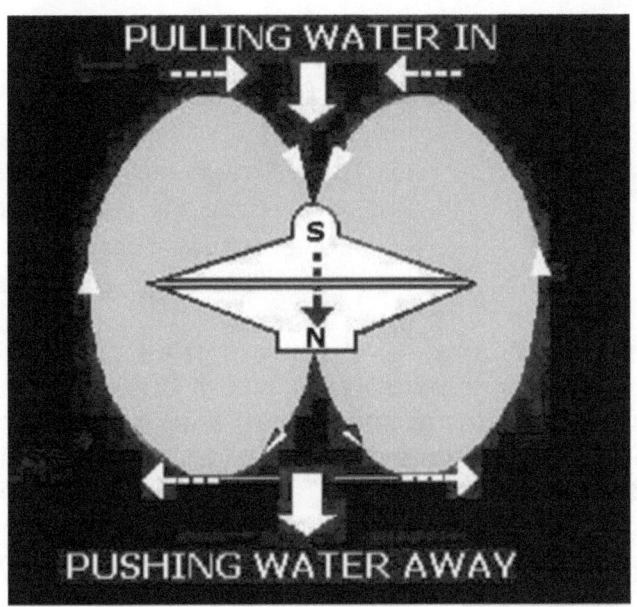

Sketch #51

This means that the field is pulling in the water close to the surface, thus affecting the water in the whole general area, which is towards the center of the craft, and thereby flattens any surface waves. In essence, the water becomes like a rubber band, with wrinkles in it, thrown on a table. Then with one end of the rubber band in a fixed position, it is pulled outward and becomes a straight line with no wrinkles, **similar to what happens to the waves as they are pulled out by the rotation of the field**. Again, the observed effects can be seen as proof of the field's motion and power.

In Chapter 20, I bring you *TERROR* and *MAYHEM* to end your education.

Chapter 20

Up Close and Personal

**"Do not turn the page!
There is a monster at the end of this book."**

—From Sesame Street*'s book* The Monster at the
End of This Book*, this is a quote by Grover,
who eventually discovers that* he *is the
monster at the end of the book.* [606]

In this final chapter are some stories of UFOs that do not exit, enter, or interact with water but are nonetheless terrifying experiences at sea or over a lake or stream. These stories might as well have occurred over land since there is no unifying connection other than that they occurred over water. One thing that comes to mind here, besides the obvious, is that these craft are quite comfortable over water and have no great concern about running out of fuel while playing with their human toys. Several of these cases, however, do contain some other attributes that show physical traces, such as heat, radar contact, and electromagnetic influences on compasses and engine performance. Therefore, in that respect, their inclusion here is appropriate. Note that the witnesses experienced extreme fear, a reaction that animals and fish exhibit as well, as discussed in Chapter 9. You have done well to get to this point, so pick a dark and dreary night with lightning and thunder, and with a soft light of a candle to read by, enjoy:

The Last Chapter

●
●
●

20.1 The Fear Factor

??-??-1959   *Lake Gualletué, Malleco, Chile*

Dr. Claudio Crocquevielle and a friend were fishing in Lake Gualletué when they saw a luminous object which was approaching circuitously. It passed by them within 30-40 cm of the **witnesses**, who **threw themselves into the water in fear and hid among the plants for about an hour**. The object, 2-3 m [in] diam., was flying in circles, rising and falling rapidly until moving away and stopping to hover for a while in the distance. The shape could not be made out because of its brilliance. A photograph appears to have been taken. [607]

10-06-1961   *Lake Maracaibo near Santa Rita, Venezuela*

In the past six months, **UFO close approaches have led to one observer's death and have badly frightened other witnesses** in the U.S. and South America.

At Santa Rita, Venezuela, **a fisherman drowned when he and panic-stricken companions jumped from their boats as a huge, glowing object approached.**

It was Oct. 6, 1961, about 11 p.m., when the enormous UFO appeared from the north, its glow lighting up the town. As it slowly crossed Lake Maracaibo, casting its glare on the fishing boats, **dozens of terrified fishermen leaped overboard. One man, Bartolomé Romero, went down in the frantic swim for safety**. The others made shore, **unnerved** by the encounter. [608]

In the next case, keep in mind what was expounded upon in Chapter 4 on heat.

05-04-1968   *Seal Island, Nova Scotia, Canada*

What Was It?

Clark's Harbour—Capt. Woodrow Atwood and the crew of the fishing boat *Which Way In* had **a frightening experience** about eight o'clock Saturday evening when about one and a half hours below Seal Island.

The captain was watching the compass when he saw a light to the nort[h] about the size of a match light. Suddenly it burst into a blood-red light and appeared to be about 50 to 75 yards away, coming towards the boat. **As he watched through the window, it became so hot he had to move away.** The light floated overhead for about five minutes then lowered and seemed to float towards Brown's Bank.

Capt. Atwood called by radio to tell of his experience and was answered by the captain of the *Racer*, who told him his crew had called to tell him that a large, red ball of light had just missed the spars when it passed over.

Crew member William Nickerson, aboard the *Which Way In*, called it a **frightening** experience. **Capt. Atwood said the heat was intense, and he expected the boat to be burned before it passed**. [609]

11-??-1972   *Grand Lake, New Brunswick, Canada*

A man and two of his children were hunting deer. They saw a very big "star" moving very slowly. It moved across the lake a ways off, and the father began shining the flashlight toward it and "signaling": 1 flash, 2, 3 flashes, 4, etc., up to 10 flashes, then stop and resume. The object turned and began coming back up the lake toward them. And it descended as it came. **The children became scared and begged their father to stop flashing the light and take them back to the cabin**. At that point the object just clicked out and was apparently no longer there. **Five to ten minutes later the air over the lake was swarming with**

Air Force jets, which criss-crossed the area for a half hour. [610]

11-03-1978 *Adriatic Sea near San Benedetto del Tronto, Italy*

TYPE: NL (Night Light)
TIME: 4 a.m.
DURATION: 2 hours
SOURCE: *Oggi* magazine (Dec. 2), *Gente* magazine (same date)

Fisherman Antonio Pallesca caught sight of a reddish light at sea following his boat. **Frightened**, he drew up his nets and directed his ship towards land. The light stayed with him, first behind him, then beside him, then in front of him. As it circled around his boat, he couldn't discern it as anything but a light. It left him shortly before he arrived in port. On Nov. 7, he had another run-in with an orange light and a dark floating object. Pallesca has been a seafarer for 30 years. [611]

All right, blow out the candle and turn on the lights. It's back to work. When witnesses unexpectedly confront the unknown, the encounter leaves an indelible impression on them. While they may be curious, amazed, or awestruck, they may also experience fear or apprehension, as we saw in the accounts above. Amazingly, the ionization of the air by a UFO's field might also be the cause of the fear itself. I remember reading an article (not UFO related) during my days of researching newspaper articles on UFOs in the Delaware papers which said that your mood is indeed geared to the weather. Sunny and bright weather produces one level of ionization, which gives that good-to-be-alive feeling, while dark and threatening clouds present another level that brings the threat of harm and fear. Rather than concentrate on an area that does not contribute any new knowledge of the craft, let us continue with a different scenario.

20.2 From Our Government Which Does Not Believe UFOs Exist

In doing my newspaper research, I came across the first case presented here from a time period when UFOs were not very active. The ship was from Nova Scotia, and I could not resist the temptation to tease Canadian UFO researchers Don Ledger and Chris Styles about it. So I sent off an e-mail to them saying, "Are you guys trying to smuggle UFOs into the U.S.?" The reply was a bit startling to me as Don said, **"I have copies of the CIA documents on that case."** However, in the news clip that I had, the CIA was not mentioned. What follows is from those CIA documents that Don copied and sent to me. Most of the blacked-out items represented by X's are names, addresses, and phone numbers of the witnesses, which is understandable to protect their privacy.

20.2 a. From the CIA

08-04-1950 *Atlantic Ocean, roughly E. of Cape May, NJ, USA*

COUNTRY	At Sea—North Atlantic	DATE DISTR. 18 Aug 1950
SUBJECT	Unidentified Airborne Object	NO. OF PAGES 2
PLACE ACQUIRED—		NO. OF ENCLS. (LISTED BELOW)
DATE ACQUIRED BY SOURCE	4 Aug '50	SUPPLEMENT TO REPORT NO.

DATE OF INFORMATION   4 Aug '50
[PARAGRAPH BLACKED OUT—XXXXXXXXXXXXXXXXXXXXXXXX]
[XXXXXXXXXXXXXXXXXXXXXXXXXXXXXXXXXXXXXXXXXXXXX]
[XXXXXXXXXXXXXXXXXXXXXXXXXXXXXXXXXXXXXXXXXXXXX]

[XXXXXXXXXXXXXXXXXXXXXXXXXXXXXXXXXXXXXXXXXXXX]
[XXXXXXXXXXXXXXXXXXXXXXXXXXXXXXXXXXXXXXXXXXXX]
[XXXXXXXXXXXXXXXXXXXXXXXXXXXXXX]
[XXXXXXXXXXXXXXXXXXXXXXXXXXXXXXX]
[XXXXXXXXXXXXXXXXXXXXXXXXXXXX] THIS IS UNEVALUATED
[XXXXXXXXXXXXXXXXXXXXXXXXXXXXXXX] INFORMATION
[XXXXXXXXXXXXXXXXXXXXXXXXXXXXXXXXXXXXXXX]

SOURCE. [PARTIALLY BLACKED OUT—XXXXXXX]

[XXXXXXXXX] reported the following observations at 10:00 a.m. EDT on 4 Aug '50 at 39° 35' North, 72° 24 ½' West.

[LINE BLACKED OUT—XXXXXXXXXXXXXXXXX]

1. On 4 Aug '50 at 10 a.m., my ship, while on a heading of 245° true, with a smooth sea and clear weather, visibility 14 miles, barometer reading 30.03, was underway from Walton, Nova Scotia, to an east coast U.S. port. I was in the chart room just aft the bridge when Third Mate, who was at mid-bridge checking the compass, shouted that there was a flying object off the starboard bow. I immediately ascended the conning tower, and by this time the object was on our starboard beam. It was traveling on a reciprocal course to ours about 50 or 100 feet above the water at an estimated speed of over 25 mph. From the conning tower I observed it with my binoculars for a period of approximately a minute and a half when it disappeared into the horizon in a northeasterly direction. I would estimate that the closest it approached my ship was one thousand feet and **it was an ovular, cylindrical-shaped object the like of which I have never seen before.** The object was quite small and I would judge that its diameter was approximately 10 feet.

It had depth but to what extent I was unable to observe. **The object made no noise**, and as it passed abeam our ship, it appeared to pick up considerable speed. **It was not flying smoothly but impressed me as having a churning or a rotary motion.** It had a shiny aluminum color and sparkled in the sunlight.

[LINE BLACKED OUT XXXXXXXXXXXXXXXXX]

2. I was on the main deck, port side, just forward of the bridge when the Third Mate shouted there was an object on our starboard bow. I looked off to the starboard and saw an object of elliptic shape looking like half an egg cut lengthwise traveling at a great rate of speed on a course reciprocal to our own. I immediately ran to the stern, port side, and with my glasses was able to observe the object disappearing into the horizon. From the time I was first alerted to its presence until it disappeared from sight, 15 seconds elapsed. **I believe that it was traveling at a tremendous rate of speed, possibly faster than 500 mph.** During the time I saw it, it was approximately 70 feet off the water and I judge it was approximately 10 miles away. I clearly saw its shadow on the water. I last observed it off the starboard quarter and it seemed to be increasing its speed and ascending. It had an elliptic shape and I could clearly see that it had three dimensions. It wobbled in the air, made no noise, and was a metallic white in color. The length was approximately six times the breath and its belly had a depth of possibly 5'.

CC-B-19864

-2-

[LINE BLACKED OUT—XXXXXXXXXXXXXXXXX]

3. At 10:00 a.m. on 4 Aug '50 as I was checking the compass at mid-bridge through a bridge porthole, I observed a flying object off the starboard bow. I immediately shouted to the Captain, who was in the chart room, and the Chief Mate, who was below on the port deck, of my observation and went out on the flying bridge myself. The object was approximately 70' above the horizon at a distance of 12 miles. It came toward us then ran on a course reciprocal to ours and turned off into the horizon in the northeast. I clearly saw its shadow on the water. My impression of the object was that it was elliptical, not unlike a Japanese diamond box kite in shape. I have no idea of its size but the length was about six times the breath and it had a depth of from two to five feet. It made no noise and was traveling at a tremendous rate of speed. As it traveled through the air, it made a spinning or wobbly motion. After it disappeared in the horizon, I saw it reappear several seconds later, ascending at an even faster speed then [sic - than] when I first observed it. I have no idea what this object was. I never saw anything comparable to it before, and **it was one of the most frightening experiences I have ever had**. I roughly estimate that the object traveled 28 miles during the 15 seconds I had it under observation.

Collector's Note: The Chief and Third Mates were interviewed on 8 August by two Intelligence

Officers. The Captain, who was absent at that time, was interviewed on 9 August by only one of the two Intelligence Officers. In describing the occurrence, the Chief and Third Mates reenacted their behavior at the time of sighting, and the period from the time the Chief Mate saw the object abeam until he reached the afterdeck and saw it disappear off the starboard quarter was timed at 15 seconds. In laying the angles of observation out on a chart and assuming the object was ten miles distant and taking the time into account, it is evident **it was certainly traveling at a very high rate of speed, which approximated 400 to 500 miles per hour**. It will be noted that there is a tremendous discrepancy between the Captain's estimate of the speed and the estimate of the two officers which could not be explained as they were very careful in making their statements and asserted that their observations had been correct. **All three men were quite evidently very much upset by the sighting.** Aside from the discrepancies, it was quite evident to the Intelligence Officers who interviewed these men that they had certainly seen some very unusual object which they could not identify but was just as certainly not any conventional type of aircraft. [LINE BLACKED OUT—XXXXXX] [612]

20.2 b. From Naval Intelligence

11-??-1954 *Strait of Florida between Cuba and Florida, USA*

Two frigates of the Cuban Navy military, *José Martí* and *Máximo Gómez*, were patrolling the north coast of Cuba between 4 to 6 miles off the coast of the province of Matanzas. Their destination was the Havana Bay. About

4:30 in the morning, they began to observe intense lights on the horizon. When something similar happens, the deck officer must inform his commander of the situation. The lights appeared in the northwest. The frigates were put on the alert. The lights came near at a tremendous speed and in seconds they were over the ships. **Three lights were forming a V, following the frigates for 4 or 6 minutes**. Then, because the two frigates were combat ready, most of the sailors were outside watching the lights. The photographer on board one of the ships activated a film camera and filmed the UFOs when they were above the ships and followed the lights until, very slowly and moving to the west, they disappeared on the horizon.

The film lasted from 7 to 10 minutes and the telephoto followed the UFOs until 2 miles off the horizon. The film, in black and white, was of superior quality. In some pictures it was possible to observe the lights and the top structure of the frigates. The camera was located next to the cannon on the bow of one of the ships.

The UFOs maintained the same distance among them, and the photographer had to move the camera to film the three lights until, in the last pictures, you can see the three lights, at slow speed, disappear on the horizon.

Ten days after the sighting, the film was shown to the Cuban military High Commands at the Cuban Navy Building. **Our witness was present and confirmed to us also that the radar of the frigates detected the 3 objects when the UFOs appeared on the horizon** until they disappeared in the west.

Analyzing the film, an expert declared that an object could be seen within the intense lights. This created a great impact on those present. **The film was sent in a Navy airplane to the U.S. Navy Intelligence in Key West, Florida.**

Most of the Cuban high-ranking officials are now living in Florida, and they wish to maintain their anonymity. This report is based on interviews with officials and sailors who were eyewitnesses. [613]

Our government's attitude reminds me, in a way, of Sgt. Schultz, a character from the TV comedy series *Hogan's Heroes*, who used to say: "I hear nothing, I see nothing, I know nothing!"

20.3 Two UFO Stories from before the UFO Era

The following comes right out of the yellowed-with-age newspaper of July 28, 1904, and happened right next to my home state of Delaware:

07-28-1904   *Coast of Delaware, USA*

ELECTRIC CLOUD ENVELOPED SHIP

Caused *Mohican*'s Sailors to Become like Animated Magnets

Compass Was Set A-Spinning and Iron Chains Could Not Be Lifted from the Deck

When the British steamer *Mohican*, from Ibraila, Roumania, which reached this port on Saturday, was making for the Delaware Breakwater, it had a most remarkable experience which terrorized the crew, played havoc with the ship's compass, and brought the vessel to a standstill for nearly a half hour.

For that length of time, the vessel was enshrouded in a strange metallic vapor, which glowed like phosphorus. The entire vessel looked as if it were afire and the sailors flitted about the deck like glowing phantoms. **The cloud had a strange magnetic effect on the vessel, for the needle of the compass revolved with the speed of an electric motor and the sailors were unable to raise pieces of steel from the magnetized decks.** Captain Urquhart described the thrilling experience and his story is vouched for by every man of the crew.

"It was shortly after the sun had gone," he said, "and we were in latitude 37 degrees 16 minutes and longitude 72 degrees 48 minutes. The sea was almost as level as a parlor carpet and scarcely a breeze ruffled the water. It was slowly growing dark when the lookout saw a strange gray cloud in the southeast. At first **it appeared as a speck on the horizon, but it rapidly came nearer and was soon as large as** [a] **balloon**."

Ship in Shining Cloud

"It had a peculiar gray tinge, and as it bore down upon us, **we saw bright glowing spots in its mass**. A mile away we perceived that it rose several hundred feet above the level of the sea and was almost that broad. It rolled over the sea toward us, **the glowing spots becoming more and more vivid**. Suddenly the cloud enveloped the ship, and the most remarkable phenomena took place. The *Mohican* suddenly blazed forth like a ship on fire, and from stem to stern and topmast to keel, everything was tinged with the strange glow. The seamen were in terror when they found themselves looking as if they had been immersed in hell fire. **Their hair stood straight on end, not from the fright so much as from the magnetic power of the cloud**.

"They rushed about the deck in consternation, and the more they rushed about, the more excited they became. I tried to calm them, but the situation was beyond me. **I looked at the needle and it was flying around like an electric fan**. I ordered several of the crew to move some iron chains that were lying on the deck, thinking that it would distract their attention. **But what was the surprise to find that the sailors could not budge the chains, although they did not weigh more than seventy-five pounds each.** Everything was magnetized, and chains, bolts, spikes and bars were as tight on the deck as if they had been riveted there."

Hair on Heads Stood Out

"The cloud was so dense that it was impossible for the vessel to proceed. I could not see beyond the decks, and it appeared as if the whole world was a mass of glowing fire. The frightened sailors fell on the decks and prayed. **I never saw anything so terrifying** in the years I have been at sea [as in the above "fear" stories]. **The hair on our heads and in our beards stuck out like bristles on a pig. After we had been in the cloud for about ten minutes, we noticed that it became difficult to move our arms and legs; in fact, all the joints of the body seemed to stiffen.**

"Then it was that my sea legs began to fail me for the first time. I've heard of phantom ships and stories about the needle running wild, but shiver me if I had ever seen the like of that. For a half hour we were enveloped in that mysterious vapor. And for nearly all that time, after the sailors' first cries of fright had subsided, there was a great silence over everything that only added to the terror. I tried to talk, but the words refused to leave my lips. The density of the cloud was so great that it would not carry sound.

"Suddenly the cloud began to lift. The phosphorescent glow of the ship and the crew began to fade. It gradually died away and, at the same time, the stiffness left the hair. In a few minutes the cloud had passed over the vessel and we saw it moving off over the sea. It loomed above the water as a great, gray mass, spotted like a leopard's back with the bright, glowing patches.

"The crew gradually regained their composure and whispered to one another. I went among them, telling that all danger was past and they slowly went about their work. When I ordered them to move the iron chains for the second time, the men had no trouble in lifting them from the deck and tossing them about. Then I took a look at the needle and it was pointing steadily toward the north, as it [sic— if] nothing had occurred. I have sailed the seas for many years, but I never encountered a cloud like that. It must have been composed of some magnetized substance, which at the same time was combined with phosphorus." [614]

On to the war years:

02-26-1942 *Timor Sea between Australia and Timor Island*

Spaceship over Dutch cruiser

The following letter to the Australian Flying Saucer Research Society, Victorian branch, is reproduced here by permission:

Dear Sir,

After listening to your programme on flying saucers on Wednesday evening, I thought I would write and tell you of an experience I had while in the Timor Sea on Thursday, February 26, 1942. **This happened while on watch for enemy aircraft that afternoon.**

I was scanning the skies with binoculars when suddenly **I saw a large, aluminium disc approaching at terrific speed** at 4,000 or 5,000 ft. above us. This proceeded to circle high above our ship, the cruiser *Tromp* of the Royal Netherlands Navy.

After reporting it to the officer on duty, he was unable to identify it as any known aircraft. **After keeping track of this object for about three to four hours, still flying in big circles and at the same height, it suddenly veered off at a tremendous speed (about 3,000 to 3,500 mph) and disappeared from sight.**

I have an account of this in my notes made the same day in a diary which I still have in my possession.

Hoping you will find this of aid to your investigations,

I remain,

Yours sincerely,

William Methorst [615]

20.4 Electromagnetic (EM) Effects after 1947

This book was conceived with the idea of a UFO interacting, in one way or another, with water. In the following accounts, however, we have some noteworthy effects that do not deal with water but rather the ship that rests thereon. What makes these cases significant is the fact that they are nothing less than a continuation of many other UFO reports both in the air and on land which encounter the same electrical anomalies.

11-12-1963   *Golfo Nuevo, Argentina*

Argentine Navy Discloses Important E-M Case

An official Argentine Navy UFO report, labeled one of the most important Argentine UFO cases, has been released to NICAP [National Investigations Committee on Aerial Phenomena] in accordance with an agreement to exchange UFO information.

This combined visual and E-M (electromagnetic) interference incident occurred late in 1963, involving the Navy transport A.R.A. *Punta Médanos*.

During the night of Nov. 12, 1963, says the official report, "A large airship (never identified) was sighted from the stern of the vessel. **The huge UFO was round-shaped, and it was moving at great speed. It displayed no lights and made no sound.**

As the unknown machine appeared, the needles of the ship's magnetic compasses "suddenly and simultaneously" swung off course, pointing toward

the UFO. The power which caused this E-M interference is indicated by the distance involved. At the time, the Navy report states, the UFO was 2,000 meters (well over a mile) away from the ship.

After the UFO was gone and the compasses returned to normal, the transport commander radioed the Commander-in-Chief of the Argentine Fleet. The Fleet Commander was so concerned he ordered a full investigation by the Hydrographic Service.

Instrument experts found no electric cause for the E-M effect. No magnets had been near the compasses, which were separately mounted on different decks. The Geologic Division determined that the sighting area was magnetically quiet and also ruled out submarines and military and civilian aircraft, leaving the UFO as definitely the cause of the E-M interference.

The official report was sent to NICAP by Lt. Cdr. O. R. Pagani, whom the Secretary of the Argentine Navy appointed to investigate UFO sightings. Pagani said that the Argentine Naval Hydrographic Service will investigate all future E-M cases, and he will send detailed reports to NICAP.

Lt. Cdr. Pagani also forwarded a summary of 1964 UFO cases registered with the Argentine Government, adding that fifteen 1965 reports already are on record with the Navy. NICAP will try to report the best cases as soon as space permits. [616]

09~10-??-1968   *Atlantic Ocean off the coast of Portugal*

Date: September/October 1968   Time: 0400 a.m.

Portuguese coast, en route from the Canary Islands to Vigo.

Francisco Fuentes, 27, officer of the watch, along with the first officer and the helmsman onboard the merchant ship *Campanario* observed at a very low altitude—approximately one meter above the surface of the ocean—a round object, the size of the full moon at its zenith which emitted a strong white glow. It was traveling at very high speed on a rectilinear trajectory. **As it paralleled the vessel, there was a complete electrical and mechanical failure onboard**. The crew decided against logging this incident in the ship's log. [617]

04~05-??-1969   *Unknown location in the Gulf of Mexico*

Arrowhead UFO Sighting, Gulf of Mexico, 1969

Although arrowhead UFOs are not the most common of UFO shapes sighted, this sighting was even more unusual because of its duration. **The craft was seen by the entire crew of the oil tanker *British Grenadier* for exactly 3 days!**

The event occurred around April/May 1969 while the *British Grenadier* was sailing through the Gulf of Mexico. The UFO appeared directly above the ship at exactly noon on the first day. It just seemed to appear, as nobody saw it arrive, and it remained above the ship for the next 3 days.

According to reports, the craft was approximately one mile above the ship and dark blue in colour during the day. At night it appeared as a silvery light. The weather conditions were good at the time, with a clear sky and calm seas.

On the first day of its arrival (one minute after midnight), **the ship's engines suddenly stopped**, and only emergency lighting and steering were available. The crew managed to restart the engines, which took some time as all the pumps had to be shut down and restarted manually.

The second day saw all the **food storage refridgerators [sic] closed down** and no reason for this event could be found.

On the third day at 11:45 p.m., one of the crew noticed that the lights were out in the ship's air conditioning room and a door leading to the crew's accomodation [sic] was opened. The engineer passed through the door to investigate and spoke to two firemen. But the next day the same door was found welded shut! Apparently, the door had been welded before the ship left dock (because of a fault), but the door was definitely open for the engineer to have walked through it.

A few minutes after this event the engines stopped again, leaving just emergency lighting and steering (again one minute after midnight!). They were restarted once more, but later it was discovered that the starter motor for the emergency diesel generator had been dismantled. All the parts of the generator were neatly placed by the machine as if someone had examined it.

At exactly noon on the third day, the craft vanished in the same manner as it had appeared.

Although all these events were recorded in the ship's log, this is the first time (to our knowledge) that this event has been published. We would be interested in hearing from any of the crew that were aboard the BP oil tanker *British Grenadier* at this time in order to obtain more information about this sighting. If you were there and would like to write to us, please send an email to: Grenadier@profindpages. com

Source: http://www.profindpages.com/news/2004/05/29/ MN132.htm [618]

NOTE: Both the e-mail address and source above are for sale. If you have any information about this sighting and are unable to reach the above URLs, please get in touch with me using my website's "Contact Me" page.

09-17-1985   The Bosphorus off Istanbul, Turkey

Now a very interesting case. What makes it interesting is not only the size of the alleged UFO, which was said to had [sic—have] been kilometers in diameter, but also that all witnesses are Turkey's well-known businessmen!

On Sept. 17, 1985, a group of businessmen were sailing in their yachts on the Bosphorus in Istanbul. According to the witnesses' testimony, at 3:30 a.m., **they saw a huge set of lights approaching them.**

It hovered over them, and everywhere was like daylight. They said that **the object was emitting a great deal of heat**. **The yachtsmen tried to talk to each other on the radio, but even though they tried 16 different channels, the radios just wouldn't work.**

The whole incident lasted about five minutes, then the object flew away. They said that even after it went away, the heat remained for awhile. **The observers were all shocked and frightened.**

When their radios started to work again, they contacted the coast guard radio and learned that the coast guard also saw the same thing. But it could never be identified. [619]

There are not many instances where witnesses have an instrument handy to measure any physical association with a UFO, and even when they do, the fear or awe factor sometimes negates its use. However, on page 70 of his book *Unconventional Flying Objects,* Paul Hill gives one of those rare instances where an instrument was used to measure radiation from a UFO.

What follows is the first account that I have come across where a UFO's presence was able to be detected and measured on a ship by means other than sonar or radar. The text is a bit lengthy, so I have clipped it to just the radiation segment:

SM-??-1986   *Atlantic Ocean off Cape Hatteras, NC, USA*

To my amazement, when I returned to the bridge after my watch, I was very pleased to learn that the conning officer and everyone else on the bridge had seen this sighting and logged it into the ship's log as a UFO sighting.

Next, after a half hour had passed since the sighting, **the radiation detection system (gamma roentgen meter) on the bridge started making a loud clicking sound.** At first, no one seemed to know what was making this sound, then a very loud bell went off notifying us as to what was going on; we were being radiated.

When the instrument stopped clicking, it indicated we had taken a hit of 385 roentgens in the period of about one minute. At this point, the captain of the ship was awoken and called to the bridge, as well as the chief in charge of the radiation metering equipment onboard ship.

The captain was not impressed with an entry of a UFO sighting being placed in the ship's log, and, at first, took the roentgen meter as being defective. However, the chief informed the captain that **the meter had been serviced and calibrated the day before and that other like meters throughout the ship had just gone off indicating the same amount of roentgens received as the bridge.** [620]

The following case is an example of a GPS navigational system affected by the proximity of a UFO:

08-23-2001 *Atlantic Ocean off the New Jersey coast, USA*

The witness reported to MUFON:

"I was out about 45 miles to sea in the Atlantic. Weather was calm and I was looking forward to a day of fishing. At about 3:00 p.m. in the afternoon on August 23, 2001, a reflection off a metallic cylinder caught my eye. The cylinder was hovering about 1000 feet over the surface and the direct spot that it was hovering over was about 1500 feet away from my boat. During the encounter, which lasted only about 45 seconds, **the object disturbed my compass and my GPS. The screen on my GPS said, 'Cannot locate signal.'** Small bait fish became excited and swam to the surface. The craft emitted no light, but **a pulsing, electrical sound came from the craft**. The craft had two sections that counter rotated around each other. I left to go see if I had a camera down below, and **when I returned, the craft had flown away**. The fish subsided, the compass returned to its bearings, and **my GPS picked up the signal.**" [621]

As in the book Grover was *not* trying to read, there is *no* monster at the end of this book...just witnesses and a strange type of craft and the realization that you and I have covered almost all the influences of UFOs on a medium other than air or land.

Conclusion

Although I began my website as an accumulation of reports on UFOs and their reaction with water, as I progressed I realized that in the books that I read, I found nothing involving the rotating field of energy, which I call the "protective field." Over the years, my work has evolved to include the study of the UFOs' physical manifestations and my theories about the observed interaction between UFOs and water which I have presented in this text. Throughout the chapters of this book, the UFO's protective field has been revealed in simplistic form, being seen not just from the exterior of the craft, through its interaction with water, but from the inside as well, by the abductees who have ridden in these craft. We have seen how this protective field has such a high energy output that its heat is felt at a distance and can melt snow and ice with impunity to its own existence. This "field" or force, as I became more familiar with it, has opened the door to understanding many of the "mysteries" of the crafts' idiosyncrasies that seemed previously unanswerable.

While UFOs maneuver in space, in gaseous atmospheres, and in liquids (water in our case), without changing design or accessories, we do not yet know the driving force that governs their movement or hovering ability. These craft operate by physical principles, and though the details behind their operations are unknown to us, they should not be considered beyond our understanding. Excusing the crafts' functions as paranormal or holographic projections or the result of some other impenetrable complication is not the way to comprehend the physical principles governing their technology. We should concentrate on the crafts' mechanics because they are clearly machines. Perhaps a craft like this is in the Earth's future just as the Wright brothers' first kite-plane in 1903 turned into the Boeing 747 of 1970. At some point in the aliens' evolution, they may have had their own version of a Wright brothers' "craft." More recently on November 16, 2004, NASA's X-43 suborbital research vehicle reached a speed of 7,546 mph and may reach its intended speed of 10,000 mph in the near future. We are making progress but it is still excelled by the "unconventional flying objects," according to data that has been collected, and we continue to look in awe at the extraterrestrials' capabilities.

Now as we also try to grasp a UFO's ability to operate in and out of water, we recognize that an understanding of the UFO-water connection may help us in the development of our own technology. While water has restricted our scientific know-how and has impeded our ability to travel at great speed or at great depths while in it, we have seen that the extraterrestrials and their craft have overcome these limitations. UFOs can function in both the air and the water and they can move swiftly between both mediums.

The aliens may be different from us biologically and mentally, but they can communicate with us, as eyewitnesses have stated. As new eyewitness accounts are collected, we learn more about what we must accomplish to approach their technology. While individual stories may not seem believable, the number of accounts reported worldwide and over such a long period of time lends credence to what has been recounted, and we are cognizant of the many common physical characteristics seen by the eyewitnesses. We owe a debt of gratitude to those all over the world who reported their sightings in spite of any repercussions which they may have experienced by doing so.

For those who wish to do further research in the area of UFOs and water, there are other areas which could be further examined to continue the UFO-water discussion:

1. Aircraft disappearances over water. I do not equate this to "missing ships" as several of these reports have

been tracked—on radar in one case and heard via radio in another. In the first of these two cases which comes to mind, dated 11-23-1953, an F-89C (Scorpion) over Lake Superior and a UFO combined into one target (blip) which then disappeared. No wreckage was found. The second report, dated 10-21-1978, pertains to Frederick Valentich's disappearance over Bass Strait during the course of his flight from Australia to King Island between the Australian mainland and Tasmania. He had been talking to Flight Control in Australia about a UFO that was moving back and forth above him when all contact was abruptly lost. Again, no wreckage was ever found. In addition to these, there are other cases of missing and mispositioned aircraft.

2. Photo cases of UFOs over water. Two important examples are the Trindade Island case of 01-16-1958 and the Lago de Cote, Costa Rica, case of 09-04-1971. The Trindade Island case involved the sighting of a UFO during which several pictures of it were taken by a photographer aboard a Brazilian Navy vessel and were analyzed and considered genuine by military personnel. However, other investigators of the case questioned the authenticity of the photos. As a result, the case has been the subject of much controversy. In the Lago de Cote, Costa Rican case, an automatic camera on a government mapping aircraft snapped an unexplained picture of an apparent UFO over the lake or possibly partially submerged in it. None of the flight crew or photographers saw the object, which was discovered only after the film was developed.

As we have seen, a UFO may not be the intended subject of a photograph but may show up unexpectedly in a picture. Remember that these craft travel at enormous speeds, so while they may not be seen in the viewfinder when the photographer clicks on his or her subject, they may end up caught on camera and be a surprise to the photographer when viewed later. Even when a UFO is caught on camera, the quality of the photos is often not good. This may be due to the ionization of the air in the immediate vicinity of a UFO which might distort the photo as the first thing received by the camera would be light from the field and then the craft itself. This initial contact with the field could affect the clarity of the photo. In addition, many people today use cell phone cameras, some of which may produce poorer quality pictures. Despite the fact that photographs are being taken daily, the main problem is verification, and there is always a concern as to whether the pictures may be hoaxed.

3. Ancient civilizations and water gods. While our *total* history is uncertain, tales passed down from various ancient civilizations have survived to the present day, telling how the gods supposedly arrived and taught agriculture, architecture, and other sciences to the indigenous populations. Among these in the water realm are Neptune of the Roman era, Poseidon of Greece, Kappa of Japan, the Nivatakavacas of India, and so on. This area might be suited to the work of professors of mythology who could deal with the cosmological and supernatural traditions of ancient peoples and their legends, keeping in mind that until as recently as the 1800s, Troy was considered a myth.

It is the hope of this author that I have left you with the desire to further explore the UFO-water connection. If this is so, please visit my website for the hundreds of other cases which were not able to find their way into this small volume. The site is frequently updated as I receive cases from all over the world. It is unfortunate that we lack a large, central computer database with which to work and share information worldwide. Consequently, the study of UFOs and their connection to water is not yet extensive. It is my hope that this book will encourage others to pursue this fascinating topic where we still have so much to learn.

Eyes on the skies? Yes.
But eyes on the water as well...

References

INTRODUCTION

001   Book: Fort, C. (1974). *The complete books of Charles Fort.* Mineola, NY: Dover Publications.

002   Ibid., pp. 642-643.

003   Ibid., p. 639.

004   Book: Condon, E. U. (1968). *Final report of the scientific study of unidentified flying objects.* New York: Bantam Books.

005   Book: Saunders, D. R., & Harkins, R. R. (1968). *UFOs? yes! Where the Condon committee went wrong.* New York: Signet Books [New American Library].

CHAPTER 1: WATER-UFO HISTORY

006   Book: Lorenzen, J., & Lorenzen, C. (1968). *UFOs over the Americas.* New York: New American Library.

007   Book: Sanderson, I. T. (1970). *Invisible residents.* Toronto, Canada: Fitzhenry & Whiteside Ltd.

008   Book: *New illustrated Webster's dictionary of the English language.* (1992). New York: Pamco Publishing.

009   Website: Force field. *The Oxford pocket dictionary of current English.* (2009). Retrieved March 30, 2010 from Encyclopedia.com: http://www.encyclopedia.com/doc/10999-forcefield.html

010   Book: Sanderson, I. T. (1970). *Invisible residents.* p. 32. Toronto, Canada: Fitzhenry & Whiteside Ltd.

011   Book: Vallee, J. (1969). *Passport to Magonia: On UFOs, folklore, and parallel worlds.* p. 6. Chicago: H. Regnery.

012   Website: Fordham University. (n.d.). *Medieval sourcebook: Christopher Columbus: Extracts from journal.* Retrieved February 26, 2001, from http://www.fordham.edu/halsall/source/columbus1.html

013   Website: Filer, G. (2000, August 28). *Scotland UFO report from the year 1767*. Retrieved August 28, 2000, from: http://www.nationalufocenter.com/files/2000/FilersFiles34.htm

014   Ibid.

015   Book: Gaddis V. (1965). *Invisible horizons*. p. 212. New York: Ace Books, a division of Charter Communications.

016   Book: Heyerdahl, T. (1950). *Kon-Tiki*. p. 118. New York: Permabook edition, published by arrangement with Rand McNally & Co.

017   Journal Article: Brill, J. M. (1978, September). Are UFOs operating from underwater bases off the coast of Argentina? *The MUFON UFO Journal*, No. 130, p. 4.

018   Letter: Name is confidential. (n.d.). Original on file with the Center for UFO Studies (CUFOS): http://www.cufos.org

019   Newspaper Article: Mystery object some underwater craft? (1965, November 18). *Otago Daily Times* (New Zealand).

020   Book: Cochrane, H. F. (1980). *Gateway to oblivion: The Great Lakes' Bermuda Triangle*. pp. 69-70. Garden City, NY: Doubleday & Co.

021   Book: Hesemann, M. (1998). *UFOs: The secret history*. pp. 415-416. New York: Marlowe & Co.

022   Newspaper Article: González, L. (1981, April 01). A luminous ball came out of the water. *Pueblo* (Madrid, Spain). Translation by R. Heiden.

023   Magazine Article: *Flying Saucer Review (FSR)*. (1991, July-August). Vol. 36, No. 4, p. 13. http://www.fsr.org.uk

024   E-Mail: B. Vike (personal communication, July 30, 2006). Director HBCC UFO Research: http://hbccuforesearch.blogspot.com

025   Case Investigation Files: Feindt, C. (Case date: UNDATED-4). Pityilu Island (Papua New Guinea) UFO sighting – late '70s. See: http://www.waterufo.net

026   Newsletter Article: (author unknown). (1964, January). Disc submerged.... *The A.P.R.O. Bulletin*, Vol. & No. unknown, p. 2.

027   Case Investigation Files: Feindt, C. (Case date: SM-??-1957). No title. Added to MUFON's CMS (Computer Management System). See: http://www.waterufo.net

028   Book: Gross, L. (1999). *The fifth horseman of the apocalypse: UFOs: A history. 1959: January-March*. pp. 11-13. Fremont, CA: Author.

029   E-Mail: M. Podell (personal communication, December 17, 2005).

030   Magazine Article: Creighton, G. (1975, March-April). Alleged kidnappings, and other matters. *Flying Saucer Review (FSR)*, Vol. 21, No. 2, pp. 21-22. http://www.fsr.org.uk

031   Newsletter Article: (author unknown). (1976, May). UFO submerged in N.Y. lake. *The A.P.R.O. Bulletin*, Vol. 24, No. 11, pp. 1, 3.

032   Case Investigation Files: Feindt, C. (Case date: 08-??-1981). UFO enters waters of Canadian harbor. See: http://www.waterufo.net

033   Catalog: Bianchini, M. (2003). *USOCAT: The Italian catalog of sightings of unidentified submerged objects*. Turin, Italy: Edizioni UPIAR. Case from updated, unpublished 2007 catalog, p. 100. The Italian Center for Ufological Studies, (CISU): http://www.cisu.org

034   Website: Case dated 04-14-2000. (2000, April 15). Retrieved May 01, 2000, from National UFO Reporting Center (NUFORC), P. Davenport, director: http://www.nuforc.org/webreports/012/S12627.html

035   Website: *Oklahoma UFO dives in pond*. (2006, June 29). Retrieved July 04, 2006, from MUFON CMS: http://www.mufon.com/mufonreports.htm

036   Website: *Little people on type of submarine*. (n.d.). Retrieved February 20, 2000, from http://www.mufor.org which is now archived at UFO Casebook: http://www.ufocasebook.com/malta1947.html

037   Magazine Article: Norman, S. (1956, June). Recent UFOs over Japan. *Fate*, Vol. & No. unknown, pp. 22-24.

038   Magazine Article: Hinfelaar, H. J. (1966, July-August). Submarine craft in Australasian waters. *Flying Saucer Review (FSR)*, Vol. 12, No. 4, p. 29. http://www.fsr.org.uk

039   Newspaper Article: Miller, K. (1968, September 18). Valley UFO 'sank in river.' *The Chronicle-Herald* (Halifax, Nova Scotia, Canada), pp. 1, ?.

040   Newsletter Article: (author unknown). (1969, May-June). Strange object sighted in Alaska. *The A.P.R.O. Bulletin*, Vol. & No. unknown, p. 6.

041   Newspaper Article: UFO was in the path of the Caioba. (1980, August 17). Translated by Joe Brill from *O Poti* (Natal, Rio Grande do Norte, Brazil).

042   Magazine Article: Keel, J. A. (1970, November). Ocean-based UFOs ring the U.S. *Male*, Vol. 20, No. 12, p. 48.

043   Letter: Gary R..... (n.d.). Copy on file with the Center for UFO Studies (CUFOS): http://www.cufos.org

044   Journal Article: Brill, J. M. (1978, September). Are UFOs operating from underwater bases off the coast of Argentina? *The MUFON UFO Journal*, No. 130, p. 4.

045   Book: Cooper, M. W. (1991). *Behold a pale horse*. pp. 17-21. Sedona, AZ: Light Technology Publications. Thanks to Doyel Shamley & Rob Houghton of: http://www.hourofthetime.com for permission to use in this book.

046   Book: Hall, R. H. (2001). *The UFO evidence: A thirty-year report*. Volume 2. p. 346. Lanham, MD: Scarecrow Press.

047   Transcription: Interview between UFO researcher R. Gribble and witness to UFO sighting. (n.d.). Transcription done by C. Feindt. CD from Wendy Connors of Faded Discs which is inactive but is archived at: http://web.archive.org/web/*/http://www.fadeddiscs.com

048   Website: Case dated 07-20-1993. (2000, February 16). Retrieved January 25, 2002, from National UFO Reporting Center (NUFORC), P. Davenport, director: http://www.nuforc.org/webreports/011/S11942.html

049   E-Mail: B. Vike (personal communication, July 03, 2007). Director HBCC UFO Research: http://hbccuforesearch.blogspot.com

CHAPTER 2: PHYSICAL INFLUENCES OF A UFO ON WATER

050   Magazine Article: Farish, L. & Titler, D. (1977, May). Mysteries of the deep: Underwater UFOs. *Official UFO Magazine*, Vol. 2, No. 3, pp. 54-55.

051   Book: Edwards, F. (1966). *Flying saucers-serious business*. p. 143. New York: Bantam Books.

052   Book: Hill, P. R. (1995). *Unconventional flying objects: A scientific analysis*. p. 182. Charlottesville, VA: Hampton Roads Publishing. © 1995 by Julie M. Hill.

053   Website: *Ionization*. (n.d.). In Encyclopædia Britannica. Retrieved March 13, 2013, from Encyclopædia Britannica Online: http://www.britannica.com/EBchecked/topic/293007/ionization

054   Book: Hill, P. R. (1995). *Unconventional flying objects: A scientific analysis*. p. 54. Charlottesville, VA: Hampton Roads Publishing. © 1995 by Julie M. Hill.

055   Book: Feschino, F. C., Jr. (2007). *Shoot them down! - The flying saucer air wars of 1952*. p. 23. Port Orange, FL: Author.

056   Book: Randle, K. D. (1999). *Scientific ufology*. p. 45. New York: Avon Books.

057   Book: Cramp, L. G. (1966). *UFOs and anti-gravity: Piece for a jig-saw*. p. 133. Kempton, IL: Adventures Unlimited Press.

058   Magazine Article: Poland: Report on the year 1959. (1961, September). *Le Courrier Interplanétaire*, Vol. 7, No. 56.

059   Catalog: Bianchini, M. (2003). *USOCAT: The Italian catalog of sightings of unidentified submerged objects*. Turin, Italy: Edizioni UPIAR. Case from updated, unpublished 2007 catalog, p. 32. The Italian Center for Ufological Studies, (CISU): http://www.cisu.org

060   Magazine Article: Farish, L. & Titler, D. (1977, May). Mysteries of the deep: Underwater UFOs. *Official UFO Magazine*, Vol. 2, No. 3, p. 52.

061   Ibid., pp. 52-53.

062   Book: Lorenzen, J., & Lorenzen, C. (1968). *UFOs over the Americas*. pp. 54-55. New York: New American Library.

063   Catalog: Ballester Olmos, V.-J. (1976, April). *A catalogue of 200 type-1 UFO events in Spain and Portugal*. Case No. 146, p. 34. Evanston, IL: Center for UFO Studies (CUFOS).

064   Website: Pratt, B. (n.d.). *The concert pianist and the "rat-men."* Retrieved 03-04-2012, from The Unreal World of UFOs by Bob Pratt, hosted by MUFON (Mutual UFO Network): http://www.mufon.com/bob_pratt/index.html Case at: http://www.mufon.com/bob_pratt/luli.html

065   Magazine Article: Earley, G. (1976, January). Encounter in the Gulf Stream. *Beyond Reality*, No. 18, pp. 37-39.

066   Videotape: Painter VideoGraphics (Producer). (1999, April 24). Ted Phillips, Session One. Presentation at *The 2nd Annual Kentucky MUFON Conference*, Western Kentucky University, Bowling Green, KY.

067   Magazine Article: *Flying Saucer Review (FSR)*. (1973, April). Case Histories, Supplement No. 14, pp. 13-15. http://www.fsr.org.uk

068   Magazine Article: Farish, L. & Titler, D. (1977, May). Mysteries of the deep: Underwater UFOs. *Official UFO Magazine*, Vol. 2, No. 3, p. 52.

069   Book: Cooper, M. W. (1991). *Behold a pale horse*. pp. 17-21. Sedona, AZ: Light Technology Publications. Thanks to Doyel Shamley and Rob

Houghton of: http://www.hourofthetime.com for permission to use in this book.

070   Book: Sanchez-Ocejo, V., & Stevens, W. C. (1982). *UFO contact from undersea: A report of the investigation*. pp. 155-156. Tucson, AZ: Privately published by W. C. Stevens.

071   Journal Article: (author unknown). (1976, March). *Phénomènes Spatiaux* (Space Phenomena), a publication of GEPA, Groupement d'étude de Phénomènes Aériens (Group for the Study of Aerial Phenomena), Vol. & No. unknown, pp. 23-26.

072   Website: Pratt, B. (n.d.). *Invisible object makes waves.* Retrieved 01-19-2014, from The Unreal World of UFOs by Bob Pratt, hosted by MUFON (Mutual UFO Network): http://www.mufon.com/bob_pratt/index.html Case at: http://www.mufon.com/bob_pratt/invisible.html

073   Book: Lorenzen, C., & Lorenzen, J. (1969). *UFOs: The whole story*. p. 160. New York: Signet Books [New American Library].

074   Book: Cramp, L. G. (1966). *UFOs and anti-gravity: Piece for a jig-saw*. p. 133. Kempton, IL: Adventures Unlimited Press.

075   Newsletter Article: *CSI Newsletter*, published by the Civilian Saucer Intelligence New York. No other information available. Thanks to the J. Allen Hynek Center for UFO Studies (CUFOS): http://www.cufos.org

076   Book: Beckley, T. G. (1992). *Strange encounters: Bizarre & eerie contact with UFO occupants*. p. 84. New Brunswick, NJ: Inner Light – Global Communications.

077   Website: Title unknown. (2000, February 19). Retrieved February 20, 2000, from website: http://www.mufor.org which is inactive. See WATER UFO: http://www.waterufo.net/item.php?id=40

078   Case Investigation Files: Feindt, C. (Case date: 05-??-1953). UFO has a column of fog or water under it. Case interview on 03-17-2003. See: http://www.waterufo.net

079   Magazine Article: Culverwell, A. (1958, April). Submerging UFO? *Fate*, Vol. & No. unknown, p. 114.

080   Book: Keyhoe, D. E. (1960). *Flying saucers: Top secret*. p. 56. New York: G.P. Putnam's Sons.

081   Magazine Article: *Flying Saucer Review (FSR)*. (1961, September-October). Vol. 7, No. 5, pp. 18-20. http://www.fsr.org.uk

082   Magazine Article: Farish, L. & Titler, D. (1977, May). Mysteries of the deep: Underwater UFOs. *Official UFO Magazine*, Vol. 2, No. 3, p. 52.

083   Journal Article: Neville, R. D. (1974, September). Witness says UFO came out of the water. *Skylook* [the original *MUFON UFO Journal*], No. 82, p. 8.

084   Book: Stonehill, P. (1998). *The Soviet UFO files*. p. 58. Surry, England: Quadrillion Publishing Ltd.

085   Book: Beckley, T. G. (1992). *Strange encounters: Bizarre & eerie contact with UFO occupants*. pp. 86-87. New Brunswick, NJ: Inner Light - Global Communications.

086   Catalog: Bianchini, M. (2003). *USOCAT: The Italian catalog of sightings of unidentified submerged objects*. Turin, Italy: Edizioni UPIAR. Case from updated, unpublished 2007 catalog, p. 43. The Italian Center for Ufological Studies, (CISU): http://www.cisu.org

087   Book: Swann, I. (1998). *Penetration: The question of extraterrestrial and human telepathy*. p. 93. Rapid City, SD: Ingo Swann Books.

088   Journal Article: (author unknown). (1979, September-October). The Adriatic triangle: Whales or submerged UFOs? *International UFO Reporter (IUR)*, Vol. 4, No. 3, p. 17. © CUFOS.

089   Book: Pratt, B. (1996). *UFO danger zone: Terror and death in Brazil— where next?* pp. 231-232. Madison, WI: Horus House Press.

090   Book: Walters, E., & Maccabee, B. (1997). *UFOs are real: Here's the proof*. pp. 56-57. New York: Avon Books.

091   Website: *1995, Widnes*. (2002). Retrieved December 08, 2004, from Para Science: http://www.parascience.org.uk/misc/ngr/ nwest.htm which is inactive but is available through Unknown Phenomena Investigation Association: http://upia.moonfruit.com/#/ upia-ufo-files-1/4516077270

092   Magazine Article: Martín, J. (2003). Something strange happens in Vieques, and the Navy of the United States is surrounded. *Enigmas of the Millennium*, Vol. & No. unknown, pp. 21-23.

093   Magazine Article: Farish, L. & Titler, D. (1977, May). Mysteries of the deep: Underwater UFOs. *Official UFO Magazine*, Vol. 2, No. 3, p. 52.

094   Ibid.

095   Ibid.

096   Journal Article: Chalker, B. (1997-98, Winter). Tully saucer nests of 1966: Part one. *International UFO Reporter (IUR)*, Vol. 22, No. 4, pp. 14-20. © CUFOS 1998.

097   Book: Maccabee, B. S. (2000). *UFO-FBI connection: The secret history of the government's cover-up*. pp. 13-14. St. Paul, MN: Llewellyn Publications, a division of Llewellyn Worldwide, Ltd.

098   Book: Phillips, T. (2002). *Delphos: A close encounter of the second kind*. p. 57. Fairfax, VA: UFO Research Coalition.

099   E-Mail: D. Ledger (personal communication, n.d.). From documents of the Canadian Department of National Defence, Training Command Headquarters, Westwin, Manitoba, dated 01 September 1967.

100   Journal Article: Swords, M. D. (2002-2003, Winter). Timmermania: A step too far into the Timmerman files? *International UFO Reporter (IUR)*, Vol. 27, No. 4, p. 9. © CUFOS 2003.

101   Book: Pratt, B. (1996). *UFO danger zone: Terror and death in Brazil— where next?* pp. 3-5. Madison, WI: Horus House Press.

102   Book: Haines, R. F. (1999). *CE-5: Close encounters of the fifth kind*. Naperville, IL: Sourcebooks.

103   Book: Todd, D. R. (1997). *The Antilles incident*. p. 106. Sun Lakes, AZ: Blue Star Productions, a Division of Book World.

104   Book: Hill, P. R. (1995). *Unconventional flying objects: A scientific analysis*. p. 146. Charlottesville, VA: Hampton Roads Publishing. © 1995 by Julie M. Hill.

CHAPTER 3: THE FIELD IN REVERSE

105   Magazine Article: Binder, O. O. (1968, February). "Oddball" saucers...that fit no pattern. *Fate*, Vol. 21, No. 2, pp. 57-58.

106   Case Investigation Files: Feindt, C. (Case date: SM-??-1957). No title. Added to MUFON's CMS (Computer Management System). See: http://www.waterufo.net

107   Book: Sanchez-Ocejo, V., & Stevens, W. C. (1982). *UFO contact from undersea: A report of the investigation*. p. 156. Tucson, AZ: Privately published by W. C. Stevens.

108   Newsletter Article: (author unknown). (1976, May). UFO submerged in N.Y. lake. *The A.P.R.O. Bulletin*, Vol. 24, No. 11, pp. 1, 3.

109   Website: *UWO – Unidentified Water Object?* (1999). Retrieved September 10, 1999, from http://www.ufobc.org/uwo.htm and now available at UFO*BC: http://www.ufobc.ca/History/1990/uwo_v2.htm

110   Website: *Oklahoma UFO dives in pond.* (2006, June 29). Retrieved July 04, 2006, from MUFON CMS: http://www.mufon.com/mufonreports.htm MUFON's case investigation report is in a file which is not accessible to the general public.

111   Website: Filer, G. (2006, September 13). *Lake Superior.* Retrieved September 13, 2006, from http://www.nationalufocenter.com which is now archived at UFOINFO: http://www.ufoinfo.com/filer/2006/ff0637.shtml

112   Book: Corrales, S. (1998). *Flashpoint: High strangeness in Puerto Rico.* pp. 68-69. Maulden, Beds, United Kingdom: Amarna Ltd.

113   Book: Edwards, F. (1966). *Flying saucers-serious business.* pp. 175-176. New York: Bantam Books.

114   Catalog: Bianchini, M. (2003). *USOCAT: The Italian catalog of sightings of unidentified submerged objects.* Turin, Italy: Edizioni UPIAR. Case from updated, unpublished 2007 catalog, p. 29. The Italian Center for Ufological Studies, (CISU): http://www.cisu.org

115   Book: Edwards, F. (1966). *Flying saucers-serious business.* p. 304. New York: Bantam Books.

116   Website: Ridge, F. (2011, April 14). *The 1997 UFO chronology.* Retrieved November 20, 2011, from National Investigations Committee on Aerial Phenomena (NICAP): http://www.nicap.org/waves/1997fullrep.htm

117   Magazine Article: Karivieri, A. (1971, September-October). The Saapunki UFO: Results of investigations. *Flying Saucer Review (FSR),* Vol. 17, No. 5, pp. 24-27. http://www.fsr.org.uk

118   Book: Randle, K.D., & Schmitt, D.R. (1991). *UFO Crash at Roswell.* p. 50. New York: Avon Books.

119   Newspaper Article: Saturday's storm: Damage by lightning and wind: Winsted: A ball of fire in Long Lake:—Remarkable phenomenon. (1888, August 6). *The Hartford Times* (Connecticut).

120   Magazine Article: Father like son. (1967). *Flying Saucer UFO Reports,* No. 4, p. 5.

121   Book: Hobana, I., & Weverbergh, J. (1975). *UFO's from behind the iron curtain.* pp. 61-63. London, United Kingdom: Bantam Books published by arrangement with Souvenir Press Ltd.

122   Newsletter Article: White, J. D. (1976, June). UFO research in Russia. *The A.P.R.O. Bulletin*, Vol. & No. unknown, p. 5.

123   Journal Article: Hall, R. H. (1976, November). Lake objects reported in Scandinavia. *The MUFON UFO Journal*, No. 108, p. 11.

124   Book: Potter, P. (2008). *Gravitational manipulation of domed craft: UFO propulsion dynamics.* pp. 250-252. Kempton, IL: Adventures Unlimited Press.

125   Book: Fowler, R. E. (1990). *The watchers: The secret design behind UFO abduction.* New York: Bantam Books.

126   Memorandum: Treadwell, L. (1982, March). Concerning Treadwell's sighting on July 26, 1956, while aboard the USS *FDR*. Copy in Carl Feindt's investigation files.

127   Letter: Treadwell, L. (1991, April). To Chester Grusinski, in reference to Treadwell's sighting on July 26, 1956, while aboard the USS *FDR*. Copy in Carl Feindt's investigation files.

128   Book: Fowler, R. E. (1990). *The watchers: The secret design behind UFO abduction.* pp. 76-80. New York: Bantam Books.

129   Website: Ananthaswamy, A. (2009). *First black hole for light created on earth.* Retrieved September 07, 2012, from: http://www.newscientist.com/article/dn17980-first-black-hole-for-light-created-on-earth.html

130   Book: Vallee, J., & Aubeck, C. (2009). *Wonders in the sky.* pp. 253-254. New York: Tarcher/Penguin.

131   Magazine Article: Farish, L., & Titler, D. M. (1977). Strange underwater wheels of light. *Bermuda Triangle Special Report 1977*, compiled by the editors of *Saga Magazine & UFO Report*, p. 66.

132   Website: Johnson, D. A. (2006). *On this day.* Retrieved September 27, 2011, from http://www.ufoinfo.com/onthisday/July29.html

133   Catalog: Bianchini, M. (2003). *USOCAT: The Italian catalog of sightings of unidentified submerged objects.* Turin, Italy: Edizioni UPIAR. Case from updated, unpublished 2007 catalog, p. 69. The Italian Center for Ufological Studies, (CISU): http://www.cisu.org

134   Letter: Name is confidential. (2008, June 29). To researcher Carl Feindt.

CHAPTER 4: HEAT AND WATER-RELATED UFOs

135   Book: *The new encyclopædia Britannica*. (1986). Vol. 5. p. 784. Chicago: Encyclopedia Britannica.

136   Book: Hill, P. R. (1995). *Unconventional flying objects: A scientific analysis*. p. 70. Charlottesville, VA: Hampton Roads Publishing. © 1995 by Julie M. Hill.

137   Book: Sider, J. (1997). *Dossier 1954 et l'imposture rationaliste*. p. 122. Villeselve, France: Editions Ramuel.

138   Book: Schuessler, J. (1998). *The Cash-Landrum UFO incident*. La Porte, TX: Geo Graphics Printing.

139   Magazine Article: Doyle, A. C. (1891). A Scandal in Bohemia. *The Strand Magazine*, Vol., No., & page unknown.

140   Newspaper Article: A meteor storms a vessel. (1893, April 11). *Manitoba Daily Free Press* (Winnipeg, Canada).

141   E-Mail: A. Rosales (personal communication, date unknown). Source: Paolo Fiorino, *UFO Universe*, October-November 1991.

142   Website: Case dated 03-??-1931. Retrieved January 7, 2013, from UFO DNA: http://thecid.com/ufo/uf11/uf7/117588.htm. Primary reference: *Phénomènes Spatiaux* (Space Phenomena), a publication of GEPA, Groupement d'étude de Phénomènes Aériens (Group for the Study of Aerial Phenomena), June 1972, Vol. unknown, No. 32, p. 15.

143   Journal Article: Brænne, O. J. (1995, January-February). Observations of unidentified submarine objects in Norway. *International UFO Reporter (IUR)*, pp. 12 & 17.

144   Catalog: Bianchini, M. (2003). *USOCAT: The Italian catalog of sightings of unidentified submerged objects*. Turin, Italy: Edizioni UPIAR. Case from updated, unpublished 2007 catalog, pp. 23-24. The Italian Center for Ufological Studies, (CISU): http://www.cisu.org

145   Newsletter Article: (author unknown). (1964, January). Disc submerged.... *The A.P.R.O. Bulletin*, Vol. & No. unknown, p. 2.

146   Magazine Article: Farish, L. & Titler, D. (1977, May). Mysteries of the deep: Underwater UFOs. *Official UFO Magazine*, Vol. 2, No. 3, p. 40.

147   Magazine Article: Creighton, G. (1964, November-December). A Brazilian sighting. *Flying Saucer Review (FSR)*, Vol. 10, No. 6, p. 18.

148   Catalog: Bianchini, M. (2003). *USOCAT: The Italian catalog of sightings of unidentified submerged objects*. Turin, Italy: Edizioni UPIAR. Case from updated, unpublished 2007 catalog, p. 29. The Italian Center for Ufological Studies, (CISU): http://www.cisu.org

149   Documents: Unclassified U.S. Air Force documents from Project Blue Book. DD Form 173 Joint Message Form, Aug. 04, 1958, 0900.

150   Website: Rosales, A. (n.d.). *A chronology of Cuban cases*. Now available at UFOINFO: http://www.ufoinfo.com/humanoid/humanoid1959.shtml

151   Newsletter Article: Moseley, J. W. (1962, September). No article title. *Saucer News*, Vol. & No. unknown, p. 24.

152   E-Mail: A. Rullán (personal communication, date unknown).

153   E-Mail: D. Ledger (personal communication, date unknown).

154   Ibid.

155   Magazine Article: Farish, L. & Titler, D. (1977, May). Mysteries of the deep: Underwater UFOs. *Official UFO Magazine*, Vol. 2, No. 3, pp. 52-53.

156   Magazine Article: Rieseberg, H. E. (1965, December). A submerged UFO? *Exploring the Unknown*, Vol. 6, No. 2, pp. 64-67.

157   Newspaper Article: Mystery object some underwater craft? (1965, November 18). *Otago Daily Times* (New Zealand).

158   Newsletter Article: (author unknown). Submerging UFO case. (1968, May-June). *The U.F.O. Investigator* [Publication of NICAP], Vol. 4, No. 6, p. 3.

159   Newspaper Article: A ship that flew almost vertically emerged from the sea in front of Arrecife. (1967, August 20). *El Universal* (Caracas, Venezuela).

160   Magazine Article: Farish, L. & Titler, D. (1977, May). Mysteries of the deep: Underwater UFOs. *Official UFO Magazine*, Vol. 2, No. 3, p. 40.

161   Magazine Article: Wolkomir, R. (1968, November). The glowing "thing" in Moore Lake. *Fate*, Vol. 21, No. 11, pp. 32-36.

162   Journal Article: Delair, J. B. (editor) (1974). *UFO Register*, Vol. 5, Parts 1 & 2, No. 380. © Contact International UFO Research (CIUFOR) (UK). See: http://contactinternationalufo.homestead.com/history.html

163   Newsletter Article: (author unknown). (1976, May). UFO submerged in N.Y. lake. *The A.P.R.O. Bulletin*, Vol. 24, No. 11, pp. 1, 3.

164   Book: Walters, E., & Maccabee, B. (1997). *UFOs are real: Here's the proof.* pp. 144-145. New York: Avon Books.

165   Catalog: Bianchini, M. (2003). *USOCAT: The Italian catalog of sightings of unidentified submerged objects.* Turin, Italy: Edizioni UPIAR. Case from updated, unpublished 2007 catalog, p. 83. The Italian Center for Ufological Studies, (CISU): http://www.cisu.org

166   Ibid., p. 86.

167   Magazine Article: Hynek, J. A. (1981, July-August). Editorial from International UFO Reporter. *Frontiers of Science*, Vol. 3, No. 5, pp. 13-14.

168   Journal Article: Schuessler, J. F. (1982, December). Oblong UFO lands on water. *The MUFON UFO Journal*, No. 178, p. 11.

169   Website: Case dated 10-12-1985. (2003, March 21). Retrieved November 04, 2007, from National UFO Reporting Center (NUFORC), P. Davenport, director: http://www.nuforc.org/webreports/012/S12627.html

170   Catalog: Bianchini, M. (2003). *USOCAT: The Italian catalog of sightings of unidentified submerged objects.* Turin, Italy: Edizioni UPIAR. Case from updated, unpublished 2007 catalog, pp. 97-98. The Italian Center for Ufological Studies, (CISU): http://www.cisu.org

171   Book: Walters, E., & Maccabee, B. (1997). *UFOs are real: Here's the proof.* pp. 132-133. New York: Avon Books.

172   Book: Corrales, S. (1998). *Flashpoint: High strangeness in Puerto Rico.* pp. 112-113. Maulden, Beds, United Kingdom: Amarna Ltd.

173   Website: Case dated 01-31-2005. (2005, February 08). Retrieved February 17, 2005, from National UFO Reporting Center (NUFORC), P. Davenport, director: http://www.nuforc.org/webreports/041/S41709.html

174   E-Mail: P. Hassall (personal communication, July 16, 2006).

175   Website: *Oklahoma UFO dives in pond.* (2006, June 29). Retrieved July 04, 2006, from MUFON CMS: http://www.mufon.com/mufonreports.htm MUFON's case investigation report is in a file which is not accessible to the general public.

176   Book: Vallee, J., & Aubeck, C. (2009). *Wonders in the sky.* pp. 216-217. New York: Tarcher/Penguin.

177   E-Mail: C. Aubeck (personal communication, n.d.). Original source was a newspaper article: (1845, August 18). *The Malta Times.* Secondary

source: (1861). Report on observations of luminous meteors, 1860-62. *Report of the British Association for the Advancement of Science,* Vol. 31, pp. 30-31.

178   Magazine Article: *Flying Saucer Review (FSR).* (1973, April). Case Histories, Supplement No. 14, pp. 13-15. http://www.fsr.org.uk

179   Journal Article: Chalker, B. (1997-98, Winter). Tully saucer nests of 1966: Part one. *International UFO Reporter (IUR),* Vol. 22, No. 4, pp. 14-20. ©CUFOS 1998.

180   Newspaper Article: What was it? (1968, May 09). *The Coast Guard* (Shelburne, Nova Scotia, Canada).

181   Journal Article: (author unknown). (1976, March). *Phénomènes Spatiaux* (Space Phenomena), a publication of GEPA, Groupement d'étude de Phénomènes Aériens (Group for the Study of Aerial Phenomena), Vol. & No. unknown, pp. 23-26.

182   Book: Randles, J. (1991). *From out of the blue.* pp. 208-209. New Brunswick, NJ: Global Communications.

183   Catalog: Bianchini, M. (2003). *USOCAT: The Italian catalog of sightings of unidentified submerged objects.* Turin, Italy: Edizioni UPIAR. Case from updated, unpublished 2007 catalog, pp. 92-93. The Italian Center for Ufological Studies, (CISU): http://www.cisu.org

184   Journal Article: Sekerkarar, E. (2005, March). History of UFO sightings in Turkey: Businessmen's sighting. *The MUFON UFO Journal,* No. 443, p. 8.

185   Catalog: Bianchini, M. (2003). *USOCAT: The Italian catalog of sightings of unidentified submerged objects.* Turin, Italy: Edizioni UPIAR. Case from updated, unpublished 2007 catalog, p. 107. The Italian Center for Ufological Studies, (CISU): http://www.cisu.org

186   Book: Ledger, D. (1998). *Maritime UFO files.* p. 111. Halifax, NS, Canada: Nimbus Publishing Ltd.

187   E-Mail: J. Randles (personal communication, date unknown).

188   Book: Michel. A. (1958). *Flying saucers and the straight-line mystery.* pp. 203-204. New York: S. G. Phillips.

189   Transcription: Willow Grove Sighting 15/2/63 (Australia). Tape-recorded interview with Mr. Charles Brew by Mr. Charles Norris, President, Victorian Flying Saucer Research Society. Thanks to the Center for UFO Studies (CUFOS): http://www.cufos.org

190   Book: Fowler, R. E. (1974). *UFOs: Interplanetary visitors*. p. 334. Jericho, NY: Exposition Press.

191   Ibid., p. 335.

192   Pamphlet: Hall, R. H. (2004). Alien invasion or human fantasy? The 1966-67 UFO Wave. p. 20. Washington, DC: Fund for UFO Research.

193   Newspaper Article: McKerron, I. (1980, August 31). We saw a UFO say 5 policemen. *Sunday Express* (London, England).

194   Journal Article: Haines, R. (2007, October). A UAP and its safety implications: O'Hare international airport, November 7, 2006. *International UFO Reporter (IUR)*, Vol. 31, No. 3, p. 3. © CUFOS 2008. Original source is Technical Report 10 on NARCAP website: http://www.narcap.org

195   Website: *Ockham's razor or Occam's razor*. (2010). In *Encyclopædia Britannica*. Retrieved March 13, 2010, from Encyclopædia Britannica Online:http://www.britannica.com/EBchecked/topic/424706/Ockhams-razor

196   Journal Article: Efishoff, K., & Lemke, L. (2007, October). A UAP and its safety implications: O'Hare international airport, November 7, 2006: Hole-in-cloud considerations. *International UFO Reporter (IUR)*, Vol. 31, No. 3, pp. 26-27. © CUFOS 2008. Original source is Technical Report 10 on NARCAP website: http://www.narcap.org

197   E-Mail: Name withheld by this author (personal communication, October 07-08, 2008).

198   Journal Article: Swiatek, R. (2010, October). Domed disk in snow listed as 'unexplained.' *The MUFON UFO Journal*, No. 510, p. 18.

199   Book: Vallee, J. (1965). *Anatomy of a phenomenon*. p. 22. Lincolnwood, IL: NTC/Contemporary Publishing.

200   Newsletter Article: (author unknown). (1957, May 1). Remarkable Norwegian near-landing case of 1954 now published. *CSI Newsletter*, published by the Civilian Saucer Intelligence New York, No. 7, p. 16.

201   Book: Stringfield, L. H. (1977). *Situation red: The UFO siege!* p. 50. Garden City, NY: Doubleday & Co.

202   Book: Keyhoe, D. E. (1960). *Flying saucers: Top secret*. p. 56. New York: G.P. Putnam's Sons.

203   Book: Vallee, J. (1969). *Passport to Magonia: A century of landings*. No. 718. p. 321. Chicago: H. Regnery.

204   Newspaper Article: Both see huge 'saucer': UFO startles telephone pals. (1967, February 24). *Baltimore News-American* (Maryland).

205   Book: Fowler, R. E. (1974). *UFOs: Interplanetary visitors*. p. 355. Jericho, NY: Exposition Press.

206   Book: Lorenzen, C., & Lorenzen, J. (1976). *Encounters with UFO occupants*. p. 338. New York: Berkley Publishing Corporation.

207   Newspaper Article: Man reports sighting UFO. (1976, January 12). *Brantford Expositor* (Ontario, Canada). Thanks to the J. Allen Hynek Center for UFO Studies (CUFOS): http://www.cufos.org

208   Newspaper Article: Mysterious fire melts deep snow. (1924, January 16). *Los Angeles Times.*

209   Magazine Article: Binder, O. O. (1968, February). "Oddball" saucers...that fit no pattern. *Fate*, Vol. 21, No. 2, pp. 57-58.

210   Book: Swords, M. D. (2005). *Grass roots UFOs: Case reports from the Timmerman files*. p. 77. Lima, OH: CSS Publishing. Thanks to the Fund for UFO Research: http://www.fufor.com for permission to use.

211   Book: Magor, J. (1977). *Our UFO visitors*. p. 211. Sannichton, BC, Canada: Hancock House Publishers Ltd.

212   Newsletter Article: (author unknown). New sightings put AF on spot. (1965, March-April). *The U.F.O. Investigator* [Publication of NICAP], Vol. 3, No. 1, p. 4.

213   Case Investigation Files: Phillips, T. (Case date: 01-29-1967). Center for Physical Trace Research: http://www.ufophysical.com. See also: http://www.waterufo.net

214   Case Investigation Files: Martiny, W. G. (Case date: 02-21-1967). NICAP report dated February 23, 1967. See: http://www.waterufo.net

215   Newspaper Article: "Lighted object" seen in Berlin. (1968, December 30). *Lewiston Daily Sun* (Maine).

216   News Clipping Service: Merchant, D. (2008, May 12). Close encounters of the second kind in Stratham: The sledding incident. UFO News Clipping Service, June 2008, No. 467. Thanks to: http://www.seacoastonline.com/apps/pbcs.dll/article?AID=/20080512/LIFE/8051203020080512/LIFE/80512030

217   Website: Case dated 02-21-2003. (2003, February 25). Retrieved June 15, 2003, from National UFO Reporting Center (NUFORC), P. Davenport, director: http://www.nuforc.org/webreports/027/S27831.html

218   Newspaper Article: 'Flying saucers' like family at Gillies, pay second visit. (1957, January/February). *Northern News* (Latchford, Canada).

219   Book: Hobana, I., & Weverbergh, J. (1975). *UFO's from behind the iron curtain*. p. 62. London, United Kingdom: Bantam Books published by arrangement with Souvenir Press Ltd.

220   Newsletter Article: White, J. D. (1976, June). UFO research in Russia. *The A.P.R.O. Bulletin*, Vol. & No. unknown, p. 5.

221   Case Investigation Files: Lutz, J. A. (Case date: 12-29-1972). Odyssey Investigations Club of Baltimore, Report No. AA-1007 dated 11/25/73. See: http://www.waterufo.net

222   Journal Article: Hall, R. H. (1976, November). Lake objects reported in Scandinavia. *The MUFON UFO Journal*, No. 108, p. 11.

223   Newspaper Article: Bakken, R. (1981, January 27). 'A ball of light.' Part of that Friday night sky display hit close to home. *Herald* (Grand Forks, North Dakota).

224   Book: Hall, R. H. (2001). *The UFO evidence: A thirty-year report*. Volume 2. pp. 53-56. Lanham, MD: Scarecrow Press.

225   E-Mail: B. Vike (personal communication, June 02, 2001). Director HBCC UFO Research: http://hbccuforesearch.blogspot.com

226   Newspaper Article: Swedish lake mystery. (1968, April 05). *The Times* (London), p. 7.

227   Newsletter Article: (author unknown). NICAP probes crashed object report: Winter cold prevents search of lake. (1971, February). *The U.F.O. Investigator* [Publication of NICAP], No Vol. or No., p. 1.

228   Book: Clark, J. (1996). High strangeness: UFOs from 1960 through 1979: Wakefield incident. In *The UFO encyclopedia*. Vol. 3. p. 544. Detroit, MI: Omnigraphics.

229   E-mail: B. McNeff (personal communication, October 01, 2008).

230   E-Mail: B. Vike (personal communication, July 31, 2007). Director HBCC UFO Research: http://hbccuforesearch.blogspot.com

231   Website: *Hole in the ice fuels speculation about UFOs*. (2007, February 15). Retrieved February 16, 2007, from IOL (Independent Online (Pty) Ltd): http://www.int.iol.co.za/index. php?set_id=1&click_id=29&art_id=iol11714865971B243

232   Newspaper Article: Vad var det somlandade på Vikerns snöiga is? Iskristaller röjer ett UFO (What was it that landed on the snowy ice of

Vikern? Ice crystals show a UFO!). (1976, March 16). *Karlskoga-Kuriren* (Karlskoga, Sweden).

233   Book: Godfrey, L. S. (2006). *Weird Michigan: Your travel guide to Michigan's local legends and best kept secrets.* p. 67. New York: Sterling Publishing Co.

234   Website: Booth, B. J. (2005, December 19). Original source: *CCCRN NEWS-E-News from the Canadian Crop Circle Research Network,* Retrieved April 15, 2011, from UFO Casebook: http://www. ufocasebook.com/icecircle.html

235   Journal Article: Rubtsov, V. V. (1991, October). Soviet ice ring. *The MUFON UFO Journal,* No. 281, p. 16.

236   Journal Article: Rosenfield, P. (1992, April). Charles River ice ring. *The MUFON UFO Journal,* No. 288, pp. 12-13.

237   Website: Charlton, B. (2004). *Mystery at Mud Lake.* Retrieved April 8, 2011, from Signs of the Times: http://www.sott.net/signs/signs_supplement_ufo2.htm

238   Journal Article: Chalker, B. (1997-98, Winter). Tully saucer nests of 1966: Part one. *International UFO Reporter (IUR),* Vol. 22, No. 4, pp. 14-20. ©CUFOS 1998.

239   Book: Lorenzen, C., & Lorenzen, J. (1969). *UFOs: The whole story.* pp. 163-164. New York: Signet Books [New American Library].

240   Catalog: Hall, R. H. (2007). *From airships to Arnold: A catalogue of UFO reports in the early 20th century (1900-1946).* Fairfax, VA: UFO Research Coalition, p. 16.

241   Transcription: Interview between UFO researcher S. Friedman and witness. (1992). Transcription done by C. Feindt from DVD. *Recollections of Roswell.* Alexandria, VA: Fund for UFO Research.

242   Book: Friedman, S. T., & Marden, K. (2007). *Captured! The Betty and Barney Hill UFO experience.* p. 118. Franklin Lakes, NJ: Career Press.

243   Journal Article: Wood, R. M. (2008, October). McDonnell Douglas studied UFOs in 1960s. *The MUFON UFO Journal,* No. 486, p. 6.

244   Website: Rosales, A. (n.d.). Retrieved December 02, 2005. Humcat 1951-12. Original source: Juan José Benitez, *Flying Saucer Review (FSR),* Vol. 24, No. 2. Archived at UFOINFO: http://ufoinfo.com/humanoid/humanoid1951.shtml

245   Book: Good, T. (2013). *Earth: An alien enterprise: The shocking truth behind the greatest cover-up in human history.* New York: Pegasus Books.

246   Website: *Carl D. Anderson - Biographical.* (1936). Retrieved June 03, 2014, from Nobelprize.org: http://www.nobelprize.org/nobel_prizes/physics/laureates/1936/anderson-bio.html

247   Website: Fraknoi, A. (2007). *How fast are you moving when you are sitting still?* Retrieved June 9, 2014, from The Universe in the Classroom (newsletter of the Astronomical Society of the Pacific), No. 71: http://www.astrosociety.org/edu/publications/tnl/71/howfast.html#2

248   Book: Good, T. (2013). *Earth: An alien enterprise: The shocking truth behind the greatest cover-up in human history.* pp. 87-92. New York: Pegasus Books.

CHAPTER 5: ABDUCTEES—WATER AND BASES

249   Book: Friedman, S. T., & Marden, K. (2007). *Captured! The Betty and Barney Hill UFO experience.* pp. 87, 141. Franklin Lakes, NJ: The Career Press.

250   Ibid., pp. 148-149.

251   Book: Hopkins, B. (1996). *Witnessed: The true story of the Brooklyn Bridge UFO abductions.* p. 4. New York: Pocket Books.

252   Ibid., pp. 20-21.

253   Ibid., p. 40.

254   Book: *Webster's new world dictionary of the American language.* (1976). Second College Edition. Englewood Cliffs, NJ: William Collins & World Publishing, Prentice-Hall.

255   Book: Fort, C. (1974). *The complete books of Charles Fort.* p, 571. Mineola, NY: Dover Publications.

256   Book: Fowler, R. E. (1993). *The Allagash abductions: Undeniable evidence of alien intervention.* p. 24. Tigard, Oregon: Wild Flower Press.

257   Ibid., p. 42.

258   Ibid., p. 74.

259   Ibid., p. 93.

260   Journal Article: Mesnard, J. (2009, July). A failed abduction at Chaville? *International UFO Reporter (IUR)*, Vol. 32, No. 3, p. 21. Originally appeared in *Lumières dans la Nuit*, No. 380, pp. 4-7.

261   Book: Webb, W. N. (1994). *Encounter at Buff Ledge: A UFO case history.* p. 10, Chicago: CUFOS, J. Allen Hynek Center for UFO Studies.

262   Book: Collings, B., & Jamerson, A. (1996). *Connections: Solving our alien abduction mystery.* p. 119. Newberg, Oregon: Wildflower Press.

263   Book: Hill, P. R. (1995). *Unconventional flying objects: A scientific analysis* p. 219. Charlottesville, VA: Hampton Roads Publishing. © 1995 by Julie M. Hill.

264   Ibid., p. 220.

265   Book: Collings, B., & Jamerson, A. (1996). *Connections: Solving our alien abduction mystery.* p. 313. Newberg, Oregon: Wildflower Press.

266   Website: Wilson, K. (1996). *Some of them were military: Abduction step-by-step.* Retrieved September 23, 2009, from The Alien Jigsaw, MILABS: Project Open Mind, Part Seven: http://www.alienjigsaw.com/Milabs/pom7.html

267   E-Mail: K. Wilson (personal communication, date unknown).

268   Book: Wilson, K. (2007). *I forgot what I wasn't supposed to remember: An expanded view of the alien abduction phenomenon* (first draft). p. 217. Portland, Oregon: Puzzle Publishing.

269   Book: Sanchez-Ocejo, V., & Stevens, W. C. (1982). *UFO contact from undersea: A report of the investigation.* Editor's note, p. 67. Tucson, AZ: Privately published by W. C. Stevens.

270   Ibid.

271   Ibid.

272   Book: Fowler, R. E. (1994). *The Andreasson affair-phase two: The continuing investigation of a woman's abduction by alien beings.* p. 90. Mill Spring, NC: Wild Flower Press.

273   Book: Walters, E., & Walters, F., (1990). *The Gulf Breeze Sightings,* p. 156. New York: Avon Books.

274   Ibid., pp. 267-268.

275   Book: Bord, J., & Bord, C. (1989). *Unexplained mysteries of the 20th century.* p. 171. Chicago: Contemporary Books.

276   Magazine Article: Creighton, G. (1975, January-February). Underwater UFO base off Venezuela? *Flying Saucer Review (FSR)*, Vol. 21, No. 1, p. 11. http://www.fsr.org.uk

277   Book: Sanchez-Ocejo, V., & Stevens, W. C. (1982). *UFO contact from undersea: A report of the investigation.* p. 178. Tucson, AZ: Privately published by W. C. Stevens.

278   Catalog: Bianchini, M. (2003). *USOCAT: The Italian catalog of sightings of unidentified submerged objects.* Turin, Italy: Edizioni UPIAR. Case from updated, unpublished 2007 catalog, p. 53. The Italian Center for Ufological Studies, (CISU): http://www.cisu.org

279   Book: Hesemann, M. (1998). *UFOs: The secret history.* pp. 415-416. New York: Marlowe.

280   Catalog: Bianchini, M. (2003). *USOCAT: The Italian catalog of sightings of unidentified submerged objects.* Turin, Italy: Edizioni UPIAR. Case from updated, unpublished 2007 catalog, pp. 95-96. The Italian Center for Ufological Studies, (CISU): http://www.cisu.org

281   Website: Trainor, J. (1999, February 22). *UFOs videotaped in the skies of Peru.* Retrieved February 01, 2000, from UFO Roundup, Vol. 4, No. 8. Now archived at UFOINFO: http://www.ufoinfo.com/roundup/v04/rnd04_08.shtml

282   Book: Sanchez-Ocejo, V., & Stevens, W. C. (1982). *UFO contact from undersea: A report of the investigation.* Editor's note, p. 67. Tucson, AZ: Privately published by W. C. Stevens.

283   Book: Sauder, R. (1996). *Underground bases and tunnels: What is the government trying to hide?* Kempton, IL: Adventures Unlimited Press.

284   Book: Sauder, R. (2001). *Underwater and underground bases: Surprising facts the government does not want you to know!* Kempton, IL: Adventures Unlimited Press.

285   Ibid., p. 184.

286   Book: Fowler, R. E. (1994). *The Andreasson affair-phase two: The continuing investigation of a woman's abduction by alien beings.* pp. 90-91. Mill Spring, NC: Wild Flower Press.

287   Book: Blundell, N., & Boar, R. (1983). *The world's greatest UFO mysteries.* pp. 67-68. London, United Kingdom: Octopus Books Ltd.

288   Book: Sanchez-Ocejo, V., & Stevens, W. C. (1982). *UFO contact from undersea: A report of the investigation*. pp. 151, 153. Tucson, AZ: Privately published by W. C. Stevens.

289   Ibid.

290   Book: Collings, B., & Jamerson, A. (1996). *Connections: Solving our alien abduction mystery*. p. 199. Newberg, Oregon: Wildflower Press.

291   Book: Wilson, K. (1993). *The alien jigsaw*. pp. 229-230. Portland, Oregon: Puzzle Publishing.

292   E-Mail: K. Wilson (personal communication, date unknown).

293   Book: Cramp, L. G. (1966). *UFOs and anti-gravity: Piece for a jig-saw*. Kempton, IL: Adventures Unlimited Press.

CHAPTER 6: RADAR AND SONAR

294   Book: *The new encyclopædia Britannica*. (1986). Vol. 9. p. 882. Chicago: Encyclopedia Britannica.

295   Newsletter Article: (author unknown). Casebook. (1970, October). *The U.F.O. Investigator* [Publication of NICAP], No Vol. or No., p. 3.

296   E-Mail: V. Sanchez-Ocejo (personal communication, November 08, 2004).

297   Book: Keyhoe, D. E. (1960). *Flying saucers: Top secret*. pp. 121-122. New York: G.P. Putnam's Sons.

298   Book: Beckley, T. G. (1992). *Strange encounters: Bizarre & eerie contact with UFO occupants*. pp. 86-87. New Brunswick, NJ: Inner Light – Global Communications.

299   Magazine Article: Mystery at sea. (1964, March-April). *Flying Saucer Review (FSR)*, Vol. 10, No. 2, p. 22. http://www.fsr.org.uk

300   Book: Teets, B. (1995). *West Virginia UFOs: Close encounters in the mountain state*. pp. 137-138. Terra Alta, WV: Headline Books.

301   Newsletter Article: (author unknown). New sightings put AF on spot. (1965, March-April). *The U.F.O. Investigator* [Publication of NICAP], Vol. 3, No. 1, p. 4.

302   Book: Sanderson, I. T. (1970). *Invisible residents*. p. 228. Toronto, Canada: Fitzhenry & Whiteside Ltd. Text included was from UFOCAT URN 76800 citing Sanderson's reference as *Rochester* (Minn.) *Post-Bulletin*, 05 August 1965.

303   Newsletter Article: (author unknown). (1967, January-February). On the international scene. *The A.P.R.O. Bulletin*, Vol. & No. unknown, p. 7.

304   Case Investigation Files: Feindt, C. (Case date: ??-??-1970). AO-97 USS *Allagash* and the UFO. See: http://www.waterufo.net

305   Newsletter Article: Todd, D. R. (1978, May). Ship's crew sees UFO. *The A.P.R.O. Bulletin,* Vol. 26, No. 11, pp. 1-2.

306   Catalog: Bianchini, M. (2003). *USOCAT: The Italian catalog of sightings of unidentified submerged objects.* Turin, Italy: Edizioni UPIAR. Case from updated, unpublished 2007 catalog, pp. 60-61. The Italian Center for Ufological Studies, (CISU): http://www.cisu.org

307   Newsletter Article: (author unknown). The strange observation of the M. V. *Dolphin*: New object appears. (1977, June). *The U.F.O. Investigator* [Publication of NICAP], No Vol. or No., pp. 1, 3.

308   Catalog: Bianchini, M. (2003). *USOCAT: The Italian catalog of sightings of unidentified submerged objects.* Turin, Italy: Edizioni UPIAR. Case from updated, unpublished 2007 catalog, pp. 74-76. The Italian Center for Ufological Studies, (CISU): http://www.cisu.org

309   Book: Swords, M. D. (2005). *Grass roots UFOs: Case reports from the Timmerman files.* pp. 120-121. Lima, OH: CSS Publishing. Thanks to the Fund for UFO Research: http://www.fufor.com for permission to use.

310   Newspaper Article: The appearance of flying saucers to increase. (1980, August 17). Translated by Joe Brill from *Tribuna do Noste* (Natal, Rio Grande Do Norte, Brazil).

311   Magazine Article: Dodd, T. (n.d.). Confrontations in the North Atlantic. *UFO Magazine* (UK), No other information. Website: http://worldgathering. net/times/atlantic.htm

312   Website: *An encounter at sea.* (2004). Retrieved December 16, 2004, from Cosmic Conspiracies: http://www.ufosaliens.pwp.blueyonder. co.uk/ufosea.htm

313   Website: Case dated 05-23-1968. (2007, February 01). Retrieved November 03, 2007, from National UFO Reporting Center (NUFORC), P. Davenport, director: http://www.nuforc.org/webreports/054/S54893. html

314   Book: Stevens, W. C., & Dong, P. (1983). *UFOs over modern China: A survey of the phenomenon*. pp. 48-49. Tucson, AZ: UFO Photo Archives.

315   Journal Article: Chalker, B. (1985, September-October). An extraordinary UFO incident off Chile. *International UFO Reporter (IUR)*, pp. 4-6.

316   Book: Stonehill, P. (1998). *The Soviet UFO files*. p. 73. Surry, England: Quadrillion Publishing Ltd.

317   Catalog: Bianchini, M. (2003). *USOCAT: The Italian catalog of sightings of unidentified submerged objects*. Turin, Italy: Edizioni UPIAR. Case from updated, unpublished 2007 catalog, p. 73. The Italian Center for Ufological Studies, (CISU): http://www.cisu.org

318   Ibid., pp. 76-77.

319   Newspaper Article: Stewardson, J. (1979, April 12). Fishermen land with tale of possible close encounter. *Standard-Times* (New Bedford, Massachusetts).

320   Newsletter Article: Gribble, B. (n.d.). Pilot sightings and radar trackings. *The A.P.R.O. Bulletin,* Vol. 31, No. 8, pp. 3-4.

321   Website: Filer, G. (2001, October 25). *Chilean chief of naval operations says, "UFOs are real."* Retrieved October 25, 2001, from http://www.filersfiles.com and now archived at UFOINFO: http://www.ufoinfo.com/filer/2001/ff_0143.shtml

322   Magazine Article: Farish, L., & Titler, D. M. (1977). Strange underwater wheels of light. *Bermuda Triangle Special Report 1977*, compiled by the editors of *Saga Magazine & UFO Report*, p. 68.

323   Ibid.

324   Book: *The new encyclopædia Britannica*. (1986). Vol. 11. p. 9. Chicago: Encyclopedia Britannica.

325   Case Investigation Files: Feindt, C. (Case date: 01~08-??-1947). No title. See: http://www.waterufo.net

326   Case Investigation Files: Feindt, C. (Case date: 04?-??-1952). No title. See: http://www.waterufo.net

327   Journal Article: Brænne, O. J. (1995, January-February). Observations of unidentified submarine objects in Norway. *International UFO Reporter (IUR)*, p. 12.

328   Magazine Article: Argentina: The Wily Whatzit? (1960, February 22). *Newsweek*, Vol. & No. unknown, p. 57.

329   Website: Gottschall, S. (2003). *Third Australian national UFO conference review*. Retrieved June 29, 2005, from UFO Research Queensland: http://www.uforq.asn.au/articles/conference03.html

330   Book: Sanderson, I. T. (1970). *Invisible residents.* pp. 20-23. Toronto, Canada: Fitzhenry & Whiteside Ltd.

331   Journal Article: (author unknown). (1972, Autumn). USOs. *INFO Journal,* a publication of the International Fortean Organization, Vol. 3, No. 1, p. 38.

332   Website: *Pacific Ocean*. (1998, January 12). Retrieved November 11, 1999, from MUFON UFO Reports: http://mufon.com/news/cases1_12_98_2. html which is inactive. See WATER UFO: http://www.waterufo.net/ item.php?id=338

333   Journal Article: Brænne, O. J. (1995, January-February). Observations of unidentified submarine objects in Norway. *International UFO Reporter (IUR)*, p. 13.

334   Catalog: Bianchini, M. (2003). *USOCAT: The Italian catalog of sightings of unidentified submerged objects*. Turin, Italy: Edizioni UPIAR. Case from updated, unpublished 2007 catalog, pp. 73-74. The Italian Center for Ufological Studies, (CISU): http://www.cisu.org

335   Magazine Article: (1996, August). *Revelacion Magazine*. No other information available. Translated by Albert S. Rosales: http://www. iraap.org/rosales

336   Journal Article: Webb, W. N. (1984, November). Radar/Sonar contact. *The MUFON UFO Journal*, No. 199, pp. 7, 8, 10.

337   Book: Cooper, M. W. (1991). *Behold a pale horse.* pp. 17-21. Sedona, AZ: Light Technology Publications. Thanks to Doyel Shamley and Rob Houghton of: http://www.hourofthetime.com for permission to use in this book.

338   Newsletter Article: Todd, D. R. (1978, May). Ship's crew sees UFO. *The A.P.R.O. Bulletin,* Vol. 26, No. 11, pp. 1-2.

339   Newsletter Article: (author unknown). NICAP studies underwater UFO case: Search for other witnesses continues. (1971, January). *The U.F.O. Investigator* [Publication of NICAP], No Vol. or No., p. 3.

340   Magazine Article: (1996, August). *Revelacion Magazine*. No other information available. Translated by Albert S. Rosales: http://www. iraap.org/rosales

CHAPTER 7: AIRCRAFT CARRIERS AND OTHER LARGE SHIPS

341   Book: Chester, K. (2007). *Strange company: Military encounters with UFOs in World War II*. San Antonio, TX: Anomalist Books.

342   Journal Article: Cerny, P. C., & Neville, R. (1983, July). U.S. Navy 1942 sighting. *The MUFON UFO Journal*, No. 185, pp. 14-15.

343   Book: Lorenzen, C., & Lorenzen, J. (1969). *UFOs: The whole story.* p. 255. New York: Signet Books [New American Library].

344   Book: Keyhoe, D. E. (1960). *Flying saucers: Top secret.* pp. 262-263. New York: G.P. Putnam's Sons.

345   Website: *USS Franklin D. Roosevelt.* (2007). Retrieved July 25, 2007, from MUFON of Ohio: http://home.columbus.rr.com/threeemusic/mufono/USS_FDR.html which is inactive. See WATER UFO: http: // www.waterufo.net/item.php?id=1078

346   Website: *USS Franklin D. Roosevelt.* (2007). Retrieved July 26, 2007, from MUFON of Ohio: http://home.columbus.rr.com/threeemusic/mufono/USS_FDR.html which is inactive. See WATER UFO: http:// www.waterufo.net/item.php?id=1081

347   Newspaper Article: Baughman, J. (1999, December 01). A moment in (recent) history: UFOs shadowed U.S. aircraft carrier. *Broad Top Bulletin* (Saxton, PA).

348   Website: Puckett, W. (2007). *Glowing orange object buzzes U.S.S. Lexington – Mid-1960s.* Retrieved July 05, 2007, from UFOs Northwest: http://www.ufosnw.com/sighting_reports/older/uslexsight/uslexsight.htm

349   Newspaper Article: Baughman, J. (2000, December 26). A moment in (recent) history: Chet Grusinski: Witness confirms UFO sighting aboard USS *FDR*, 1958: Letter from Jordan. *Broad Top Bulletin* (Saxton, PA).

350   Website: Case dated SM~FF-??-1965. (2007, February 24). Retrieved July 06, 2007, from National UFO Reporting Center (NUFORC), P. Davenport, director: http://www.nuforc.org/webreports/055/S55295.html

351   Website: Rense, J. (2000, November 03). *U.S. Naval officer's UFO experience.* Retrieved July 29, 2007, from: http://www.rense.com/general5/ufpo.htm

352   Website: Penre, W. (2005, May 05). *UFO sighting from U.S.S.* Constellation *aircraft carrier, Spring-Summer 1994.* Retrieved June 15, 2007, from Aliens & UFOs, Top Secret: http://www.illuminati-news.com/ufos-and-aliens/html/u.s.s.-constellation-aircraft-carrier.htm

353   E-Mail: B. Vike (personal communication, April 09, 2002). Director HBCC UFO Research: http://hbccuforesearch.blogspot.com

354   Website: Dingle, P. (n.d.). USS Yorktown *and the UFO.* Retrieved July 28, 2007, from Life of a Radarman on the USS *Yorktown*: http://www.yorktownsailor.com/yorktown/dingle.htm

355   Website: Case dated 10-02-2007. (2007, October 08). Retrieved November 08, 2007, from National UFO Reporting Center (NUFORC), P. Davenport, director: http://www.nuforc.org/webreports/059/S59141.html

356   Website: Filer, G. (2007, May 30). *U.S. carrier* Roosevelt *spots three UFOs.* Retrieved June 24, 2007, from unknown URL, but now archived at UFOINFO: http://www.ufoinfo.com/filer/2007/ff0722.shtml

357   Website: Ridge, F. (n.d.). *U.S. aircraft carrier stopped by UFO.* Retrieved July 28, 2007, from National Investigations Committee on Aerial Phenomena (NICAP): http://www.nicap.org/jfk.htm

358   Case Investigation Files: Feindt, C. (Case date: WW-??-1979). No title. See: http://www.waterufo.net

359   Case Investigation Files: Feindt, C. (Case date: 09~10-??-1974). No title. See: http://www.waterufo.net

360   E-Mail: D. Ledger (personal communication, February 19, 2008).

CHAPTER 8: SECRECY ABOUT UFOs

361   Book: Hall, R. H. (2001). *The UFO evidence: A thirty-year report.* Volume 2. pp. 94-95. Lanham, MD: Scarecrow Press.

362   Journal Article: Cerny, P. C., & Neville, R. (1983, July). U.S. Navy 1942 sighting. *The MUFON UFO Journal*, No. 185, pp. 14-15.

363   Newsletter Article: (author unknown). Casebook. (1970, October). *The U.F.O. Investigator* [Publication of NICAP], No Vol. or No., p. 3.

364   Website: Creighton, G. (2004). *UFOs seen by crew of an American aircraft carrier (1952-1958).* Retrieved April 07, 2009, from http://www.fsr.org.uk/fsrart4.htm

365   Website: Goudie, D. (n.d.). *JANAP 146 (C) communication instructions for reporting vital intelligence sightings from airborne and waterborne sources - 10 March 1954*. Retrieved November 20, 2009, from the Computer UFO Network (CUFON): http://www.cufon.org/cufon/janp146c.htm

366   Website: Feindt, C. (n.d.). *Blue book UFO reports by ships at sea*. Retrieved December 20, 2002, from: http://www.waterufo.net/bluebook/bbpdf.pdf

367   Website: *USS Franklin D. Roosevelt*. (2007). Retrieved July 26, 2007, from MUFON of Ohio: http://home.columbus.rr.com/threeemusic/mufono/USS_FDR.html which is inactive. See WATER UFO: http://www.waterufo.net/item.php?id=1081

368   Newspaper Article: Baughman, J. (1999, December 01). A moment in (recent) history: UFOs shadowed U.S. aircraft carrier. *Broad Top Bulletin* (Saxton, PA).

369   Website: Puckett, W. (2007). *Glowing orange object buzzes U.S.S. Lexington – Mid-1960s*. Retrieved July 05, 2007, from UFOs Northwest: http://www.ufosnw.com/sighting_reports/older/uslexsight/uslexsight.htm

370   Newspaper Article: Baughman, J. (2000, December 26). A moment in (recent) history: Chet Grusinski: Witness confirms UFO sighting aboard USS *FDR*, 1958: Letter from Jordan. *Broad Top Bulletin* (Saxton, PA).

371   Newsletter Article: (author unknown). New sightings put AF on spot. (1965, March-April). *The U.F.O. Investigator* [Publication of NICAP], Vol. 3, No. 1, p. 4.

372   Website: Case dated SM~FF-??-1965. (2007, February 24). Retrieved July 06, 2007, from National UFO Reporting Center (NUFORC), P. Davenport, director: http://www.nuforc.org/webreports/055/S55295.html

373   Book: Cooper, M. W. (1991). *Behold a pale horse*. pp. 17-21. Sedona, AZ: Light Technology Publications. Thanks to Doyel Shamley & Rob Houghton of: http://www.hourofthetime.com for permission to use in this book.

374   Book: Swords, M. D. (2005). *Grass roots UFOs: Case reports from the Timmerman files*. p. 30. Lima, OH: CSS Publishing. Thanks to the Fund for UFO Research: http://www.fufor.com for permission to use.

375   Website: Ridge, F. (n.d.). *U.S. aircraft carrier stopped by UFO.* Retrieved July 28, 2007, from National Investigations Committee on Aerial Phenomena (NICAP): http://www.nicap.org/jfk.htm

376   Case Investigation Files: Feindt, C. (Case date: 09~10-??-1974). No title. See: http://www.waterufo.net

377   E-Mail: N. Burns (personal communication, February 08, 2009). © 2008. http://www.trueghosttales.com. All rights reserved, republished here with permission.

378   Newsletter Article: Todd, D. R. (1978, May). Ship's crew sees UFO. *The A.P.R.O. Bulletin,* Vol. 26, No. 11, pp. 1-2.

379   Book: Swords, M. D. (2005). *Grass roots UFOs: Case reports from the Timmerman files.* pp. 120-121. Lima, OH: CSS Publishing. Thanks to the Fund for UFO Research: http://www.fufor.com for permission to use.

380   Case Investigation Files: Feindt, C. (Case date: WW-??-1979). No title. See: http://www.waterufo.net

381   Website: Greenewald, J., Jr. (2007, April 25). *USS* Midway, *USS* Tuscaloosa *UFO encounter.* Retrieved July 12, 2008, from The Black Vault: http://www.theblackvault.com/ftopic-62192-next.html and available at WATER UFO: http://www.waterufo.net/item.php?id=1141

382   Website: Penre, W. (2005, May 05). *UFO sighting from U.S.S.* Constellation *aircraft carrier, spring-summer 1994.* Retrieved June 15, 2007, from Aliens & UFOs, Top Secret: http://www.illuminati-news.com/ufos-and-aliens/html/u.s.s.constellation-aircraft-carrier.htm

383   Newsletter Article: (author unknown). A British naval sighting. (1956, 2nd quarter). *Australian Saucer Record,* Vol. 2, No. 2, pp. 20-21. Copy on file with the Center for UFO Studies (CUFOS): http://www.cufos.org

384   Book: Lorenzen, C. E. (1970). *The shadow of the unknown.* p. 84. New York: Signet Books.

385   Journal Article: (author unknown). (2005, February-March). Seeing is believing: HMAS *Voyager* incident (1962) verification, reported January 2005. *UFO Encounter,* No. 222, Journal of UFO Research Queensland, Australia: http://www.uforq.asn.au Copy on file with the Center for UFO Studies (CUFOS): http://www.cufos.org

386   Website: Gottschall, S. (2003). *Third Australian national UFO conference review.* Retrieved June 29, 2005, from UFO Research Queensland: http://www.uforq.asn.au/articles/conference03.html

387   Journal Article: Webb, W. N. (1984, November). Radar/Sonar contact. *The MUFON UFO Journal*, No. 199, p. 9.

388   Journal Article: Chalker, B. (1997-98, Winter). Tully saucer nests of 1966: Part one. *International UFO Reporter (IUR)*, Vol. 22, No. 4, p. 19. ©CUFOS 1998.

389   Book: Swords, M. D. (2005). *Grass roots UFOs: Case reports from the Timmerman files*. p. 91. Lima, OH: CSS Publishing. Thanks to the Fund for UFO Research: http://www.fufor.com for permission to use.

390   Book: Vallee, J. (1992). *UFO chronicles of the Soviet Union: A cosmic samizdat*. pp. 29-30. New York: Ballantine Books.

391   Book: Good, T. (1991). *Alien contact: Top-secret UFO files revealed*. pp. 106-107. New York: William Morrow & Co.

392   Website: Svahn, C. (n.d.). *Swedish military search for fallen UFO in lake.* Translated by Eileen Fletcher. Retrieved June 21, 2006, from UFO-Sweden: http://www.ufo.se/english/news/index.html

393   Magazine Article: Handler, E. H. (1964, September). The flying submarine. *Proceedings*. Vol. & No. unknown, pp. 144-146. Copyright 1964 by the U.S. Naval Institute: www.usni.org

CHAPTER 9: FISH AND ANIMAL REACTIONS TO UFOs

394   E-Mail: J. Woodward (personal communication, August 30 through September 4, 2006).

395   Book: Vallee, J., & Aubeck, C. (2009). *Wonders in the sky*. p. 217. New York: Tarcher/Penguin.

396   Book: Heyerdahl, T. (1950). *Kon-Tiki*. p. 118. New York: Permabook edition, published by arrangement with Rand McNally & Co.

397   Book: Stringfield, L. H. (1977). *Situation red: The UFO siege!* p. 108. Garden City, NY: Doubleday & Co.

398   Magazine Article: (1975, July-August). *Flying Saucer Review (FSR)*. Vol. 21, No. 4, p. 22. Translated from Philippe Piet van Putten's *Fenomenos Aeroespaciais (Aerospatial Phenomena)* by Gordon Creighton.

399   Newsletter Article: (author unknown). (1975, January-February). UFOs over Arizona. *The A.P.R.O. Bulletin,* Vol. 23, No. 4, pp. 3-4.

400   Book: Corrales, S. (1998). *Flashpoint: High strangeness in Puerto Rico*. pp. 68-69. Maulden, Beds, United Kingdom: Amarna Ltd.

401   Website: *Cylinders hovers above fisherman in the ocean, New Jersey, USA.* (2006). Retrieved August 26, 2006, from UFOs at Close Sight: http://www.ufologie.net/htm/2001aug.htm Original source is MUFON CMS case #1709 submitted June 9, 2002, to: http://mufon.com/mufonreports.htm

402   E-Mail: J. Woodward (personal communication, August 30 through September 04, 2006).

403   Magazine Article: Keel, J. A. (1970, November). Ocean-based UFOs ring the U.S. *Male*, Vol. 20, No. 12, p. 48.

404   Magazine Article: Ghana: UFO 2 nights running. (1958, September-October). *Flying Saucer Review (FSR)*, Vol. 4, No. 5, p. 6. http://www.fsr.org.uk

405   Letter: Oswald, J. P. (1969, November 03). To NICAP investigator Raymond Fowler. Copy on file with the Center for UFO Studies (CUFOS): http://www.cufos.org

406   Magazine Article: Malthaner, H. (1972, July-August). Mystery flying object rolls along a German road. *Flying Saucer Review (FSR)*, Vol. 18, No. 4, pp. 16-17. http://www.fsr.org.uk

407   Website: Vike, B. (n.d.). *Zanesville, Ohio unseen huge craft rises out of the water*. Retrieved September 04, 2006, from HBCC UFO Research: http://www.hbccufo.org and now available at UFOINFO: http://www.ufoinfo.com/sightings/usa/990500.shtml

408   Newsletter Article: Svahn, C., & Liljegren. A. (1984, January-December). The Kölmjärv ghost rocket crash revisited. *AFU Newsletter,* No. 27, pp. 1-5. AFU's website: http://www.afu.info

409   Journal Article: (author unknown). (1961, March). UFO-Pilots in Cabo Frio (Brazil). *UFO Nachrichten (News)*, Vol., No., & page unknown. Thanks to Anders Liljegren of Archives for UFO Research (AFU) for supplying the text to me: http://www.afu.info

410   Newspaper Article: Flying saucers reported seen in North Carolina July of last year. (1966, March 30). *Cherryville Eagle* (North Carolina).

411   Journal Article: Chalker, B. (1997-98, Winter). Tully saucer nests of 1966: Part one. *International UFO Reporter (IUR)*, Vol. 22, No. 4, pp. 14-20. ©CUFOS 1998.

412   Newspaper Article: Mysterious unidentified object sighted in south country. (1967, January 19). *Phillips County News* (Malta, Montana).

413   Document: NICAP report by Wayne G. Martiny, dated February 23, 1967. Copy on file with the Center for UFO Studies (CUFOS): http://www.cufos.org

414   Document: Notes taken by Mr. Gordon Lore in a telephone interview with witness, Mrs. Weston. (n.d.). Thanks to the Center for UFO Studies (CUFOS): http://www.cufos.org

415   Book: Lorenzen, C., & Lorenzen, J. (1976). *Encounters with UFO occupants*. pp. 337-339. New York: Berkley Publishing Corporation.

416   Catalog: Bianchini, M. (2003). *USOCAT: The Italian catalog of sightings of unidentified submerged objects*. Turin, Italy: Edizioni UPIAR. Case from updated, unpublished 2007 catalog, p. 54. The Italian Center for Ufological Studies, (CISU): http://www.cisu.org

417   Ibid., pp. 86-87.

418   Book: Stringfield, L. H. (1977). *Situation red: The UFO siege!* pp. 52-53. Garden City, NY: Doubleday & Company.

419   Website: Case dated 08-15-1971. (2003, January 26). Retrieved November 04, 2007, from National UFO Reporting Center (NUFORC), P. Davenport, director: http://www.nuforc.org/webreports/023/S23213.html

420   Book: Rutkowski, C. (1989). *Visitations? Manitoba UFO experiences*. p. 19. Winnipeg, MB, Canada: Gateway Publishing.

421   E-Mail: J. Woodward (personal communication, August 30 through September 04, 2006).

422   Website: Campbell, E. (n.d.). *Lightning and the diver*. Retrieved June 03, 2009, from Scubadoc's Diving Medicine: http://scuba-doc.com/lightdive.htm

423   Newspaper Article: An electric monster. (1893, July 03). *The Tacoma Daily Ledger* (Washington).

424   E-Mail: R. H. Hall (personal communication, January 19, 2007).

425   Journal Article: Brænne, O. J. (1995, January-February). Observations of unidentified submarine objects in Norway. *International UFO Reporter (IUR)*, p. 12.

426   Newspaper Article: Flying saucers reported seen in North Carolina July of last year. (1966, March 30). *Cherryville Eagle* (North Carolina).

427   Book: Beckley, T. G. (1992). *Strange encounters: Bizarre & eerie contact with UFO occupants*. pp. 86-87. New Brunswick, NJ: Inner Light - Global Communications.

428   Newsletter Article: (author unknown). (1967, September-October). UAO dives into sea at Nova Scotia. *The A.P.R.O. Bulletin,* Vol. & No. unknown, p.7.

429   Magazine Article: Wolkomir, R. (1968, November). The glowing "thing" in Moore Lake. *Fate*, Vol. 21, No. 11, pp. 32-36.

430   Website: Case dated 07-05-1969. (2003, October 31). Retrieved November 03, 2007, from National UFO Reporting Center (NUFORC), P. Davenport, director: http://www.nuforc.org/webreports/032/S32366. html

431   Book: Stringfield, L. H. (1977). *Situation red: The UFO siege!* p. 53. Garden City, NY: Doubleday & Company.

432   Magazine Article: Chionetti, A. H. (n.d.). A very important case: The accident of Lake Lacar (Argentina). *Lumières dans la Nuit*, Issue 241, pp. 10-14. Translated by Henri Julien.

433   Website: Case dated 04-15-1985. (2003, April 22). Retrieved November 04, 2007, from National UFO Reporting Center (NUFORC), P. Davenport, director: http://www.nuforc.org/webreports/028/S28494.html

434   Website: Case dated 06-09-2004. (2004, June 18). Retrieved June 26, 2004, from National UFO Reporting Center (NUFORC), P. Davenport, director: http://www.nuforc.org/webreports/037/S37494.html

435   Website: Trainor, J. (1998, January 11). *UFOs sighted in Rockland, Massachusetts*. Retrieved February 01, 2000, from UFO Roundup, Vol. 3, No. 2. Now available at UFOINFO: http://www.ufoinfo.com/roundup/v03/rnd03_02.shtml

436   Magazine Article: (1975, July-August). *Flying Saucer Review (FSR)*. Vol. 21, No. 4, p. 22. Translated from Philippe Piet van Putten's *Fenomenos Aeroespaciais (Aerospatial Phenomena)* by Gordon Creighton.

CHAPTER 10: NEGATIVE ASPECTS OF WATER UFOLOGY

437   Newspaper Article: Puzzle picture from seabed. (1964, December 05). *New Zealand Herald* (Auckland, New Zealand).

438   Magazine Article: Steiger, B. & Whritenour, J. (1968, June). Unidentified underwater saucers. *Saga Magazine,* Vol. & No. unknown, pp. 36-37.

439   Book: Cathie, B. (1968). *Harmonic 33*. p. 32. Wellington, Sydney, London: A. H. and A. W. Reed.

440   Book: Pauwels, L., & Bergier, J. (1960-original French). *Le matin des magiciens [Morning of the magicians]*. New York: Avon (1968-in English).

441   Journal Article: (author unknown). (1970, Winter). UFOs and USOs. *INFO Journal*, a publication of the International Fortean Organization, Vol. 11, No. 3, p. 14.

442   Website: Hatch, L. (2003, October 22). *The Eltanin antenna identified.* Archived at: http://web.archive.org/web/20060717153647/http://www.larryhatch.net/ELTANIN.html

443   Journal Article: Hendrickson, R. (1987, March). Marine light wheels. *The MUFON UFO Journal*, No. 227, p. 6.

444   Book: Wilkins, H. T. (1954). *Flying saucers on the attack.* p. 78. New York: Citadel Press.

445   Book: Corliss, W. R. (2001). *Remarkable luminous phenomena in nature: A catalog of geophysical anomalies.* p. 346. Glen Arm, MD: The Sourcebook Project.

446   Book: Fort, C. (1974). *The complete books of Charles Fort.* p. 278. Mineola, NY: Dover Publications.

447   Ibid., pp. 276-277.

448   Ibid., p. 277.

449   Magazine Article: Farish, L., & Titler, D. M. (1977). Strange underwater wheels of light. *Bermuda Triangle Special Report 1977*, compiled by the editors of *Saga Magazine & UFO Report*, p. 66.

450   Ibid.

451   Ibid.

452   Ibid., p. 22

453   Magazine Article: Bodler, J. R. (1952, September). An unexplained phenomenon of the sea. (U.S. Naval Institute) *Proceedings.* Vol. 78, No. 66, pp. 66-67.

454   Book: Sanderson, I. T. (1970). *Invisible residents.* p. 99. Toronto, Canada: Fitzhenry & Whiteside Ltd.

455   Newspaper Article: A shower of meteors around the St. Andrew. (1906, November 05). *New York Times*, p. 1.

456   E-Mail: C. Aubeck (personal communication, n.d.).

457   Newspaper Article: Balloon 'saucer' mystifies Morro Bay area residents. (1969, July 26). *Fresno Bee* (California).

458   Catalog: Bianchini, M. (2003). *USOCAT: The Italian catalog of sightings of unidentified submerged objects.* Turin, Italy: Edizioni UPIAR. Case from updated, unpublished 2007 catalog, p. 62. The Italian Center for Ufological Studies, (CISU): http://www.cisu.org

459   Newspaper Article: Sailor says he saw UFO with yellow pilots floating on the sea. (1968, September 23). Unknown newspaper from Spain & translated into English by unknown person. Thanks to the J. Allen Hynek Center for UFO Studies (CUFOS): http://www.cufos.org

460   E-Mail: V.-J. Ballester Olmos (personal communication, September 16, 2001).

461   Catalog: Ballester Olmos, V.-J. (1976, April). *A catalogue of 200 type-1 UFO events in Spain and Portugal.* Case No. 114, p. 26. Evanston, IL: Center for UFO Studies (CUFOS).

462   E-Mail: V.-J. Ballester Olmos (personal communication, September 16, 2001).

463   Catalog: Bianchini, M. (2003). *USOCAT: The Italian catalog of sightings of unidentified submerged objects.* Turin, Italy: Edizioni UPIAR. Case from updated, unpublished 2007 catalog, p. 115. The Italian Center for Ufological Studies, (CISU): http://www.cisu.org

464   Ibid., p. 50.

465   Ibid., p. 89.

466   Ibid., pp. 82-83.

467   Ibid., p. 59.

468   Ibid., p. 107.

469   Ibid., p. 78.

470   Ibid., pp. 37-38.

471   Ibid., pp. 17-18.

472   Ibid., pp. 84-85.

473   Ibid., pp. 42-43.

474   Ibid., p. 57.

475   Ibid., pp. 89-90.

476   Ibid., p. 118.

477   Ibid., pp. 64-65.

478   Ibid., pp. 105-106.

479   Book: Clark, J. (1992). The emergence of a phenomenon: UFOs from the beginning through 1959. In *The UFO encyclopedia.* Vol. 2. p. 244. Detroit, MI: Omnigraphics.

480   Catalog: Bianchini, M. (2003). *USOCAT: The Italian catalog of sightings of unidentified submerged objects*. Turin, Italy: Edizioni UPIAR. Case from updated, unpublished 2007 catalog, p. 16. The Italian Center for Ufological Studies, (CISU): http://www.cisu.org

481   Book: Clark, J. (1992). The emergence of a phenomenon: UFOs from the beginning through 1959. In *The UFO encyclopedia*. Vol. 2. p. 194. Detroit, MI: Omnigraphics.

482   Catalog: Bianchini, M. (2003). *USOCAT: The Italian catalog of sightings of unidentified submerged objects*. Turin, Italy: Edizioni UPIAR. Case from updated, unpublished 2007 catalog, p. 31. The Italian Center for Ufological Studies, (CISU): http://www.cisu.org

483   Newspaper Article: Mates, R. (1999, November 07). Chronicles of a UFO sighting. *The Sunday Times* (Scranton, Pennsylvania).

484   Catalog: Bianchini, M. (2003). *USOCAT: The Italian catalog of sightings of unidentified submerged objects*. Turin, Italy: Edizioni UPIAR. Case from updated, unpublished 2007 catalog. pp. 62-63. The Italian Center for Ufological Studies, (CISU): http://www.cisu.org

485   Ibid., p. 84.

486   Ibid., p. 122.

487   Ibid., pp. 123-124.

CHAPTER 11: UFOs – TAKING ON WATER BY CONVENTIONAL MEANS

488   Journal Article: Magin, U. (2006, May). USO reports from Germany. *Journal für UFO-Forschung* (Journal for UFO Research), a publication of GEP, the Society for the Study of UFO Phenomenon, Lüenscheid, Vol., No., & page unknown.

489   Book: Vallee, J. (1969). *Passport to Magonia: On UFOs, folklore, and parallel worlds*. pp. 144-146. Chicago: H. Regnery.

490   Ibid., pp. 142-143.

491   E-Mail: C. Aubeck (personal communication, date unknown). Humcat 1906-2. Source was *Phenomena Research Reporter*, No. 3.

492   Website: Rosales, A. (n.d.). Retrieved December 02, 2005. Humcat 1951-12. Original source: Juan José Benitez, *Flying Saucer Review*

(FSR), Vol. 24, No. 2. Archived at UFOINFO: http://ufoinfo.com/ humanoid/humanoid1951.shtml

493   Website: Rosales, A. (n.d.). Original source: *Revista Brasileira de Ufologia*, Vol., No., & page unknown. Retrieved September 27, 2009, from UFO Casebook: http://www.ufocasebook.com/1978brazil.html

494   E-Mail: A. Rosales (personal communication, May 26, 2007). Original source: Grupo O.S.I.F.E., Argentina at: http://www.angelfire.com/ scifi/etdelsol/archivos/Merkabah/caso_las_prusianas.htm Translation by Albert S. Rosales. See WATER UFO: http://www.waterufo.net/item. php?id=1062

495   Website: Williamson, S. G. (n.d.). *Udo Wartena letter—html transcription of original* (West Linn, Oregon, 1980). Retrieved September 15, 2007, from S. Gill Williamson, Professor Emeritus, CSE, UCSD: http:// cseweb.ucsd.edu/~gill/UdoWartLetter.htm

496   Website: Rosales, A. (n.d.). Retrieved August 12, 2004, from UFOINFO: http://www.ufoinfo.com/humanoid/humanoid1969.shtml

497   Website: Chalker, B., Basterfield, K. (n.d.). *Rosedale, Victoria: A close encounter*. Retrieved February 21, 2011, from: http://www.auforn. com/Bill_Chalker_12.htm

498   Journal Article: (author unknown). (1978, June). Foreign forum. *International UFO Reporter (IUR)*, Vol. 3, No. 6, p. 2. Original source is New Zealand *Spaceview*, 1st Quarter, 1978.

499   Booklet: Hind, C. (1992, January). Case no. 28 (C) Namibia. *UFO Afrinews*, No. 5, pp. 19-21. Also available at: http://www.ufoafrinews.com/pdfs/ UFO_AFRINEWS05-150.pdf

500   Website: Case dated 08-19-1973. (2005, April 16). Retrieved December 22, 2011, from National UFO Reporting Center (NUFORC), P. Davenport, director: http://www.nuforc.org/webreports/042/S42332.html

501   Magazine Article: Encounter by the pool. (1982, January-February). *Frontiers of Science*, Vol. 3, No. 6: Including the *International UFO Reporter*, pp. 11-12.

502   Journal Article: Hill, J. M. (1982, January). South Texas vehicle effects cases. *The MUFON UFO Journal*, No. 167, p. 8.

503   Website: LaFata, P., & translated by S. Corrales. (2010, September 07). *Argentina: UFO stole water from military base*. Retrieved July 24, 2011, from Inexplicata—The Journal of Hispanic Ufology: http://

inexplicata.blogspot.com/2010/09/argentina-ufo-stolewater-from-military.html

504   Website: Burgos, L., & translated by S. Corrales. (2011, January 10). *Argentina: Another water tank mysteriously emptied.* Retrieved August 9, 2011, from Inexplicata—The Journal of Hispanic Ufology: http://inexplicata.blogspot.com/2011_01_01_archive.html

505   Magazine Article: Keel, J. A. (1970, November). Ocean-based UFOs ring the U.S. *Male*, Vol. 20, No. 12, p. 48.

506   Journal Article: Taylor, H. S. (2006, May). Cloud cigars: A further look. *International UFO Reporter (IUR)*, Vol. 30, No. 3, pp. 10-12. Taken from *Flying Saucers and the Straight-Line Mystery* by Aimé Michel, p. 23, © 1958.

507   E-Mail: R. H. Hall (personal communication, January 19, 2007).

508   Book: Lorenzen, J., & Lorenzen, C. (1968). *UFOs over the Americas.* p. 53. New York: New American Library.

509   Magazine Article: Thanks to "Post." (1965, May). *Australian Flying Saucer Review*, No. 3.

510   Book: Fowler, R. E. (1974). *UFOs: Interplanetary visitors.* p. 349. Jericho, NY: Exposition Press.

511   Newspaper Article: Williams, M. J. (1971, January 23). Plummeting airplane? Flashing light, noise upset dogs, stir search in Adams. *The Idaho Statesman* (Boise), Second Section, p. 13.

512   Journal Article: Frederick, C. (1974, January). Man observes UFO landing. *Skylook* [the original *MUFON UFO Journal*], No. 74, pp. 3-4.

513   Website: Ridge, F. (2007, September 15). *The 1978 UFO chronology.* Retrieved October 03, 2012, from National Investigations Committee on Aerial Phenomena (NICAP): http://www.nicap.org/waves/1978fullrep.htm

514   Journal Article: Gordon, S. (1982, June). Pennsylvania low-level UFO sightings. *The MUFON UFO Journal*, No. 172, p. 3.

515   Journal Article: Hamilton, B. (1992, February). Light show over Lancaster. *The MUFON UFO Journal*, No. 286, p. 3.

516   Website: Ridge, F. (2008, November 15). *The 1992 UFO chronology.* Retrieved January 29, 2012, from National Investigations Committee on Aerial Phenomena (NICAP): http://www.nicap.org/waves/1992fullrep.htm

517   Website: Case dated 07-07-2000. (2003, March 21). Retrieved November 25, 2011, from National UFO Reporting Center (NUFORC), P. Davenport, director: http://www.nuforc.org/webreports/013/S13324.html

518   Catalog: Bianchini, M. (2003). *USOCAT: The Italian catalog of sightings of unidentified submerged objects.* Turin, Italy: Edizioni UPIAR. Case from updated, unpublished 2007 catalog, p. 131. The Italian Center for Ufological Studies, (CISU): http://www.cisu.org

519   Website: Directive. (n.d.). In *Wikipedia.* Retrieved July 10, 2014, from: http://en.wikipedia.org/wiki/Prime_Directive

CHAPTER 12: BEAM OF LIGHT...INTO A BODY OF WATER

520   Book: Pratt, B. (1996). *UFO danger zone: Terror and death in Brazil— where next?* pp. 247-248. Madison, WI: Horus House Press.

521   Ibid., pp. 246-247.

522   Book: Magor, J. (1977). *Our UFO visitors.* pp. 37-38. Sannichton, BC, Canada: Hancock House Publishers Ltd.

523   Magazine Article: I see by the papers. (1971, April). *Fate,* Vol. & No. unknown, pp. 19-20.

524   Website: Voss, N. (2003, June 29). Retrieved July 15, 2003, from UFO Wisconsin: http://www.ufowisconsin.com/county/reports2003/ r2003_0629_marathon.html

525   Book: Rutkowski, C. (1989). *Visitations? Manitoba UFO experiences.* p. 18. Winnipeg, MB, Canada: Gateway Publishing.

526   E-Mail: B. Vike (personal communication, April 08, 2007). Director HBCC UFO Research: http://hbccuforesearch.blogspot.com

527   Journal Article: Puckett, B. (2007, November). Washington disc about 35 feet. *The MUFON UFO Journal* (Filer's Files), No. 475, p. 20. Original source: UFOs Northwest: http://ufosnw.com

528   Website: *Noctiluca scintillans* (2010). Retrieved March 12, 2010, from Oceana: Protecting the World's Oceans: http://na.oceana.org/en/ explore/creatures/noctiluca-scintillans

CHAPTER 13: BURNING OBJECTS OR WATER-UFOs?

529   Book: Magor, J. (1977). *Our UFO visitors*. p. 24. Sannichton, BC, Canada: Hancock House Publishers Ltd.

530   Book: Sanderson, I. T. (1970). *Invisible residents*. p. 226. Toronto, Canada: Fitzhenry & Whiteside Ltd.

531   Ibid.

532   Ibid.

533   Newspaper Article: Sea search launched: 'Fiery finger' seen. (1956, January 18). *The Register* (Santa Ana, California).

534   Newspaper Article: Couple spot 'fiery ball' on lake. (1958, October 13). *Times-Union* (Rochester, New York).

535   Book: Walton, T. (1979, 1996). *Fire in the sky: The Walton experience*. New York: Marlowe & Co.

536   Newspaper Article: Sparks, G. (1987, July 02). The unexplained: Pulsating UFO lands in Florida swamp. *Republican Journal* (Belfast, Maine).

537   Book: Magor, J. (1977). *Our UFO visitors*. pp. 25-27. Sannichton, BC, Canada: Hancock House Publishers Ltd.

538   Newspaper Article: Lady sees space ship hover over North East Saturday and Sunday before it zooms away. (1954, December 05). *The Breeze* (North East, Pennsylvania).

539   Magazine Article: Gregory, J. (1957, November-December). UFOs ahoy! *Flying Saucer Review (FSR)*, Vol. 3, No. 6, p. 9. http://www.fsr.org.uk

540   Newspaper Article: Boatman reports phantom vessel. (1960, August 29). *Long Island Daily Press* (New York).

541   Magazine Article: *Flying Saucer Review (FSR)*. (1973, April). Case Histories, Supplement No. 14, pp. 13-15. http://www.fsr.org.uk

542   Catalog: Bianchini, M. (2003). *USOCAT: The Italian catalog of sightings of unidentified submerged objects*. Turin, Italy: Edizioni UPIAR. Case from updated, unpublished 2007 catalog, pp. 81-82. The Italian Center for Ufological Studies, (CISU): http://www.cisu.org

543   Newsletter Article: (author unknown). (1976, May). UFO submerged in N.Y. lake. *The A.P.R.O. Bulletin*, Vol. 24, No. 11, pp. 1, 3.

544   Website: Filer, G. (2000, August 28). *Scotland UFO report from the year 1767*. Retrieved August 28, 2000, from: http://www.nationalufocenter.com/files/2000/FilersFiles34.htm

545   Magazine Article: Norman, S. (1956, June). Recent UFOs over Japan. *Fate*, Vol. & No. unknown, pp. 22-24.

CHAPTER 14: SEA SERPENTS OR WATER-UFOs?

546   Newspaper Article: An electric monster. (1893, July 03). *The Tacoma Daily Ledger* (Washington).

547   Newspaper Article: Monster was 'all lit up.' (1908, March 28). *Oakland Tribune* (California).

548   Newspaper Article: Deep-sea monster with headlights seen on record descent in ocean. (1934, August 16). *Port Arthur News* (Texas).

CHAPTER 15: PHYSICAL INFLUENCES OF A UFO ON A BOAT'S MOTION

549   Book: Vallee, J., & Aubeck, C. (2009). *Wonders in the sky*. pp. 190-191. New York: Tarcher/Penguin.

550   Book: Hall, R. H. (2001). *The UFO evidence: A thirty-year report*. Volume 2. p. 222. Lanham, MD: Scarecrow Press.

CHAPTER 16: TOTALLY SUBMERGED

551   Newspaper Article: Trawler snags atom sub, is nearly pulled to bottom. (1956, May 03). *Wilmington Morning News* (Delaware).

552   Magazine Article: I see by the papers: "Something." (1961, December). *Fate*, Vol. & No. unknown, p. 27.

553   Book: Sanderson, I. T. (1970). *Invisible residents*. p. 52. Toronto, Canada: Fitzhenry & Whiteside Ltd.

554   Newspaper Article: 'Object' snags dragger, tows it backward. (1963, August 10). *The Boston Globe* (Massachusetts).

555   Magazine Article: UFOs in two worlds. (1971, August). *Flying Saucer Review (FSR)*, Special Issue No. 4, p. 55. http://www.fsr.org.uk

556   Newspaper Article: Navy hunts mystery sub sighted off Florida shore. (1955, February 17). *Wilmington Morning News* (Delaware).

557   Magazine Article: Case of the speedy submarine. (1962, October). *Fate*, Vol. & No. unknown, p. 10.

558   Magazine   Article: Sweden: Unidentified   submarine.   (1962, November-December). *Flying Saucer Review (FSR)*, Vol. 8, No. 6, pp. 23-24. http://www.fsr.org.uk

559   Magazine Article: Sweden: USOs reported: USO outside Norrtälje. (1970, March-April). *Flying Saucer Review (FSR)*, Vol. 16, No. 2, pp. 32-33. http://www.fsr.org.uk

560   Newspaper Article: Danes believe unknown sub is deep in Greenland fiords. (1972, December 13). *New York Times*, p. 13.

561   Newsletter Article: (author unknown). (1964, May). Object sinks yacht. *The A.P.R.O. Bulletin,* Vol. & No. unknown, p. 2.

562   Magazine Article: Sweden: USOs reported: Unknown submarine theory. (1970, March-April). *Flying Saucer Review (FSR),* Vol. 16, No. 2, pp. 32-33. http://www.fsr.org.uk

563   Magazine Article: Sweden: USOs reported. (1970, March-April). *Flying Saucer Review (FSR),* Vol. 16, No. 2, pp. 32-33. http://www.fsr.org.uk

564   Magazine Article: Sweden: USOs reported: USO in Bottenhavet. (1970, March-April). *Flying Saucer Review (FSR),* Vol. 16, No. 2, pp. 32-33. http://www.fsr.org.uk

565   Newspaper Article: Submarine hits whale: Collision of nuclear vessel causes some damage. (1959, October 07). *New York Times*, p. 84.

566   Magazine Article: Robertson, W. S. (1965, May-June). UFOs and the Scottish seas. *Flying Saucer Review (FSR)*, Vol. 11, No. 3, pp. 36 & iii. http://www.fsr.org.uk

567   Website: *UFO Sightings: Indonesia*. (2000, February 05). Retrieved March 20, 2002, from Mystical Universe: http://www.mysticaluniverse.com/ ufos_aliens/ufosightings/australia/scotland/malaysia/indonesia.html

568   Newspaper Article: Hunt fails to find strange subs reported off N. C. (1955, February 23). *Wilmington Morning News* (Delaware), pp. 1, 10.

569   Magazine Article: Bord, J., & Bord, C. (1982). Unidentified submarine objects: An unfathomable mystery. *The unexplained: Mysteries of mind, space & time*. Vol. 7, No. 82, p. 1623. London, UK: Orbis Publishing Ltd.

570   Journal Article: Themaras, P. (2002, December). Underwater UFO in 1929. *The MUFON UFO Journal*, No. 416, p. 15.

571   Newsletter Article: (author unknown). "Hidden" UFO reports given NICAP. (1960, July-August). *The U.F.O. Investigator* [Publication of NICAP], Vol. 1, No. 10, p. 7.

572   Book: Lorenzen, J., & Lorenzen, C. (1968). *UFOs over the Americas.* pp. 52-53. New York: New American Library.

573   Book: Cathie, B. (1968). *Harmonic 33.* pp. 17-19. Wellington, Sydney, London: A. H. and A. W. Reed.

574   Book: Naud, Y. (1978). *U.F.O.s and extra-terrestrials.* Vol. 4. pp. 108-109. Geneva: Ferni.

575   Catalog: Bianchini, M. (2003). *USOCAT: The Italian catalog of sightings of unidentified submerged objects.* Turin, Italy: Edizioni UPIAR. Case from updated, unpublished 2007 catalog. pp. 108-109. The Italian Center for Ufological Studies, (CISU): http://www.cisu.org

576   Newspaper Article: Speeding object seen underwater off Florida. (1965, July 06). *Los Angeles Times.*

CHAPTER 17: UNDERWATER "SOLID" LIGHTS

577   Magazine Article: Ribera, A. (1965, November-December). More about UFOs and the sea: Cardiganshire sightings. *Flying Saucer Review (FSR)*, Vol. 11, No. 6, p. 17. http://www.fsr.org.uk

578   Newspaper Article: Youth reports UFO sighting. (1965, December 03). *News-Pilot* (San Pedro, California).

579   Newsletter Article: Todd, D. R. (1978, April). Underwater UFO with "mother ship." *The A.P.R.O. Bulletin,* Vol. 26, No. 10, pp. 5-6.

580   Book: Randles, J. (1991). *From out of the blue.* pp. 122-123. New Brunswick, NJ: Global Communications.

581   Website: Case dated 09-24-1998. (1998, September 26). Retrieved July 31, 2000, from National UFO Reporting Center (NUFORC), P. Davenport, director: http://www.nuforc.org/webreports/004/S04004.html

582   Magazine Article: Earley, G. (1976, January). Encounter in the Gulf Stream. *Beyond Reality*, No. 18, pp. 37-39.

583   Website: Cashman, M. (n.d.). Retrieved June 17, 2006, from The Temporal Doorway: http://www.temporaldoorway.com/ufo/catalog/emeffect/radio.htm which is inactive. See WATER UFO: http://www.waterufo.net/item.php?id=539

584   E-Mail: B. Vike (personal communication, April 26, 2008). Director HBCC UFO Research: http://hbccuforesearch.blogspot.com

585   Letter: Gary R..... (n.d.). Copy on file with the Center for UFO Studies (CUFOS): http://www.cufos.org

586   Book: Sanchez-Ocejo, V., & Stevens, W. C. (1982). *UFO contact from undersea: A report of the investigation*. p. 156. Tucson, AZ: Privately published by W. C. Stevens.

587   E-Mail: N. Burns (personal communication, February 08, 2009). © 2008. http://www.trueghosttales.com. All rights reserved, republished here with permission.

588   Book: Hayter, A. (1959). *Sheila in the wind*. pp. 106-107. London: Hodder & Stoughton Ltd.

589   Website: Chalker, B. (1996). *The "first" official "unknown" — the electromagnetic light wheel near Groote Eylandt*. Retrieved September 22, 2000, from Project 1947 Forum: http://www.project1947.com/forum/bcoz5.htm#groote

590   Newsletter Article: (author unknown). The question of submerging. (1968, March). *The U.F.O. Investigator* [Publication of NICAP], Vol. 4, No. 5, p. 5.

591   Website: Case dated 05-23-1968. (2007, February 01). Retrieved November 03, 2007, from National UFO Reporting Center (NUFORC), P. Davenport, director: http://www.nuforc.org/webreports/054/S54893.html

592   Website: Booth, B. J. (n.d.). *Sea captain recalls 1968 USO sighting*. Retrieved March 06, 2008, from UFO Casebook: http://www.ufocasebook.com/1968usosighting.html

593   Website: Voss, N. (2002, May 16). *U.S. marine witnesses USO*. Retrieved June 22, 2005, from UFOWisconsin: Pancake Perspectives, Issue 03: http://www.ufowisconsin.com/pancakeperspectives/pp2002_0516.html

594   Case Investigation Files: Feindt, C. (Case date: 02~09-??-1982). No title. See: http://www.waterufo.net

595   Magazine Article: (1996, August). *Revelacion Magazine*. No other information available. Translated by Albert S. Rosales: http://www.iraap.org/rosales

596   Catalog: Bianchini, M. (2003). *USOCAT: The Italian catalog of sightings of unidentified submerged objects*. Turin, Italy: Edizioni UPIAR. Case from updated, unpublished 2007 catalog, p. 117. The Italian Center for Ufological Studies, (CISU): http://www.cisu.org

597   Website: *An encounter at sea.* (2004). Retrieved December 16, 2004, from Cosmic Conspiracies: http://www.ufosaliens.pwp.blueyonder.co.uk/ufosea.htm

598   Case Investigation Files: Feindt, C. (Case date: 10-05-2005). No title. See: http://www.waterufo.net

599   E-Mail: T. Phillips (personal communication March 15, 2007). Thanks to Ted Phillips of the Center for Physical Trace Research: http://www.ufophysical.com

600   E-Mail: M. Bianchini (personal communication, June 16, 2008). The Italian Center for Ufological Studies, (CISU): http://www.cisu.org

CHAPTER 18: SHAG HARBOUR: THE "WATER ROSWELL"

601   Newsletter Article: (author unknown). UAO dives into sea at Nova Scotia. (1967, September-October). *The A.P.R.O. Bulletin,* Vol. & No. unknown, p. 7.

602   E-Mail: C. Styles (personal communication, June 30, 2009).

CHAPTER 19: THE SEA WAS LIKE A SHEET OF GLASS

603   Case Investigation Files: Feindt, C. (Case date: WW-??-1979). No title. See: http://www.waterufo.net

604   Website: *Oklahoma UFO dives in pond*. (2006, June 29). Retrieved July 04, 2006, from MUFON CMS: http://www.mufon.com/mufonreports.htm MUFON's case investigation report is in a file which is not accessible to the general public.

605   Website: Filer, G. (2006, September 13). *Lake Superior.* Retrieved September 13, 2006, from http://www.nationalufocenter.com which is now archived at UFOINFO: http://www.ufoinfo.com/filer/2006/ff0637.shtml

CHAPTER 20: UP CLOSE AND PERSONAL

606   Book: Stone, J. (1971). *The monster at the end of this book*. p. 4. New York: Golden Books, division of Random House.

607   Newsletter Article: Rogerson, P. (1979, Summer). INTCAT case 867, published in John Rimmer's *MUFOB* [*Metempirical UFO Bulletin*], Part 20, News series No. 15 [whole No. 49].

608   Newsletter Article: (author unknown). UFO's cause panic, one death. (1962, January-February). *The U.F.O. Investigator* [Publication of NICAP], Vol. 2, No. 3, p. 1.

609   Newspaper Article: What was it? (1968, May 09). *The Coast Guard* (Shelburne, Nova Scotia, Canada).

610   Book: Swords, M. D. (2005). *Grass roots UFOs: Case reports from the Timmerman files*. p. 26. Lima, OH: CSS Publishing. Thanks to the Fund for UFO Research: http://www.fufor.com for permission to use.

611   Journal Article: (author unknown). (1979, October). Foreign forum. *International UFO Reporter (IUR)*, Vol. 4, No. 3, p. 18.

612   Documents: CIA documents. (1950, August 18). Forwarded to this author by Don Ledger.

613   E-Mail: V. Sanchez-Ocejo (personal communication, November 08, 2004).

614   Newspaper Article: Electric cloud enveloped ship. (1904, August 01). *The Philadelphia Inquirer* (Pennsylvania).

615   Magazine Article: Australasia. (1957, November-December). *Flying Saucer Review (FSR)*, Vol. 3, No. 6, p. 6. http://www.fsr.org.uk

616   Newsletter Article: (author unknown). Argentine Navy discloses important E-M case. (1965, August-September). *The U.F.O. Investigator* [Publication of NICAP], Vol. 3, No. 4, p. 6.

617   E-Mail: A. Rosales (personal communication, June 20, 2008).

618   Website: Booth, B. J. (n.d.). *Arrowhead UFO sighting, Gulf of Mexico, 1969*. Retrieved March 07, 2008, from UFO Casebook: http://www.ufocasebook.com/arrowheadufo.html

619   Journal Article: Sekerkarar, E. (2005, March). History of UFO sightings in Turkey: Businessmen's sighting. *The MUFON UFO Journal*, No. 443, pp. 7-8.

620   Website: Booth, B. J. (n.d.). *1986-UFO seen from bridge of USS* Edenton. Retrieved August 18, 2006, from UFO Casebook: http://www. ufocasebook.com/1986ussedenton.html

621   Website: *Cylinders hovers above fisherman in the ocean, New Jersey, USA.* (2006). Retrieved August 26, 2006, from UFOs at Close Sight: http://www.ufologie.net/htm/2001aug.htm Original source is MUFON CMS case #1709 submitted June 9, 2002, to: http://mufon.com/ mufonreports.htm

Index of Cases in Chronological Order

DATE	LOCATION	PAGE

UNDATED

UNDATED—4	Bismarck Sea off Pityilu Island, Admiralty Islands	16, 428
UNDATED—8	Hartwell Lake near Anderson, SC, USA	84
UNDATED—12	Gulf of Mexico - USS *Lexington* (CVS-16)	241-242, 275
UNDATED	Unknown location – abductee Anna Jamerson	187-188, 190, 203, 208, 209, 374
UNDATED	Unknown location – abductee Beth Collings	187, 208-209
UNDATED	Unknown location – abductee Ed Walters	195-199

PRIOR TO 1800

??-??-1067	North Sea off northeastern England, UK	xx-xxi, 7-8
??-??-1361	Sea of Japan off western Japan	8
10-11-1492	Atlantic Ocean off San Salvador	8-9
03-??-1638	Muddy River, Massachusetts, USA	410-411
08-15-1663	Ozero (Lake) Zarobozero, Sëmkino, Russia	127, 293-294
02-23~24-1740	Mediterranean Sea off Toulon, France	111
09-08-1767	River Isla near Coupar Angus, Scotland, UK	10, 395

1800-1899

06-18-1845	Mediterranean Sea near Malta	128
04-??-1875	Veracruz, Mexico	328
05-15-1879	In the Persian Gulf	328-329
06-11-1881	Between Sydney & Melbourne, Australia	11
08-04-1888	Highland Lake, Connecticut, USA	95
02-21-1893	Atlantic Ocean off the coast of the United States	119
07-02-1893	Puget Sound, Washington, USA	306, 399-405
04-22-1897	Josserand, Texas, USA	354-355
05-06-1897	Hot Springs, Arkansas, USA	355-356

1900-1949

10-28-1902   Gulf of Guinea, Africa – SS *Fort Salisbury* xxvii
07-28-1904   Coast of Delaware, USA 462-465
06-02-1906   Gulf of Oman (between Oman and Iran) 329
10-30-1906   Atlantic, 600 miles N.E. of Cape Race, Canada 333
??-??-1906   Mitchell, South Dakota, USA 356
03-12-1908   Atlantic Ocean heading north from Jamaica 405-407
07-EE-1910   Coast off Normandy, France xxviii
08-LL-1914   Near Georgia Bay, Canada 60
08-27-1917   Outside Grand Harbour, Malta 61
02-22-1922   Hubbell, Nebraska, USA 141
10-14-1923   Indian Ocean south of Sri Lanka 111, 330, 331
01-15-1924   Neillsville, Wisconsin, USA 142
08-23-1925   Indian Ocean southwest of Sri Lanka 111, 330, 331
08-25-1925   Arabian Sea, 150 miles south of Oman 331
10-31-1926   Indian Ocean southwest of Sri Lanka 111, 331
??-??-1927   Po River near Corbola, Italy 119
??-??-1929   Tarpon Springs, Florida, USA 424-425
03-??-1931   Indian Ocean north of the Maldive Islands 120
SM-??-1933   Cherryville, Pennsylvania, USA 161
08-16-1934   Atlantic Ocean near Bermuda 407-408
02-11-1937   Near Kvalsvik off the coast of Norway 375
03-29-1938   Morecambe Bay, England, UK 24, 297
05-EE-1940   Stream near Townsend, Broadwater Cty, Montana, USA 361
02-26-1942   Timor Sea between Australia and Timor Island 465-466
08-05-1942   Pacific near Guadalcanal - USS *Helm* 235-237, 265-266
??-??-1943   Unknown position in the Persian Gulf 425
SM-??-1945   Unknown position in the Pacific Ocean 237-238
04-25-1946   Anima-Nipissing Lake, Ontario, Canada 146-147
07-19-1946   Lake Kölmjärv, Sweden 299
01~08-??-1947 Unknown position in the Pacific Ocean 225
06-21-1947   Puget Sound off Maury Island, Washington, USA 346
06~08-??-1947 Mediterranean Sea, 20 miles south of Malta 21
07-EE-1947   Roswell, New Mexico, USA .... vii, 94-95, 112, 161-162,
              266, 440, 444
08-13-1947   Snake River Canyon at Twin Falls, Idaho, USA 73

??-??-1947   Pacific, Humboldt Current off South America 11, 294
10-14-1948   Into Lake Ontario, Bear Creek Harbor, NY, USA 388
11-14-1949   Passed the Strait of Hormuz heading for India 332

1950-1959

03-31-1950   Riccione, Italy 347
06-EE-1950   Atlantic Ocean off several Argentinean cities........................ 13
07-02-1950   Steep Rock Lake, Ontario, Canada............................347-348
08-04-1950   Atlantic Ocean, roughly E. of Cape May, NJ, USA 456-460
08-25-1950   Cervia, Italy..336, 342-343
09-14-1950   Ontario, Canada 326
FF-??-1950   Unknown location – abductee Betty Andreasson Luca ... 194-195,
                206-207
02-10-1951   Atlantic Ocean off Newfoundland, Canada 213-214, 266-267
04-07~18-1951 Red Sea near Djibouti.. 434
11-30-1951   Unknown position Persian Gulf.................................222-223
SP-??-1951   Paarl, South Africa 164, 357-358
FF-??-1951   Unknown waters off the Korean Peninsula 239-240
SM-??-1951/2 Valley Forge, Pennsylvania, USA 57-58
02-??-1952   Atlantic Ocean 400 miles off New Jersey, USA...............252-253
04?-??-1952   Atlantic Ocean, Cape Chidley, Labrador, Canada 225
07-14-1952   Chesapeake Bay, off Virginia, USA................................. 41-42
07-19-1952   Langley Air Force Base, Virginia, USA 42
10-??-1952   River Lågen, Norway 120
02-09-1953   Malabar Coast off Mt. Delby, India............................223-224
05-??-1953   Loch Raven Reservoir, Towson, Maryland, USA 61
??-??-1953   Near Guantánamo Bay Naval Base, Cuba 240
??-??-1953   Lake of the Ozarks, MO, USA... 89
09-14-1954   Several towns southwest of Paris, France 375
09-LL-1954   Off the coast of Georgia, USA 29, 50-51, 431
10-20-1954   Lusigny Forest near Troyes, France............................132-133
10-27-1954   Plougasnou, Finistère, France 115
11-12-1954   Unnamed river near Arquà Polesine, Italy.......................... 120
11-23-1954   Torpo, Norway... 141
11-??-1954   Strait of Florida between Cuba and Florida, USA.....214, 460-462
11-??-1954   South of Lundy Island, United Kingdom 280

12-05-1954   Lake Erie, towns in PA, NY, & Ohio, USA 392-393

12-07-1954   Isla del Francés, Uruguay 308, 375

??-??-1954   Unnamed location - USS *FDR* 255-256

02-16-1955   Atlantic Ocean off Fort Pierce, Florida, USA 416-417

02-22-1955   Atlantic Ocean off North Carolina, USA 423-424

02-??-1955   Ocean Beach near San Diego, CA, USA 61

03-24-1955   Rhoslefain, Wales, UK 430

07-LL-1955   Lake Ontario near Niagara Falls, NY, USA 13

01-15-1956   Pusan, South Korea 21-22, 396

01-17-1956   Pacific off Newport & Laguna Beaches, CA, USA 389

04-22-1956   Off the coast of New Jersey, USA 413-414

07-26-1956   Port at Rio de Janeiro, Brazil – USS *FDR* 98-99, 100-101,
              240, 274

09-??-1956   Cabo Frio, Brazil 299-300

12-13-1956   Off the coast of Venezuela 17, 120

03-08-1957   Baudette, Minnesota, USA 61, 141

04-19-1957   Pacific Ocean south of Yokohama, Japan 120

07-29-1957   Oldsmar, Florida, USA 111

09-01-1957   Ocean off Porthcawl, Wales, UK 393

10-10-1957   Mariaville, New York, USA 73

11-05-1957   Gulf of Mexico, south of New Orleans, LA, USA 214

SM-??-1957   Penobscot Bay off Camden, Maine, USA 17, 84

??-??-1957   Atlantic Ocean off São Sebastião, Brazil 62

01-10-1958   Atlantic Ocean off the coast of Curitiba, Brazil 200

01-??-1958   Niagara Falls/Depew, NY, USA 143

04-09~10-1958 Atlantic Ocean off Keta, Ghana 297

04-??-1958   Atlantic Ocean off Saúde, Brazil 57, 120

06-01-1958   Altafjord, Norway 226, 308

06-12-1958   Mediterranean Sea off Le Brusc, France 53, 128

06-??-1958   Adriatic Sea off Casal Borsetti, Italy 90, 121

07-20-1958   Private lake in Glennie, Michigan, USA 121

07-23-1958   Cook Strait, New Zealand 280

10-13-1958   Lake Ontario off Sea Breeze, New York, USA 389-390

10-??-1958   Caribbean Sea near Cuba - USS *FDR* 98, 240, 241, 242, 252,
              267, 274-275

??-??-1958   Las Vegas, Nevada, USA 163

??-??-1958   Ocean off the west coast of South Korea 379-380
01-08-1959   Derwent River off Risdon, Tasmania, Australia...................... 17
03-LL-1959   Baltic Sea off Kołobrzeg, Poland 46-47
04-04-1959   Indian Ocean.. 214
06-??-1959   Buenos Aires, Argentina... 425
08-??-1959   Unnamed creek near Tulsa, Oklahoma, USA 95
10-05-1959   Atlantic Ocean off Portsmouth, NH, USA............................ 421
??-??-1959   Off the coast of Havana, Cuba 121
??-??-1959   Lake Gualletué, Malleco, Chile 453

1960-1969

02-14~21-1960 Golfo Nuevo, Argentina.. 226
08-28-1960   Oyster Bay Harbor, Long Island, NY, USA...................... 393-394
02-??-1961   Unnamed lake in the Karelia region of Russia............ 95-96, 147
04-27-1961   Lake Onega, Russia.. 97, 147
04-??-1961   Montesilvano, Italy.. 348
06-03-1961   Ligurian Sea off Savona, Italy................................. 47, 70, 71
07-EE-1961   Port Aransas, Texas, USA 414-415
09-19-1961   New Hampshire, USA 163, 177
09-24-1961   Baltic Sea off Łeba, Poland 47, 54, 62, 68-69, 70, 71
10-06-1961   Lake Maracaibo near Santa Rita, Venezuela 453
11-14-1961   Off Sydney Heads, Australia 417
05-30-1962   Gulf of Siam - SS *Telemachus* 332
07-08-1962   Unknown body of water, Nashville, NC, USA 300, 309
07-19-1962   Unnamed creek, Asheboro, NC, USA................................ 121
08-24-1962   Ocean off Gottland, Sweden 417-418
??-??-1962   Bass Strait, Australia.. 226, 280
??-??-1962   Unnamed river, assumed in USA 90
02-15-1963   Moe, Victoria, Australia.. 133
02-28?-1963   30-50 miles off Spitsbergen, Norway 229-230, 281
04-08-1963   Montauk Point, New York, USA... 415
04-11-1963   Atlantic Ocean near the Puerto Rico Trench................... 48, 124
08-09-1963   Georges Bank off Portland, Maine, USA.......................415-416
09-12-1963   Atlantic Ocean off Cape Cod, MA, USA 375
10-31-1963   Peropava River, Iguape, Brazil 124
11-12-1963   Golfo Nuevo, Argentina...466-467

11-20-1963	Off the eastern coast of Scotland, UK	214-215
SM-??-1963	Atlantic Ocean between Puerto Rico & Key West, FL, USA	227
LL-??-1963	Off the coast of Sardinia, Italy - USS *FDR*	242-248, 275
??-??-1963	Atlantic Ocean off the coast of Puerto Rico	226-227
01-23-1964	Northeast point of Groote Eylandt, Australia	434-435
01-27-1964	Into Lake Winnebago, Oshkosh, WI, USA	388
02-05-1964	Off Cape Mendocino, California, USA	419
08-29-1964	1000 mi. W. of Cape Horn, Chile – U.S. ship *Eltanin*	315-324
08-MM-1964	Upper Nemahbin Lake off Delafield, WI, USA	24-25, 432
09-09-1964	New River, Radford, Virginia, USA	388
09-20-1964	Golfo de San Jorge, Argentina	25
??-??-1964	Casitas Reservoir near Ventura, CA, USA	62
??-??-1964	625 miles east of Bermuda	215
01-12-1965	Near Custer, Washington, USA	143
01-12-1965	Near Lynden, Washington, USA	215, 275-276
02-03-1965	Sea of the Hebrides off Barra, Scotland, UK	421-422
03-10-1965	St. Lawrence River, Baie-St.-Paul, QC, Canada	424
03-12-1965	Kaipara Harbour, New Zealand	425-426
03-14-1965	Everglades Swamp, Florida, USA	390-392
05-??-1965	Tasman Sea, near Sydney, NSW, Australia	375-376
06-06-1965	Fraser Island, Australia	22
07-05-1965	Gulf Stream off the coast of Florida, USA	336, 428
08-04-1965	St. Louis Bay, Duluth, Minnesota, USA	215
08-??-1965	The Red Sea	62
09-10-1965	Mediterranean Sea off Le Brusc, France	394
09-15-1965	Sudbury, Massachusetts, USA	133
11-13-1965	Off Rugged Islands, New Zealand	14, 124
11-16-1965	Salem, Massachusetts, USA	133
11-29-1965	Springhill, Nova Scotia, Canada	142
12-02-1965	Pacific Ocean off San Pedro, CA, USA	430
12-12-1965	Tyrrhenian Sea, Isle of Capri, Italy	336, 342
SM~FF-??-1965	Atlantic Ocean between the Azores and Spain	248, 276
01-19-1966	Tully Lagoon, Queensland, Australia	71-72, 128, 156-157, 281, 300
03-26-1966	Atlantic Ocean off Florida Keys, USA	63, 309
03-??-1966	Gulf of Tonkin, China-Vietnam	227

04-07-1966 Lincoln, Nebraska, USA... 134

07-MM-1966 Pacific between Seattle, WA and HI, USA ...25, 54, 230-231, 276

12-14-1966 Porsangerfjorden, Norway 216

??-??-1966 Pacific Ocean - USS *Yorktown* 253-254

??-??-1966 Unknown location in the Atlantic Ocean 435

FF-??-1966/7 Pacific Ocean off San Juan Capistrano, CA, USA 18

01-09-1967 Malta, Montana, USA 300-301

01-29-1967 Knox City, Missouri, USA 144

02-21-1967 Sheboygan, Wisconsin, USA 144, 301

02-23-1967 Severna Park, Maryland, USA 142, 301

05-20-1967 Falcon Lake, Manitoba, Canada 73

05-29-1967 Trout Brook Lake, New Brunswick, Canada 124

06-24-1967 Waters off Trenton, Maine, USA 376

08-04-1967 Atlantic Ocean off Arrecife, Venezuela 124

08-08-1967 Salina, Venezuela 48

10-04-1967 Shag Harbour, Nova Scotia, Canada .... xvi, xx, 18, 48, 122, 191,
     309, 440-447

12-??-1967 Unknown location in the Gulf of Mexico 124

SP-??-1967 Atlantic Ocean south of Bermuda 282

SP-??-1967 Golfo de Penas, Chile 221-222

??-??-1967 Gibsons & Keats Island, BC, Canada 381-382

??-??-1967 Unknown location – abductee Betty Andreasson Luca 103-108

01-??-1968 Baraga, Michigan, USA 141

02-18-1968 Pond on Vashon Island, Washington, USA...... 156, 158-159, 161

03-15/16-1968 East China Sea, southwest of Okinawa 232

04-EE-1968 Upprämen Lake, Sweden 150

05-04-1968 Seal Island, Nova Scotia, Canada 128, 454

05-20-1968 Moore Lake, Littleton, NH, USA 125, 310

05-23-1968 Atlantic Ocean near the Azores 219, 435

08-07-1968 Buff Ledge Camp, N. of Burlington, Vermont, USA 184-186

09-15-1968 Cornwallis River, Nova Scotia, Canada 22-23

09-21-1968 La Escala (Gerona), Spain 336, 338

09~10-??-1968 Atlantic Ocean off the coast of Portugal 467-468

12-15/16-1968 Pacific Ocean near Hawk Inlet, Alaska, USA 23

12-26-1968 Berlin, New Hampshire, USA 144-145

EE-??-1968 Unknown waters off Lüda/Dalian, China 219

??-??-1968   Atlantic Ocean off Palm Beach, Florida, USA 435-436

02-10-1969   Dartmouth, Massachusetts, USA 142

04~05-??-1969 Unknown location in the Gulf of Mexico 468-469

07-04-1969   Duck Pond (a lake), Ossipee, NH, USA 297-298

07-05-1969   Valdez Channel, Alaska, USA 310

07-LL-1969   Morro Bay off Cayucos, California, USA 336, 337

08-23-1969   Mataró (Barcelona), Spain.. 336, 339

08-??-1969   Kama River near Voronov, Russia 361-362

09-13-1969   Waters off Norrtälje, Sweden 418

10-06-1969   South China Sea 420

10-11-1969   Mediterranean Sea off Cabo Cope (Murcia), Spain 416

10-24-1969   Pacific Ocean 350 miles south of Valparaíso, Chile 219

11-19-1969   Between Domsjö and Bureå, Sweden.............................. 420

11-20-1969   Waters off Hälsingland, Sweden 420-421

WW-??-196.   Swift Current, Saskatchewan, Canada............................. 143

1970-1979

04-09-1970   Aufhofen, Baden-Württemberg, Germany 298

07-??-1970   Off San Benedetto del Tronto, Italy 337, 343-344

11-05-1970   Cholla Bay, Sonora, Mexico 382-383

??-??-1970   Atlantic Ocean near the Bermuda Triangle 216

01-03-1971   Saapunki, Finland 90, 92-94

01-07-1971   Scargo Lake, Dennis, Massachusetts, USA......................... 150

01-21-1971   Brush Mountain, Idaho, USA... 376

02~03-??-1971 Near the Azores – USS *Saratoga* 277

07-02-1971   Atlantic Ocean near the Bahamas – USS *JFK*..........256-259, 277

08-04-1971   Tyrrhenian Sea off Milazzo, Italy 63

08-15-1971   Unknown location along New River, Guyana 303-304

09-??-1971   Unnamed lake in Arroyo de la Miel, Spain............................. 49

11-21-1972   Hermansverk, Norway 228

11-??-1972   Grand Lake, New Brunswick, Canada......................... 454-455

12-07-1972   Disko Bugt (Bay), Greenland 418-419

12-29-1972   Saranac Lake, New York, USA... 148

12-30-1972   Small stream near Tres Arroyos, Argentina.... 303, 307, 310-311

SM-??-1972   Lake Erie close to Buffalo, New York, USA 14

LL-??-1972   Unknown waters off Puerto Rico...............54, 84, 207-208, 433

03-25/26-1973 Caribbean Sea off the coast of Venezuela 200

04-??-1973 Gulf Stream off Bimini and Miama, FL, USA 426-427

07-12-1973 Stanley Draper Lake, Oklahoma, USA 125

08-19-1973 Yellville, Arkansas, USA 366, 368-369

09-30-1973 Giles County, Tennessee, USA.. 376

10-18-1973 Mansfield, Ohio, USA.. 81

11-06-1973 Pascagoula River, Pascagoula, Mississippi, USA................... 294

WW-??-1973? Stratham, New Hampshire, USA.. 145

02-??-1974 Sullivan, Missouri, USA 387-388

03-29-1974 Sea off Togo, Africa 54-55, 129

04-15-1974 Between Ceuta and Algeciras, Spain 25-26

05-01-1974 Cowichan Lake, British Columbia, Canada 392

06-18-1974 Navegantes Beach, Santa Catarina, Brazil 18

07-02-1974 Navegantes Beach, Santa Catarina, Brazil 294, 312-313

07-19-1974 Cannizzaro, Italy 336, 339-340

09~10-??-1974 Pacific Ocean 15 miles off Guam 260-261, 277

11-09-1974 Pond in Carbondale, Pennsylvania, USA.......................348-349

EE-??-1974 Waters near Vietnam... 433

LL-??-1974 Near Vietnam - USS Reeves (DLG/CG-24) 278

01-18-1975 Puget Sound off Anderson Island, WA, USA 26

02-26-1975 San Carlos Reservoir, Gila, Arizona, USA 295

03-02-1975 San Matías Gulf off Valdés Pen., Argentina..................... 200

03-13-1975 Near Mellen, Wisconsin, USA 142, 302

05-16-1975 Stephenfield Lake, Manitoba, Canada 383-384

08-01-1975 Trinity Lake, New York, USA 18-19, 84, 125, 395

01-12-1976 Brantford, Ontario, Canada 142

01-21-1976 Lake Vikern near Gyttorp, Sweden 152

03-28-1976 Adriatic Sea off Porto San Giorgio, Italy 200-201

04-23-1976 Atlantic Ocean 700 miles S.W. of Bermuda 216, 231, 278

04-30-1976 Lake Siljan, Dalarna Province, Sweden..........................97, 148

06-??-1976 St. Johns River, Jacksonville, Florida, USA 125

08-12-1976 Ligurian Sea off Calignaia, Italy 302

08-24-1976 Allagash Wilderness Waterway, Maine, USA 180-183

09-19-1976 Near Tehran, Iran 263-264

01-10-1977 Wakefield, New Hampshire, USA 150

01-21-1977 Canal in St. Bernard Parish, Louisiana, USA...................... 411

02-15-1977   Swamp in Sundown, Manitoba, Canada............................ 304

02-EE-1977   Gulf of Mexico off Ft. De Soto Park, Florida, USA 55

04-10-1977   Mediterranean Sea off Israel.. 217

07-20-1977   Off Sorso, Sardinia, Italy 337, 344

07-??-1977   Lake in Alaska near the Arctic Circle 63, 336

08-10-1977   Francavilla al Mare, Italy.. 336, 341

09-30-1977   Day's Corner, Prince Edward Island, Canada 131

10-07-1977   Unnamed sea, Russian submarine repair ship 282

12-06/07-1977 Near Tatapouri, New Zealand.................................366-367

12-22-1977   Atlantic Ocean near Boston, MA, USA 430

12-??-1977   South Georgia Island near the Antarctic........................... 220

??-??-1977   Unknown location in the Caribbean Sea 436

03-23-1978   Atlantic Ocean off Guaiúba, Brazil 358-359

05-02-1978   Escazu, Costa Rica 376

06-24-1978   Adriatic Sea off Civitanova Marche, Italy 336, 337-338

07-09-1978   Lake Paluda, Jesolo, Italy.......................................349-350

09-16-1978   Off Naples, Italy 337, 345

10-18-1978   Off Hardys Bay, New South Wales, Australia 432

10-23-1978   Adriatic Sea off Pedaso, Italy.. 63-64

10-??-1978   Zone of the Channels, Chile.................................... 228, 233

11-03-1978   Adriatic Sea near San Benedetto del Tronto, Italy................ 455

11-07-1978   Adriatic Sea off Silvi Marina, Italy 220

11-08-1978   Adriatic Sea off Porto d'Ascoli, Italy 228

11-09-1978   Adriatic Sea off Silvi Marina, Italy 217

11-13-1978   Adriatic Sea off Giulianova, Italy 111, 220

12-14-1978   Unnamed lake in Avigliana, Italy 336, 337, 340

12-20-1978   Adriatic Sea off Bellaria, Italy 394-395

12-28-1978   Ionian Sea off Santa Tecla, Italy 125

12-??-1978   Venice, Italy.. 336, 342

??-??-1978   Adriatic Sea off San Benedetto del Tronto, Italy 216-217

??-??-1978/9 Gulf of Oman near Iran.. 218, 278

01-03-1979   Atlantic Ocean off Hialeah, Florida, USA 203, 208

01-04-1979   Favignana, Italy... 336, 343

03-05-1979   Atlantic Ocean near the Canary Islands.................. 14-15, 201

04-10-1979   Atlantic Ocean off New Bedford, MA, USA.......................... 220

06-22-1979   Ligurian Sea off Gorgona Island, Italy.............................. 302

06-??-1979   Ligurian Sea off Cinque Terre, Italy 125
07-05-1979   Gulf of Alaska, south of Seward, AK, USA..................... 220-221
08-28-1979   Rimini, Italy 336, 340
10-15-1979   Atlantic coast north of Rio de Janeiro, Brazil .... vii, 49
WW-??-1979   South China Sea................................. 259, 278-279, 448-449
??-??-1979   Off Roseto degli Abbruzzi, Italy... 350
EE-??-197.   Furnace Creek, California, USA 73-74

1980-1989

01-??-1980   Tapajós River, Amazon Basin, Brazil............................ 380-381
02-13-1980   Stream near Rafaela, Argentina 359-360
04-16-1980   Venice, Italy... 337, 334
07-26-1980   Atlantic Ocean off the coast of Brazil....................... 23-24, 218
08-30-1980   Dumfries, Scotland, United Kingdom 134
09-30-1980   Near Rosedale, Victoria, Australia................. 362, 363-364, 366
11-28-1980   West Yorkshire, England, UK.. 132
11-??-1980   Atlantic Ocean off Rio de Janeiro, Brazil 207
12-26-1980   Atlantic Ocean off Paco de Arcos, Portugal........................ 129
12-29-1980   Near Dayton, Texas, USA 115-116
01-23-1981   Grand Forks, North Dakota, USA 148-149
02-08-1981   Isla Cristina in the Gulf of Cadiz, Spain 15
05-16-1981   Thompson River at Kamloops, BC, Canada 125
06-12-1981   Alice, Texas, USA 366, 371
07-11-1981   Port Byron, Illinois, USA... 366, 370
07-15-1981   Newberry, South Carolina, USA....................................... 126
08-LL-1981   Hamilton Harbour, Ontario, Canada 19
10-24-1981   Mediterranean Sea off Licata, Italy 129
02-19-1982   Lake Lacar, Argentina 307, 308, 311
02~09-??-1982 Atlantic Ocean off the coast of Virginia, USA 436
04-01-1982   Petrolia, Pennsylvania, USA.. 377
06-23-1983   Sea of Japan - USS *Midway* (CV-41) 279
10-15-1983   Sizewell, N. of Orford Ness, England, UK............................ 431
12-05-1984   Off the island of Elba, Italy... 201
04-15-1985   Unnamed pond in Granville Summit, PA, USA 312
04-26-1985   Gomaub, Namibia, Africa 366, 367-368
09-17-1985   The Bosphorus off Istanbul, Turkey 129, 469-470

10-12-1985   White Oak River near Swansboro, NC, USA.......................... 126

02-20~28-1986 Tyrrhenian Sea off Milazzo, Italy 126

06-??-1986   Atlantic Ocean off the coast of New Jersey, USA.................. 126

11-03-1986   Lake Trasimeno off Castiglione del Lago, Italy...................... 19

SM-??-1986   Atlantic Ocean off Cape Hatteras, NC, USA 470-471

12-25-1987   Lake Elmo, Minnesota, USA.. 151

03-04-1988   Lake Erie, several cities, USA.. 149

06-??-1988   Pacific Ocean off Antofagasta, Chile................................ 437

08-23-1988   Atlantic Ocean northeast of the Antilles........................ 75-76

07-??-1989   Aguadilla, Puerto Rico......................85-87, 112, 295, 428

09-03-1989   Caspian Sea off Sari, Iran 15

11-25-1989   Mediterranean Sea off Licata, Italy......................... 337, 345

11-30-1989   New York City, New York, USA 177-179

SP-??-1989   Sea of Okhotsk, Russia 283

??-??-1989   Indian Ocean steaming toward the Gulf of Oman.......... 249-251

??-??-198.   Lake Constance off Stein am Rhein, Switzerland 353

1990-1999

01-07-1990   Mzha River in Ukraine...............................153-155, 156

04-21-1990   Coast of Lazio, Italy 336, 341-342

07-06-1990   Off the shore of Palominos Island, Puerto Rico 126

07-24-1990   Tyrrhenian Sea off Amantea, Italy................................ 129

01-15-1991   Charles River, Waltham, Massachusetts, USA 155-156

02-06~08-1991 Chaville, France..183-184

05-14-1991   Lancaster, California, USA 377

05-??-1991   Unnamed location in Brazil.. 74

04-12-1992   Near Deming, New Mexico, USA.................................... 377

04-29-1992   Unknown location – abductee K. Wilson 209

06-17-1992   Unknown location – abductee K. Wilson 210

07-17-1992   Atlantic Ocean near Iguape, Brazil................................. 64

07-??-1992   Ligurian Sea off island of Gorgona, Italy.......................... 427

07-20-1993   Pacific Ocean off Oahu, Hawaii, USA 27

04-26-1994   Santa Rosa Sound, Gulf Breeze, Florida, USA.................. 64-65

SP~SM-??-1994 Pacific Ocean - USS *Constellation* (CV-64) 251-252, 279

02-27-1995   Mersey River near Fiddlers Ferry, England, UK 65

07-??-1995   Unknown location – abductee K. Wilson 191

02-18-1996   Sea off Iceland 218
07-20-1996   Off Porto Empedocle, Italy 336, 339
LL-??-1996   La Esperanza Bay, Vieques Island, Puerto Rico 65
07-05-1997   St. Margarets Bay, Nova Scotia, Canada 90
07-06-1997   Ionian Sea off Fiumefreddo di Sicilia, Italy 437
12-31-1997   Pond in Rockland, Massachusetts, USA 312
03-07-1998   Pantano Basso - Termoli, Italy 337, 344-345
09-24-1998   East River, New York City, NY, USA 431
MM-??-1998   Southeast of Coronado Island, CA, USA 218-219
LL-??-1998   Unknown position in the Pacific Ocean 437
02-09-1999   Pacific Ocean 180 miles south of Lima, Peru 201-202
04-17-1999   Lake near Invermere, BC, Canada 84-85
05-??-1999   Dillon Dam, Zanesville, Ohio, USA 298-299
07-27-1999   Lake Backsjön, Sweden 283
11-24-1999   Santa Fe, Argentina 366, 372-373

2000-2013
01-15-2000   Waters off Indonesia 422-423
04-01-2000   Off the coast of Rimini, Italy 350
04-14-2000   Montauk, Long Island, New York, USA 19-20
07-07-2000   Red Rock Canyon, Nevada, USA 377
04-11-2001   Torremezzo, Italy 350-351
05-23-2001   Ottawa River, Ontario, Canada 149
08-23-2001   Atlantic Ocean off New Jersey, USA 295-296, 471
SM-??-2002   Mosinee, Wisconsin, USA 383
02-21-2003   Wellington, Prince Edward Island, Canada 145
12-28-2003   Mud Lake, Liberty Township, Michigan, USA................. 153, 156
06-09-2004   Tasman Bay near Nelson, New Zealand 312
08~09-??-2004 Pacific Ocean 200 miles off China's coast........................... 112
01-31-2005   Farm pond in Columbia City, Indiana, USA........................ 126
10-05-2005   Atlantic Ocean off Myrtle Beach, SC, USA 437-438
11-10-2005   Ligurian Sea off Genoa, Italy 377
12-09-2005   Sudbury, Ontario, Canada 153
05-04-2006   Caspian Sea off Baku, Azerbaijan 127
06-27-2006   A pond near Flagler, Oklahoma, USA............... 20, 85, 127, 449
07-25-2006   Waterhen Lake, Saskatchewan, Canada 16

08-09-2006	Lake Springfield, Springfield, MO, USA	438
08-??-2006	Unknown location – abductee K. Wilson	192
09-03-2006	White Fish Bay, Michigan, USA	85, 450
11-07-2006	Chicago, Illinois, USA	134-139
01-21-2007	Unnamed pond near Anson, Maine, USA	151
02-15-2007	Unnamed pond near Karki, Latvia	151-152
04-06-2007	North Park Lake near Pittsburgh, PA, USA	384
06-03-2007	Tyrrhenian Sea between Paola & Stromboli, Italy	438-439
07-02-2007	Okanagan Lake, Westbank, BC, Canada	27
10-02-2007	Atlantic Ocean off Virginia, USA	254-255
10-10-2007	Lake Easton, Washington, USA	385
04-02-2008	Lake Ontario off Hamilton, Ontario, Canada	432
10-07-2008	St. Pete Beach, Florida, USA	139
02-10-2010	Bayonne, New Jersey, USA	139-140
01-09-2011	Salitral de la Vidriera, Argentina	366, 373-374

450 cases total:

Many cases are used more than once because they fit into multiple categories.

About the Author

Carl Feindt's interest in water-related UFOs was preceded by his fascination with aircraft. As a boy growing up in New York City during World War II, he carved balsa replicas of aircraft and in his teens flew model aircraft. He followed his passion and became a cadet in the Civil Air Patrol. After graduating from high school, he went on to study aircraft engineering at the Academy of Aeronautics in New York City. Following two years of service in the Air Force, he was forced to return to civilian life due to a family hardship. He joined a major airline where he worked in customer service for over three decades.

Upon retirement he was interested in becoming active in ufology, about which he had been reading since the early 1960s. When Jan Aldrich, a UFO researcher, asked for volunteers to do local newspaper searches, Carl began an eight year study of microfilm rolls at the University of Delaware, exhaustively covering the years between 1923 and 1967. To his amazement, he found approximately 750 UFO-related articles in the Delaware newspapers, all this from the second-smallest state in the United States!

Intrigued by the fact that aircraft, as we know them, cannot emerge from water or submerge into water, he started assembling water-related UFO cases in 1998. He continues to do research into this aspect of the UFO mystery, which has grown into his increasingly large database and website, and he is a recognized authority in the field of water-related UFO research.

Carl has appeared on the History Channel documentaries "Deep-Sea UFOs" and "Deep-Sea UFOs: Red Alert" and has been a guest on numerous radio talk shows nationally. He is a well-received lecturer in the United States and Canada. His 2006 award-winning paper "Physical Influences of a UFO on Water" is available on his website and on DVD, through ufocongresstore.com.